RECENT ADVANCES IN CARBON CAPTURE AND STORAGE

Edited by **Yongseung Yun**

Recent Advances in Carbon Capture and Storage
http://dx.doi.org/10.5772/62966
Edited by Yongseung Yun

Contributors

Tomohiro Asai, Yasuhiro Akiyama, Satoschi Dodo, Quan Zhuang, Bruce Clements, Bingyun Li, Wonjun Cho, Heijin Yu, Yonggi Mo, John Bergstrom, Dyna Ty, Ming Zhao, Guozhao Ji, Rouzbeh G.Moghanloo, Xu Yan, Gregory Law, Soheil Roshani, Garrett Babb, Wesley Herron, Susan E. Crow, Jon Wells, Norman Meki, Carlos A. Sierra, Kimberly M. Carlson, Adel Youkhana, Daniel T. Richardson, Lauren Deem, Gonçalo Carrera, Luis C. Branco, Manuel Nunes Da Ponte, Kenneth Noll, Dongqi Wen, Wenjuan Zhai, Dumitru Cebrucean, Viorica Cebrucean, Ioana Ionel

Published by InTech

Janeza Trdine 9, 51000 Rijeka, Croatia

Publishing Process Manager Martina Usljebrka

Technical Editor SPi Global

Cover InTech Design team

First published March, 2017

Printed in Croatia

Legal deposit, Croatia: National and University Library in Zagreb

Additional hard copies can be obtained from orders@intechopen.com

Recent Advances in Carbon Capture and Storage, Edited by Yongseung Yun

p. cm.

Print ISBN 978-953-51-3005-5

Online ISBN 978-953-51-3006-2

For EU product safety concerns :
IN TECH d . o . o . , Prolaz Marije Krucifikse Kozulić 3 , 51000 Rijeka , Croatia ,
info@intechopen . com or visit our website at intechopen . com .

Meet the editor

Dr. Yongseung Yun majored in chemical engineering and received his PhD degree at the University of Utah, USA, in 1990; MS from KAIST, Korea, in 1981; and BS at the Yonsei University, Korea, in 1979. He currently works as the director at Plant Engineering Division, IAE, in Korea. He had participated as a Korean delegate at the preparation workshops on CCS from 2007 to 2008 to reach the G8 Hokkaido Toyako Summit Leaders Declaration. He has concentrated on CO2 capture technologies from 2005, mainly on the pre-combustion dry sorbent process development. He works as the president in KAWET from 2013 and vice president in Korea DME Association from 2008. Also he had served as the editor for Korean Industrial Chemistry News of KSIEC from 2009 to 2016.

Contents

Preface

Carbon capture and storage (CCS) has been highlighted during the last decade as a practical way of dealing with anthropogenic CO_2 that should be removed from the atmosphere at least to the level of 450 ppm. Also CCS is considered as the only practical way in sequestering the huge CO_2 amount with a reasonable cost at this moment. But, CCS has not reached the full commercial level yet due to the high cost involved as well as due to many uncertain environmental and legal limitations. Hopefully some revolutionary CO_2 utilization methods that can replace the storing of the CO_2 underground can solve all the cost and huge volume issues. Unfortunately, the CCS technology itself had not attained the acceptable cost goal that most technical milestones in many developed countries have targeted as less than US $20–30 per ton in capturing CO_2. Most importantly, cost incentives for CCS by carbon tax or other similar systems have not been implemented globally yet. The CCS process typically requires a heavy instrumentation with a high energy penalty in capturing and storing facilities. After the Paris Agreement on reducing global warming in December 2015, reducing CO_2 in every industrial sector becomes a key task that can guarantee the continuation of business in the long run. Industry waits the reliable and reasonable cost methods in eliminating CO_2 that can continue their business, and this is the right time to provide the technical solution in CO_2 problems.

Scientific evidence from IPCC, etc., clearly indicates that carbon dioxide is a major contributing source to climate change. Among the CO_2-generating sources, fossil fuel power generation produces almost a third of the global CO_2 emission. Since the social infrastructure has already evolved to use fossil fuels that can provide a relatively cheap energy, it is not an easy task to eliminate suddenly the use of them. Especially in developing countries, coal will remain as a cheap and reliable energy source, probably till the 2050s at least. In the long run, renewable energy should replace the fossil fuel and open the era of no-CO_2 emission. The goal is very clear. We should replace the energy source that can minimize the generation of CO_2 or store CO_2 underground until a more pragmatic solution appears. Unfortunately, it requires a heavy burden in cost for individuals and investment for the newer infrastructure. Without technical breakthroughs in CO_2 reducing, sequestering, storing, and converting methods, any enforced endeavor will yield a futile resonance in each person and companies even with morally justified causes. The CCS can work as a bridge before the no-CO_2 era of the future by applying to large-scale CO_2-emitting facilities. Already there are many success stories that exhibited profitable CCS operation by connecting to enhanced oil recovery (EOR).

There is still the question whether CO_2 really needs to be captured and sequestered. Even today many people insist that the recent climate change is just one of the climate cycles that happened throughout the earth's life span. In Cambrian period of about 500 million years ago, CO_2 concentration had reached 7,000 ppm compared to the current CO_2 target of 450 ppm that IPCC intends to control. But, most land life forms had started about 400 million years ago which was even before the highest CO_2 level at ca. 500 million years ago. Thus the high CO_2 concentration of 7,000 ppm appears not to be applicable in the argument. The important thing is that the CO_2 concentration has never reached above 300 ppm during the last 400 million years in that life forms flourish. In the year 2012–2013, the earth's CO_2 level has passed 400 ppm level and continues to rise. Many reports say that the 400 ppm level is the highest during the last 3–20 million years. With or without consenting the impact of CO_2 level on climate change, if a very faint probability of the fatal climate consequences exists as many data forecast, we have to do something to prevent the worst scenario; otherwise, the world ecosystem might fall apart to the level we cannot do anything anymore.

The 1 ppm CO_2 in global atmosphere means 2.13 gigatons of carbon. The current largest CCS plant can sequester 3 million CO_2 tons per year, which is the amount of CO_2 emitted from just one 500 MW coal-fired power plant. This suggests that there is a long way for CCS to yield a meaningful impact on the global CO_2 issues, unless the technology achieves some kind of standardized form in reducing cost and applies at least hundreds of it a year. Considering the immense volume of CO_2 amount that CCS has to deal with, most of the CCS technologies still contribute too small portion of the CO_2 problem.

The CCS was a big issue in major economies of the world during the years 2005–2008, precipitating by the recommendation (G8 Hokkaido Toyako Summit Leaders Declaration) at the G8 summit meeting held on July 8, 2008, at Hokkaido in Japan. At the Declaration, CCS was mentioned as "launching of 20 large-scale CCS demonstration projects globally by 2010, taking into account various national circumstances, with a view to beginning broad deployment of CCS by 2020." I had participated personally in the preparation of workshops to reach this Declaration from the IEA/CSLF Assessment Workshop in Oslo 2007 to the 2008 Major Economies Meeting in Hawaii. The CCS had been highlighted as a key solution for tackling CO_2 issues in the early 2000s, but it appears to lose some passion by the lack of progress in technical developments and in commercial success stories other than EOR.

Renewable energies without any pollutants or CO_2 emission should be the way of the future, but it might take several decades with current steps of technical and political stagnation in dealing with climate change. Then, we have to go back to basics, starting from finding a solution in small steps. Soil carbon sequestration that is included as a section of the book is a good example even though it will never give a big impact solution in CCS, but it can give a grass root impact that every individual can contribute a small token in tackling the CO_2 issue. Since the CCS processes involve many energy consuming steps like CO_2 separation to higher concentrations, CO_2 compression, etc., there are many potentials in reducing the energy penalty by advanced technologies. The CCS technology desperately needs far newer ideas and breakthroughs that can separate earlier attempts to capture and sequester CO_2 through improving, modifying, and switching the known principles. This book tries to give some insight into developing an urgently needed technical breakthrough through the recent advances in CCS research, in addition to the available small steps like soil carbon sequestration. Another recent direction dealing with climate change focuses on carbon utilization rather than the direct carbon capture and storage. Conceptually, converting CO_2 to chemicals or fuels should be more beneficial to environment because it can substitute the fossil fuels like oil, natural gas, or coal. Carbon utilization is considered as a continuation of CCS, and one chapter is included in the book.

The book consists of five sections: CCS in coal power plants, carbon capture methods, soil carbon sequestration, carbon storage and utilization, and economics of CCS. The first section tells about some of the recent advances in dealing with CO_2 issues by the coal power industry. The second section illustrates the most critical components in CCS, which are CO_2-capturing methods involving absorbents, membranes, and biomaterials. The third section deals about the soil carbon sequestration that can be viewed as a meaningful CO_2 study that starts with very minute step but can end with huge impact on agricultural cultivation techniques. The fourth section contains two chapters, one on the carbon utilization by conversion to chemicals and fuels and the other on the CO_2 underground storage with actual site data. Finally, the fifth section talks about the basic economic principles that should be discussed in CCS.

Personally, I had read all the details of each chapter and acted as a critical reviewer to make a better quality book during the last nine months. I hope this book can serve as a small cornerstone in finding new concepts and more reliable technologies in CCS and CO_2 utilization.

I thank all participating authors for contributing their chapters and for helping in revising where needed. Also, I really thank Ms. Martina Usljebrka who had helped me in every editing step throughout the whole nine months in 2016.

Yongseung Yun
Institute for Advanced Engineering, Yongin,
Republic of Korea

Section 1

CCS in Coal Power Plants

Development of a State-of-the-Art Dry Low NOx Gas Turbine Combustor for IGCC with CCS

Tomohiro Asai, Yasuhiro Akiyama and
Satoschi Dodo

Additional information is available at the end of the chapter

http://dx.doi.org/10.5772/66742

Abstract

The successful development of the coal-based integrated gasification combined cycle (IGCC) with carbon capture and storage (CCS) requires gas turbines capable of achieving dry low nitrogen oxide (NOx) combustion of hydrogen-rich syngas fuels for low emissions and high plant efficiency. This chapter describes the development of a "multi-cluster combustor" as a state-of-the-art dry low NOx combustor for hydrogen-rich syngas fuels. The combustor consists of multiple clusters of pairs of one fuel nozzle and one air hole that are installed coaxially. The essence of the design concept is the integration of two key technologies: rapid mixing of fuel and air for low NOx and flame lifting for flashback-resistant combustion. The combustor has been developed in three steps: burner development, combustor development, and feasibility demonstration for practical plants. The combustor was tested with a practical syngas fuel in a multi-can combustor configuration in an IGCC pilot plant in the final step. The combustor achieved the dry low NOx combustion of the syngas fuel in the pilot plant and the test results demonstrated the feasibility for achieving dry low NOx combustion of the syngas fuel in practical plants.

Keywords: integrated gasification combined cycle (IGCC), carbon capture and storage (CCS), gas turbine, dry low NOx combustor (DLNC), multi-cluster combustor, hydrogen-rich syngas fuels

1. Introduction

Coal is a vital energy source for power generation with over 40% of the electricity produced worldwide stemming from coal [1]. Coal is able to ensure energy supply stability and security due to its low cost, abundant reserves, and worldwide availability. However, conventional pulverized coal–fired power plants are the most carbon dioxide (CO_2)-intensive source of power generation. An effective method for cutting CO_2 emissions from coal-fired power plants is to employ a coal-based integrated gasification combined cycle (IGCC). IGCC plants release less CO_2 than conventional pulverized coal–fired power plants because of their higher plant efficiency. IGCC also possesses the capability to capture and store CO_2 before combustion [precombustion carbon capture and storage (CCS)]. CCS technology suppresses the release of CO_2 into the atmosphere by capturing and storing CO_2 emissions from thermal power plants. A report by the Intergovernmental Panel on Climate Change (IPCC) estimates that an IGCC plant with CCS might cut CO_2 emissions by about 80–90% compared with an IGCC plant without CCS [2]. However, the major technical hurdle with CCS is that CCS decreases plant efficiency because of the additional energy for capture and storage. The report by the IPCC estimates that a CCS-equipped IGCC plant might need 14–25% more energy than an IGCC plant of equivalent output without CCS [2]. Improving the efficiency of CCS-equipped IGCC plants is a key to the successful combination of the two technologies. In order to achieve high plant efficiency and low emissions, a gas turbine in IGCC plants requires a diluent-free ("dry"), low NOx combustor. This chapter describes the development of a state-of-the-art dry low NOx combustor intended for CCS-equipped IGCC plants.

2. CCS-equipped oxygen-blown IGCC technology and technical hurdles with gas turbines

2.1. Overview

Coal-based IGCC technology with CCS converts coal to syngas, removes CO_2 from the syngas, and generates electric power in the combined cycle by utilizing the produced hydrogen-rich syngas as gas turbine fuel. An oxygen-blown IGCC plant with a precombustion CCS system is composed of five key components: an air separation unit (ASU), a gasifier, a syngas cleanup unit, a CCS system, and a combined cycle unit. A schematic diagram of the plant is shown in **Figure 1**.

The plant generates electric power through the process as follows. The ASU separates air into oxygen (O_2) and nitrogen (N_2). The gasifier converts coal to raw syngas by reacting it with oxidant (oxygen). The gasifier employs oxygen as the oxidant, and this type of gasification is referred to as "oxygen-blown." This chapter addresses oxygen-blown IGCC technology. The syngas cleanup unit removes impurities, such as particulate matter, sulfur, and ammonia from the raw syngas, producing a clean syngas consisting mainly of carbon monoxide (CO) and hydrogen (H_2). A shift reactor of the CCS system converts CO in the clean syngas to H_2 and CO_2 by a water-gas shift reaction, producing a shifted syngas. A CO_2 capture unit removes CO_2 from the shifted syngas, thus producing a hydrogen-rich syngas. The hydrogen-rich syngas is supplied to a gas turbine as fuel. A gas turbine combustor burns the syngas, and the combustion gas operates a turbine, which, in turn, generates electric power. A heat recovery steam generator (HRSG) produces steam using the waste heat of exhaust gas from the gas turbine, and sends the steam to the gasifier in order to produce the raw syngas and to a steam turbine in order to generate additional power. The gas turbine combustors are required to operate on oil fuel as the startup fuel at ignition and during part load in order to provide the HRSG with the waste heat until the syngas fuel is supplied to the gas turbine [3].

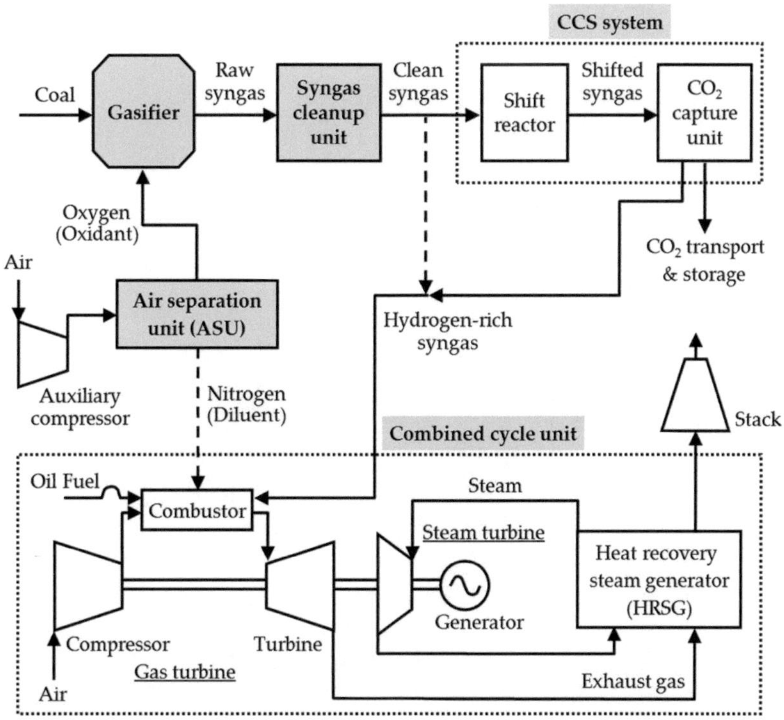

Figure 1. Schematic diagram of oxygen-blown IGCC plant with precombustion CCS system.

2.2. Technical hurdles with gas turbines

The implementation of IGCC technology with CCS poses significant challenges to gas turbine combustors owing to properties of hydrogen-rich syngas fuels. **Figure 2** shows the variation in fuel compositions against the carbon capture rate for typical syngas fuels [4]. The fuel compositions vary widely depending on the carbon capture rate. As the carbon capture rate increases from 0 to 90%, as a result of the conversion of CO to H_2 and CO_2 in the shift reactor, H_2 concentration increases widely from approximately 25 to over 80 vol%. Some properties of hydrogen, specifically its higher flame speed, lower ignition energy, and broader flammability limits compared with conventional gas turbine fuels (e.g., natural gas), increase the risk of flashback and autoignition [5].

The challenges posed by these properties of hydrogen require advanced combustion technologies for hydrogen-rich syngas fuels. Conventional gas turbine combustors are incapable of achieving low NOx emissions and high plant efficiency for hydrogen-rich syngas fuels. **Figure 3** summarizes technical hurdles with their use. The combustors are broadly classified into two types: premixed combustors and diffusion-flame combustors. Conventional premixed combustors are capable of achieving low NOx by supplying premixed fuel-air mixtures because they maintain low local flame temperatures. However, premixed combustors burning hydrogen-rich fuels are prone to flashback into their large premixing section because they are highly tuned to operate on low-flame-speed fuels like natural gas. This flashback tendency characteristic hinders the application of premixed combustion technology to

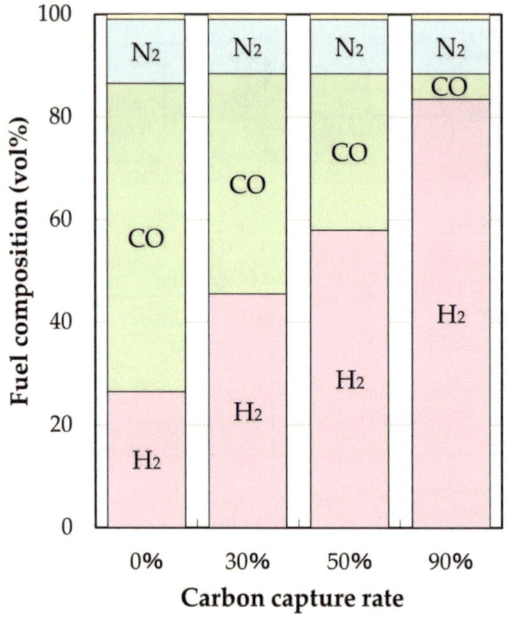

Figure 2. Fuel compositions of typical syngas fuels in oxygen-blown IGCC with CCS.

	Premixed combustor	Diffusion-flame combustor
Combustor type	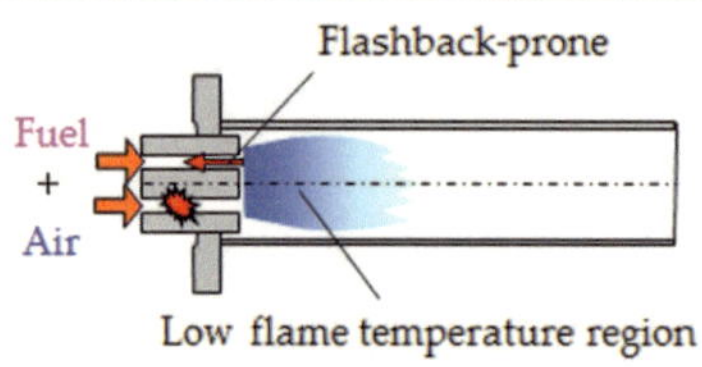	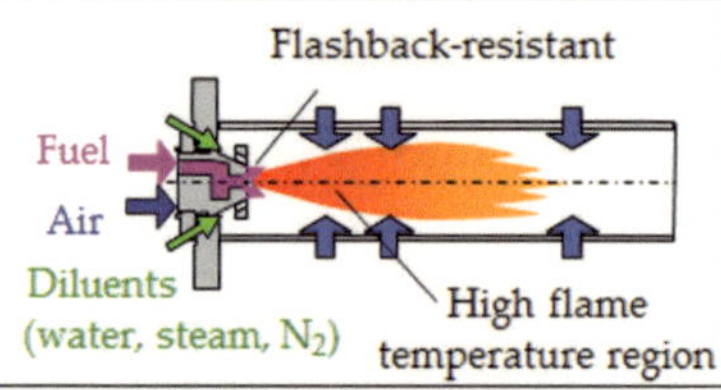
Merit	Low NOx	Flashback-resistant
Demerit	Flashback-prone	Plant efficiency decrease (due to injection of diluents for NOx decrease)
Employed for IGCC	No	Yes

Figure 3. Technical hurdles with conventional combustors.

hydrogen-rich syngas fuels in IGCC. In contrast, conventional diffusion-flame combustors are capable of achieving flashback-resistant combustion of hydrogen-rich fuels by supplying fuel and air separately into their combustion chamber. However, diffusion-flame combustors are incapable of achieving high plant efficiency because they require additional energy to inject a diluent, such as water, steam, or nitrogen, into the combustion zone in order to suppress the increased NOx emissions due to high local flame temperatures. IGCC plants have thus far employed diffusion-flame combustors at the expense of decreased plant efficiency in order to achieve flashback-resistant combustion of hydrogen-rich fuels.

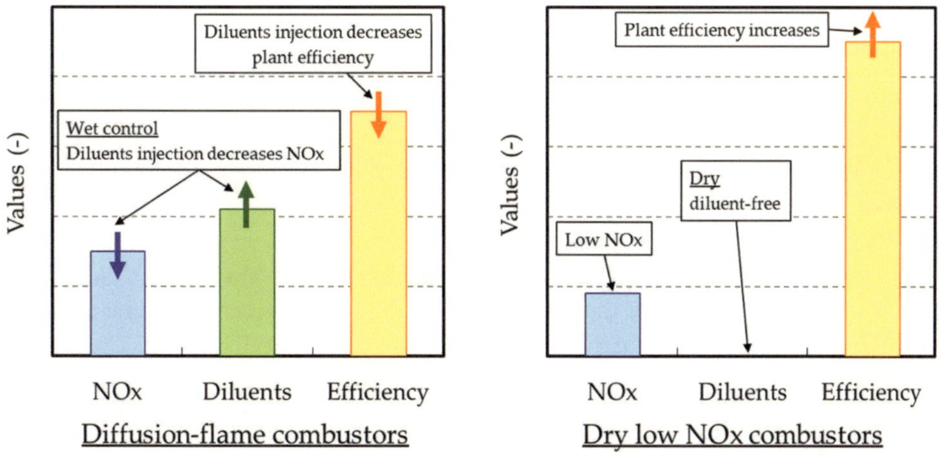

Figure 4. Advantages of dry low NOx combustor.

A solution to these hurdles is to develop state-of-the-art technologies for diluent-free (dry), low NOx combustion of hydrogen-rich fuels. **Figure 4** compares the advantages of dry low NOx combustors (DLNC) with those of diffusion-flame combustors. Diffusion-flame combustors decrease NOx by injecting diluents. This method is referred to as "wet control." However, injection of diluents decreases plant efficiency. In contrast, dry low NOx combustors achieve low NOx diluent-free (dry) combustion, thereby increasing plant efficiency. The successful implementation of IGCC technology with CCS requires state-of-the-art technologies for the dry low NOx combustion of hydrogen-rich syngas fuels that can achieve both lower NOx emissions and higher plant efficiency.

Many research groups and gas turbine manufacturers have been developing dry low NOx combustion technologies for hydrogen-rich fuels [6–18]. Technologies described in the literature include a multi-tube mixer fuel nozzle [7], a triple-fuel syngas burner [8], a MBtu EV burner [9], a low-swirl injector [10], a flameless-oxidation burner [11], a micro-mixing lean-premix injector [12], a DLN micromix burner [13, 14], a DLE combustor with supplemental burners [15], a lean premixed swirl-stabilized hydrogen burner with axial air injection [16], a rich catalytic hydrogen injector [17], and a rich/lean staged burner [18]. This chapter describes the development of a state-of-the-art dry low NOx combustor for hydrogen-rich syngas fuels in CCS-equipped oxygen-blown IGCC plants.

3. A state-of-the-art dry low NOx combustor

3.1. Combustor configuration

Figure 5 shows the configuration of the state-of-the-art dry low NOx combustor [19]. The combustor consists of multiple fuel nozzles and multiple air holes. The key elements of the combustor each consist of one fuel nozzle and one air hole that are installed coaxially. A cluster of key elements constitutes one burner, which forms one flame. The air holes are embedded in one plate. Multiple burners constitute a can combustor, and several can combustors are installed on a gas turbine. The combustor is classified as a multi-can type [20]. Hereafter, this burner is referred to as a "cluster burner," and this combustor is referred to as a "multi-cluster combustor."

An individual multi-cluster combustor consists of multiple cluster burners, a cylindrical liner, a cylindrical casing, and an end cover. **Figure 6** illustrates a cross-sectional diagram of an individual multi-cluster combustor. The cluster burners are installed on the end cover equipped with fuel supplying systems. The liner is mounted concentrically inside the casing. **Figure 7** illustrates a detailed diagram of the cluster burners. The burners consist of one pilot burner at the center and several identical main burners surrounding the pilot burner. The combustor forms seven individual flames, each of which is anchored to the corresponding burner. The combustor assigns the function of maintaining operational stability to the pilot burner and the function of maintaining low NOx operation to the main burners.

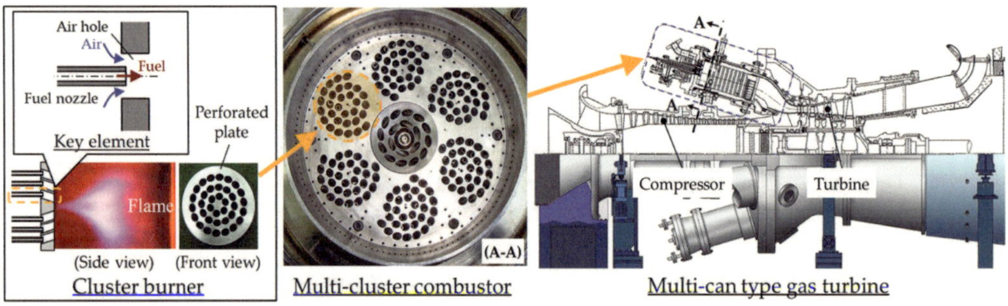

Figure 5. Configuration of the state-of-the-art dry low NOx combustor for hydrogen-rich syngas fuels.

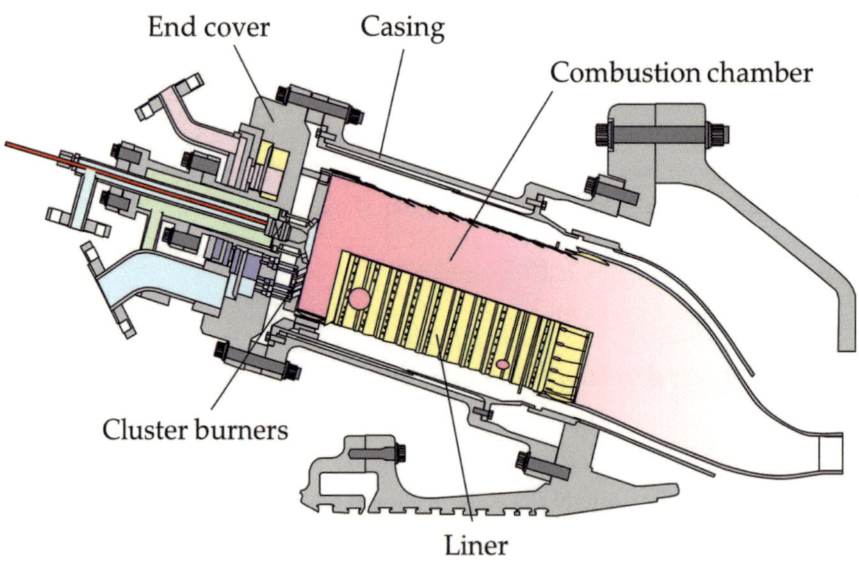

Figure 6. Cross-sectional schematic diagram of individual multi-cluster combustor.

http://dx.doi.org/10.5772/66742

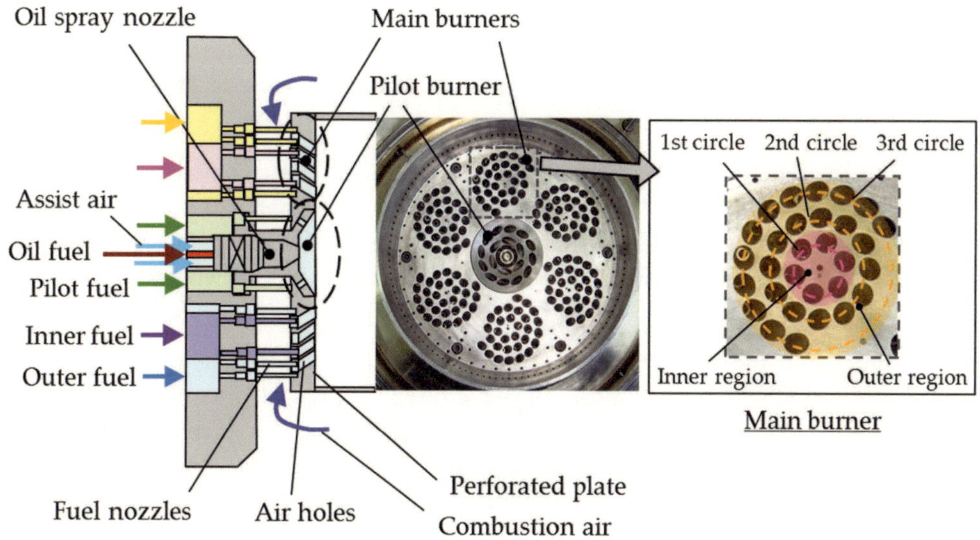

Figure 7. Detailed diagram of cluster burners.

The pilot burner can ensure combustion stability over the operating range by forming a well-stabilized flame in the center. The pilot burner is equipped with an air-assisted oil spray nozzle at the center. The oil spray nozzle operates on oil fuel at ignition and during part load before syngas fuel is supplied to the gas turbine in IGCC plants [19].

The main burners can achieve homogeneous fuel-air mixing for low NOx combustion by dispersing fuel to multiple injection points. The injection points are arranged in three circles on each main burner: six points on the first circle with the smallest diameter, 12 points on the second circle with the intermediate diameter, and 12 points on the third circle with the largest diameter. Here, the region within each first circle on the perforated plate is referred to as the "inner region," and the region outside each first circle is referred to as the "outer region." The gaseous fuel injected from six fuel nozzles on each first circle is referred to as "inner fuel," and the gaseous fuel injected from 24 fuel nozzles on each of the second and third circles is referred to as "outer fuel." The main burners characterize the low NOx performance of the combustor [19].

3.2. Burner concept

The next subsections describe the concept of the cluster burner for hydrogen-rich fuels. The essence of this burner concept is the integration of two key technologies: rapid mixing of fuel and air for low NOx combustion and flame lifting for flashback-resistant combustion. The cluster burner provides both the advantage of the premixed combustor of low NOx combustion and the advantage of the diffusion-flame combustor of flashback-resistant combustion.

3.2.1. Rapid mixing for low NOx *combustion*

Rapid mixing achieves low NOx combustion. Thermal NOx from atmospheric air is formed extensively at high temperatures [6, 21]. As a result, NOx emissions are decreased to low levels by eliminating high-temperature regions. Such high-temperature and NOx-generating regions are removed by the formation of homogeneous fuel-air mixtures before combustion, because of the rapid mixing of fuel and air within a short distance. Here, rapid mixing achieves low NOx combustion.

The cluster burner mixes fuel and air rapidly by producing multiple coaxial fuel-air jets, each of which consists of a central fuel jet surrounded by an annular air jet. The burner is equipped with multiple injection points. The burner installs the fuel nozzles in separate air holes coaxially at each injection point, thereby producing multiple coaxial fuel-air jets [19].

The coaxial jets mix fuel and air rapidly within a short distance by enhancing turbulence through contracting and expanding air passages. **Figure 8** shows the fuel concentration distribution in the mixing process of a coaxial jet analyzed by large eddy simulation (LES). The simulation results show that turbulence increases the amplitude of a wavelike disturbance at the boundary between fuel and air jets downstream of the air hole exit, thus mixing fuel and air rapidly. The burner disperses fuel by multiplying the coaxial jet, thereby enhancing mixing of fuel and air [19].

Conventional premixed combustors can mix fuel and air almost completely. However, premixed combustors burning hydrogen-rich fuels are prone to flashback into their large premixing section because of their higher flame speeds. Thus, this flashback characteristic hinders the application of premixed combustion technology to hydrogen-rich fuel combustion.

3.2.2. Flame lifting for flashback-resistant combustion

Flame lifting achieves flashback-resistant combustion. Flame lifting means that a flame is stably held at a point away from the burner. As a result, flame lifting suppresses the occurrence of flashback into the burner.

The burner can lift a flame by generating converging and diverging swirl flows downstream from itself. **Figure 9** illustrates a cross-sectional diagram of the main burner to describe the operating principle of this flame-lifting technology. The air holes cause the combustion air passing through them to swirl because the central axis of each air hole is inclined in the direction of a tangent to each circle. The swirling flow issuing from the air holes first converges toward and then diverges from an axial position (flame-anchoring point) located away from the burner. As shown in the figure, the converging-diverging swirl flows induce a pressure profile in the

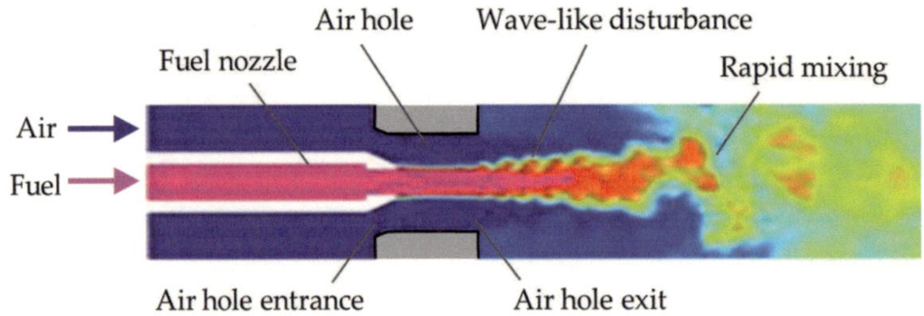

Figure 8. Mixing process of a coaxial jet.

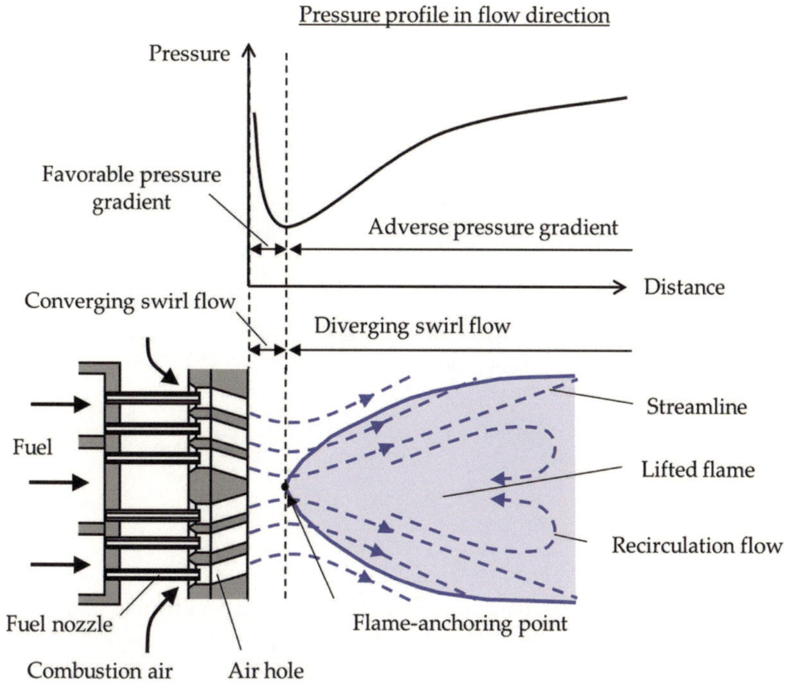

Figure 9. Operating principle of flame-lifting technology for the main burner.

flow direction. The converging flow induces a favorable pressure gradient due to the decrease in pressure downstream with increasing swirl velocity, whereas the diverging flow induces an adverse pressure gradient due to the increase in pressure downstream with decreasing swirl velocity. The adverse pressure gradient causes a vortex breakdown at the boundary between the converging and diverging swirl flows, thereby generating a recirculation flow. The recirculation flow stabilizes the flame by providing a stable heat source of combustion gas for the continuous ignition of fresh reactants. The reverse flow of the combustion gas from the boundary can be suppressed by the favorable pressure gradient in the converging flow. As a result, the flame is stabilized at the flame-anchoring point on the boundary. According to this operating principle, the flame is lifted from the burner and thus can suppress the flashback into the burner [19].

3.3. Combustor operability

The combustors are required to operate stably from ignition through part load to base load in practical IGCC plants. The next subsections describe the fuel supply system and fuel staging.

3.3.1. Fuel supplying system

The fuel supply system supplies syngas and oil fuels to the multi-cluster combustor. **Figure 10** shows the fuel supplying system for the combustor [19]. The six main burners are divided into two groups (F2 and F3) consisting of three burners each, and arranged alternately around the pilot burner (F1). The syngas fuel is distributed into five fuel circuits: F1 fuel to the F1 pilot

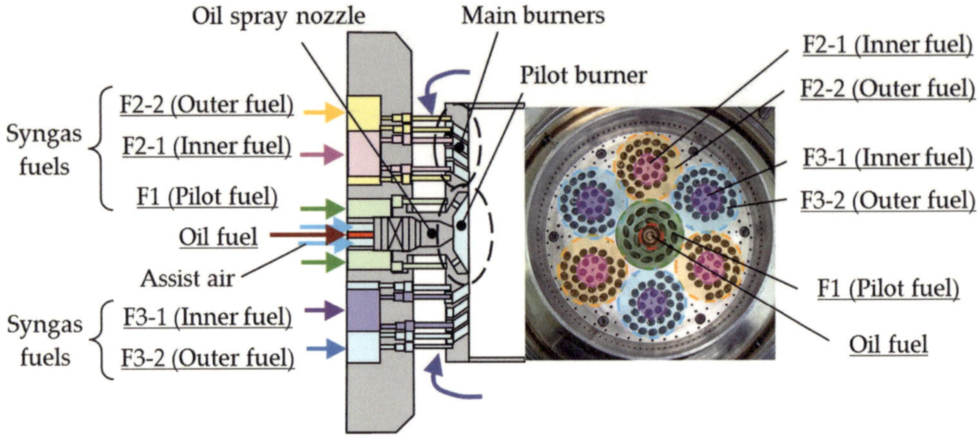

Figure 10. Fuel supplying system.

burner, F2-1 fuel to the inner region of the F2 main burners, F2-2 fuel to the outer region of the F2 main burners, F3-1 fuel to the inner region of the F3 main burners, and F3-2 fuel to the outer region of the F3 main burners. The oil fuel is supplied to the oil spray nozzle.

The fuel distribution ratios (F1 ratio and outer-fuel ratio) are important test parameters influencing combustion performance. The F1 ratio is defined as the ratio of the mass flow rates of F1 fuel to all fuel. The F1 ratio is expressed as follows:

$$\text{F1 ratio (\%)} = \frac{Gf_{F1}}{Gf_{all}} = \frac{Gf_{F1}}{Gf_{F1} + Gf_{F2-1} + Gf_{F2-2} + Gf_{F3-1} + Gf_{F3-2}} \tag{1}$$

where Gf denotes the mass flow rate, and subscripts "all," "F1," "F2-1," "F2-2," "F3-1," and "F3-2" denote all the fuel, F1 fuel, F2-1 fuel, F2-2 fuel, F3-1 fuel, and F3-2 fuel, respectively. The outer-fuel ratio is defined as the ratio of the mass flow rates of the outer fuel to all fuel supplied to the main burners. The outer-fuel ratio is expressed by Eq. (2).

$$\text{Outer} - \text{fuel ratio (\%)} = \frac{Gf_{F2-2} + Gf_{F3-2}}{Gf_{F2-1} + Gf_{F2-2} + Gf_{F3-1} + Gf_{F3-2}} \tag{2}$$

3.3.2. Fuel staging

The multi-cluster combustor can achieve low emissions and high operability over the entire operating range by switching combustion modes according to operating conditions. The combustor switches the modes by manipulating the combination of operating burners for which the fuel circuit is fueled. The fuel staging with combustion modes and switching loads hinges on such factors as operating conditions, operational requirement, and environmental regulation for each practical plant. This chapter shows an instance of the fuel staging sequences and combustion modes. **Figure 11** shows the fuel staging [22]. In this figure, colored regions shown on the burner pictures indicate operating burners. This fuel staging consists of three distinct combustion modes: oil mode, partial mode, and final mode. The combustor operates on oil fuel in the oil mode and on syngas fuel in the partial and the final

Operating range	Turbine speed (%) | Low ←	Gas turbine load (%)	→ High
Fuel	Oil	Syngas	
Mode	Oil mode	Partial mode	Final mode
Operating burners	Oil spray nozzle	F1 + F2-1 + F2-2 + F3-1	F1 + F2-1 + F2-2 + F3-1 + F3-2

Figure 11. Fuel staging.

modes. The combustor switches from the oil mode, through the partial mode, to the final mode between ignition and base load.

In the oil mode, the combustor operates on oil fuel with the oil spray nozzle. The oil mode is used to ignite, accelerate, and operate the combustor over low loads. The oil fuel operation requires injection of diluent into the combustion zone to lower NOx emissions to the level required by environmental regulations, because NOx from oil fueled combustion increase owing to the local high-temperature regions compared with syngas fueled combustion. The combustor injects diluent nitrogen from the fuel nozzles in the outer regions of the main burners. The diluent nitrogen is separated from air by the ASU in IGCC plants. In the partial mode, the combustor operates on syngas fuel with the pilot burner (F1), inner regions of the F2 and F3 main burners (F2-1 and F3-1), and outer regions of the F2 main burners (F2-2). The partial mode is employed during part load from a low load to a middle load. At a low part load, the combustor switches from oil to syngas fuels. The combustor achieves stable combustion over low to middle loads by combusting the pilot fuel and inner fuel of the main burners associated with flame stabilization. In the final mode, the combustor operates on syngas fuel with the pilot burner and all the main burners. The final mode is used from a middle part load to base load. The combustor achieves low NOx combustion by distributing syngas fuel to all the burners [22].

4. Combustor development

4.1. Development approach

This section describes the development work for the multi-cluster combustors intended for hydrogen-rich syngas fuels in CCS-equipped oxygen-blown IGCC plants. **Figure 12** shows the development approach for the multi-cluster combustors. The development approach consists of three steps: burner development; combustor development; and feasibility demonstration for practical plants.

	Step 1 Burner Development	Step 2 Combustor Development	Step 3 Feasibility demonstration for practical plants
Purpose	Optimization of burner configurations	Optimization of combustor configurations	Demonstration of combustor performance in practical plants
Configuration	- Pair of a fuel nozzle and an air hole - Single burner	Single-can combustor	Multi-can combustors
Test	- Single-nozzle mixing test - Single-burner combustion test	Single-can combustor test	Real gas turbine test in multi-can combustor configuration
Pressure	Atmospheric pressure	Medium and high pressures	Practical pressure
Fuel	Test fuels (mixtures of H_2, CH_4, & N_2)	Test fuels (mixtures of H_2, CH_4, & N_2)	Practical syngas fuels
Evaluation	- Emissions (NOx, CO, UHC) - Stability (Pressure fluctuations)	- Emissions (NOx, CO, UHC) - Stability (Pressure fluctuations) - Reliability (Metal temperatures)	- Emissions (NOx, CO, UHC) - Stability (Pressure fluctuations) - Reliability (Metal temperatures) - Operability (Dynamic characteristics)

Figure 12. Development approach for multi-cluster combustors.

In Step 1, the purpose is to optimize configurations of single burners with pairs of a fuel nozzle and an air hole by performing single-nozzle mixing test and single-burner combustion test at atmospheric pressure with test fuels comprised of H_2, methane (CH_4), and N_2. Step 1 evaluates performance of single burners in terms of emissions of NOx, CO, and unburned hydrocarbons (UHC), and stability, which is related to pressure fluctuations due to combustion oscillation. The single-burner combustion test showed that the operating range of stable low NOx combustion was restricted by the occurrence of combustion oscillation, and it was probably triggered by the attachment of the flame to the perforated plate due to the ignition of flammable mixtures in the wake behind the plate [3]. In order to suppress the combustion oscillation, a convex burner was suggested. The convex burner was equipped with a convex perforated plate, of which the center projected into the combustion chamber and the surface was inclined. The combustion test showed that the convex burner was effective in suppressing the combustion oscillation and it expanded the operating range of stable low NOx combustion [23].

In Step 2, the purpose is to optimize configurations of single-can combustors by performing the single-can combustor test at medium and high pressures with test fuels that were mixtures of H_2, CH_4, and N_2 on the basis of the burner configurations optimized in Step 1. Step 2 evaluates performance of single-can combustors in terms of emissions, stability, and reliability. The performance for the reliability is related to metal temperatures of burners and liners. On the basis of the findings from the single-burner combustion tests, multi-cluster combustors equipped with the flat burner and the convex burner were developed.

In Step 3, the purpose is to demonstrate combustor performance in practical plants by real gas turbine test in a multi-can combustor configuration at practical pressure with practical syngas fuel. Step 3 evaluates performance of multi-can combustors in terms of emissions, stability, reliability, and operability. The performance for the operability is related to dynamic characteristics of the combustors during their operation.

The next subsections describe the development work in Step 2 and Step 3.

http://dx.doi.org/10.5772/66742

4.2. Single-can combustor test

Step 1 evaluated performance of the flat burner and the convex burner by performing the single-burner combustion test in order to optimize the burner configuration. This subsection describes the development of multi-cluster combustors equipped with the flat burner and the convex burner [24, 25].

Figure 13 shows the configurations of two types of prototype multi-cluster combustors: a flat multi-cluster combustor and a convex multi-cluster combustor. The two combustors differed in terms of the main burner configurations. The flat combustor was equipped with one concave pilot burner at the center and six flat main burners surrounding the pilot burner. The convex combustor was equipped with one concave pilot burner at the center and six convex main burners surrounding the pilot burner. The combustors were tested at a medium pressure under the base load condition.

Figure 14 shows a schematic diagram of the single-can combustor test facility. A single-can combustor was assembled into the test stand. An air compressor supplied combustion air to the combustor through a preheater, and the pressure in the combustion chamber was adjusted with a back pressure valve downstream. The test fuels were mixtures of H_2, CH_4, and N_2. The fuel supplying system independently supplied the following gases to a gas mixer: H_2 from H_2-cylinder-loaded trailers; CH_4 from CH_4-cylinder-loaded trailers; and N_2 from a liquefied nitrogen tank. The gas mixer produced a gas mixture with certain volume percentages (vol%) of the three gases as a test fuel. The compositions of the gas mixture were varied by changing the flow rates of the constituents independently. The gas mixture was separated into five fuel circuits. The measuring equipment consisted of a gas analyzer and a fluctuating-pressure-measuring system. The gas analyzer measured gas concentrations in the exhaust gas at a measuring duct downstream in the test stand. The fluctuating-pressure-measuring system measured pressure fluctuations inside the combustion chamber.

Combustor		Flat multi-cluster combustor	Convex multi-cluster combustor
Surface shape	Pilot burner	Concave	Concave
	Main burners	Flat	Convex
Photo			

Figure 13. Configurations of prototype multi-cluster combustors.

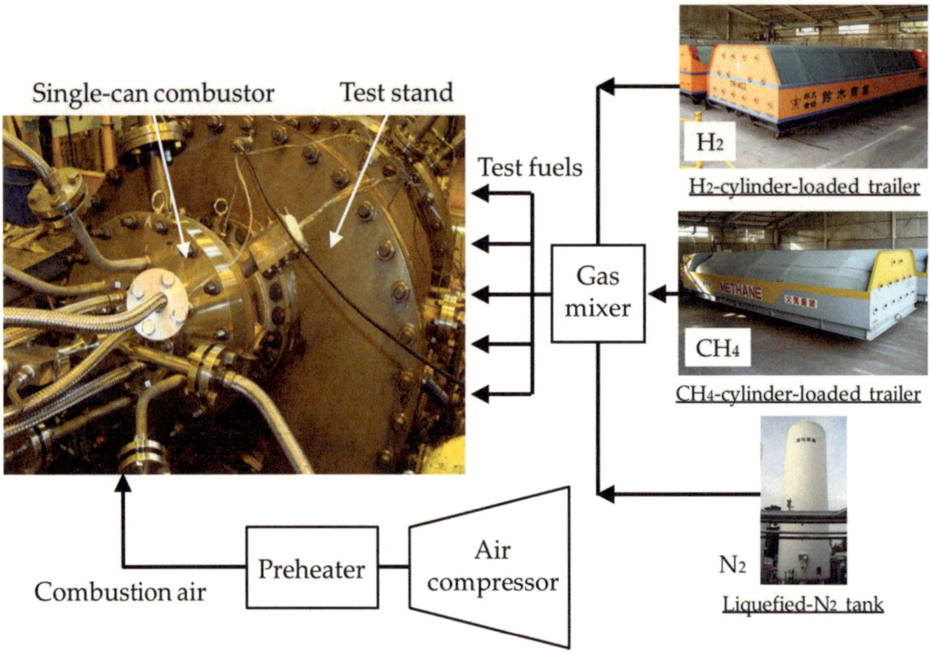

Figure 14. Single-can combustor test facility.

The practical syngas fuels used in IGCC plants include a large amount of CO. However, the road traffic law in Japan prohibits the transport of a large amount of CO required for combustor tests mainly for safety reasons. This practical restriction requires use of CO-free test fuels for the combustor tests. **Table 1** lists properties of three mixtures of test fuels used. The test fuels contained 40, 55, and 65 vol% H_2, simulating the practical hydrogen-rich syngas fuels at carbon capture rates of 0, 30, and 50% for CCS-equipped oxygen-blown IGCC, respectively. Hereafter, the test fuels are referred to as "CCS-0% fuel," "CCS-30% fuel," and "CCS-50% fuel."

Minimization of NOx requires homogeneous lean combustion. Homogeneous lean combustion was achieved by supplying syngas fuel to each fuel nozzle of the main burners at the

Test fuels		**CCS-0% fuel**	**CCS-30% fuel**	**CCS-50% fuel**
Constituents:				
H_2	vol%	40.0	55.0	65.0
CH_4	vol%	18.0	15.7	6.3
N_2	vol%	42.0	29.3	28.7
Density	kg/m^{3*}	0.640	0.490	0.429
Lower heating value	MJ/m^{3*}	10.0	10.8	8.6
	MJ/kg	15.7	22.0	20.1

*At 273.15 K, and 0.1013 MPa.

Table 1. Properties of test fuels.

http://dx.doi.org/10.5772/66742

same flow rate. This uniform fuel supply yielded an outer-fuel ratio of 80%. This ratio equaled the proportion of the number of fuel nozzles in the outer region (24 nozzles) to the total number of fuel nozzles (30 nozzles) of each main burner. This study set the target outer-fuel ratio at 80% for minimum NOx. This study defines a certain value of the maximum design amplitude of pressure fluctuations for safely operating the combustors. The combustors are required to be developed so that they can maintain the pressure fluctuation amplitudes below the maximum design value. The maximum design value is referred to as the criterion of combustion oscillation here. The combustion oscillation with an amplitude above the criterion may increase the risk of damage to the combustors.

Figures 15 and **16** show variations in pressure fluctuation amplitude and NOx emissions, respectively, for the flat combustor as a function of the outer-fuel ratio. In **Figure 15**, the amplitude of pressure fluctuation was normalized by the maximum design value. For CCS-0% fuel, the flat multi-cluster combustor could increase the outer-fuel ratio to the target ratio with the pressure fluctuation amplitude below the criterion, and thus achieved the minimum NOx at the target ratio. For CCS-30% and CCS-50% fuels, however, the flat combustor could not increase the outer-fuel ratio to the target ratio because the pressure fluctuation amplitudes increased abruptly above the criterion before the outer-fuel ratio reached the target ratio. Consequently, the NOx minimization was restricted by the abrupt increase in pressure fluctuation amplitude above the criterion.

In contrast, **Figure 17** shows that the convex multi-cluster combustor could increase the outer-fuel ratio to the target ratio with the pressure fluctuation amplitude below the criterion for CCS-0%, CCS-30%, and CCS-50% fuels. **Figure 18** shows that the convex combustor achieved the minimum NOx at the target ratio for all the test fuels. The test results demonstrated that the convex combustor was effective in suppressing the occurrence of combustion oscillation for hydrogen-rich fuels.

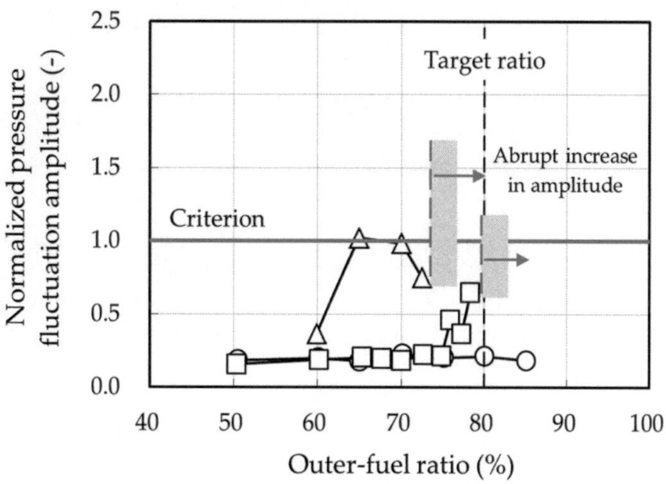

Figure 15. Pressure fluctuation amplitude in single-can combustor test for flat multi-cluster combustor. Symbols: circles, CCS-0% (H_2 = 40 vol%); squares, CCS-30% (H_2 = 55 vol%); triangles, CCS-50% (H_2 = 65 vol%).

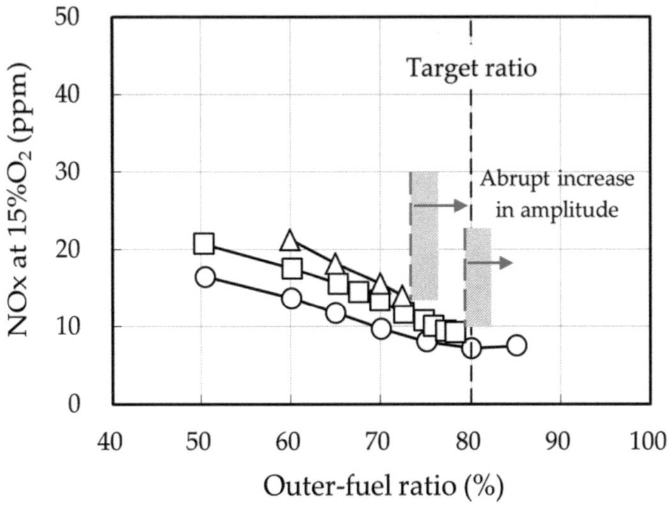

Figure 16. NOx emissions in single-can combustor test for flat multi-cluster combustor. Symbols: circles, CCS-0% (H_2 = 40 vol%); squares, CCS-30% (H_2 = 55 vol%); triangles, CCS-50% (H_2 = 65 vol%).

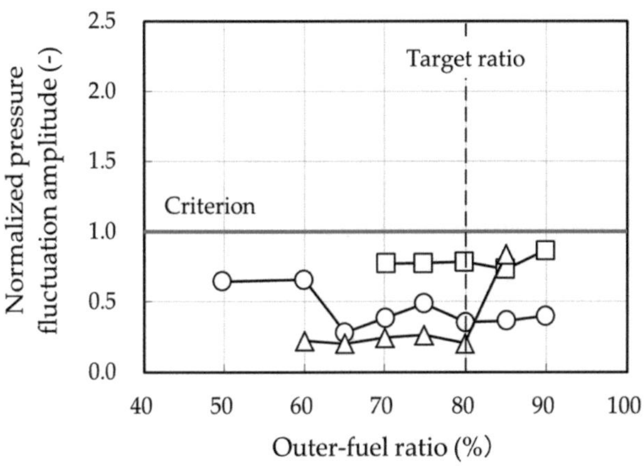

Figure 17. Pressure fluctuation amplitude in single-can combustor test for convex multi-cluster combustor. Symbols: circles, CCS-0% (H_2 = 40 vol%); squares, CCS-30% (H_2 = 55 vol%); triangles, CCS-50% (H_2 = 65 vol%).

Both multi-cluster combustors achieved flashback-free combustion throughout the tests. The test results demonstrated that the multi-cluster combustors could feasibly achieve the dry low NOx combustion of hydrogen-rich surrogate fuels with hydrogen content to 65 vol%.

4.3. Pilot plant test

In order to demonstrate the feasibility for practical IGCC plants, the multi-cluster combustor was tested on a real gas turbine in a multi-can combustor configuration in an IGCC pilot plant at practical pressure with practical syngas fuel [19, 22, 26, 27].

4.3.1. Pilot plant EAGLE and test conditions

The pilot plant was an oxygen-blown integrated coal gasification power generation pilot plant "EAGLE" ("coal Energy Application for Gas, Liquid and Electricity") at the Wakamatsu Research Institute of the Electric Power Development Co., Ltd. (J-POWER) (Japan). This pilot plant (**Figure 19**) was a test facility for developing coal gasification technologies with innovative

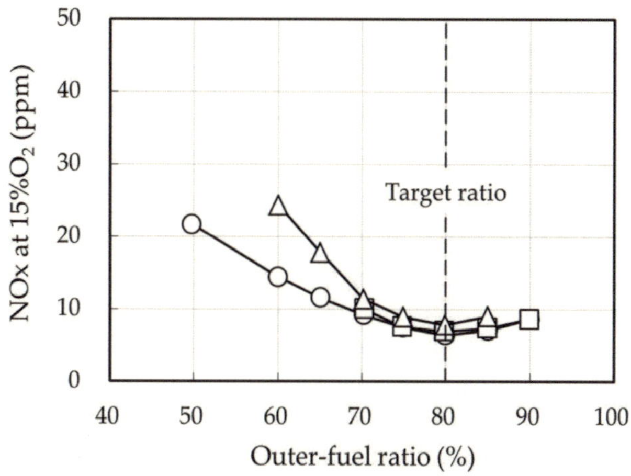

Figure 18. NOx emissions in single-can combustor test for convex multi-cluster combustor. Symbols: circles, CCS-0% (H_2 = 40 vol%); squares, CCS-30% (H_2 = 55 vol%); triangles, CCS-50% (H_2 = 65 vol%).

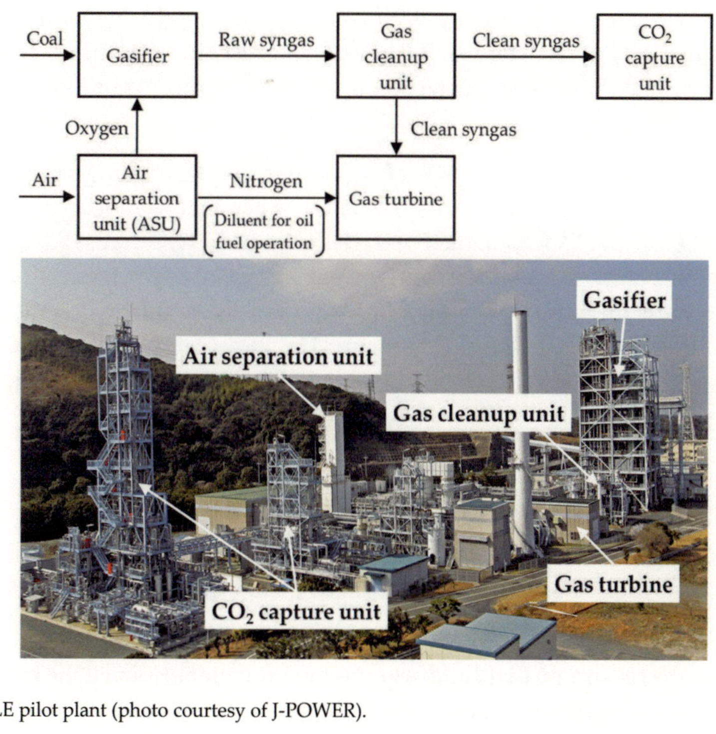

Figure 19. EAGLE pilot plant (photo courtesy of J-POWER).

CO_2 capture [28–32]. The five main components of the EAGLE plant were an ASU, a gasifier, a gas cleanup unit, a gas turbine, and a CO_2 capture unit. The ASU separated air into oxygen and nitrogen. Oxygen was supplied to the gasifier as an oxidant for the gasification process. Nitrogen was supplied to the gas turbine as a diluent for oil fuel operation. The gasifier converted coal to raw syngas by reacting it with oxygen. The gasifier employed an oxygen-blown, single-chamber, two-stage, swirling flow entrained bed gasification method. The gas cleanup unit removed impurities from the raw syngas, thus producing a clean syngas consisting mainly of CO, H_2, and N_2. The clean syngas was supplied separately to the gas turbine and the CO_2 capture unit. This separate syngas supply was employed because of the plant's operational circumstance that the test of the gas turbine combustor proceeded with the test of the CO_2 capture individually in the test series [32].

The gas turbine was an open simple-cycle/single-shaft type. It was originally equipped with a conventional diffusion-flame combustor with one oil fuel supplying system and one syngas fuel supplying system. The diffusion-flame combustor needed to inject diluent nitrogen into the combustion chamber to decrease NOx emissions. The diffusion-flame combustor on the gas turbine was replaced by the multi-cluster combustor with four additional syngas fuel supplying systems for the present tests. The multi-cluster combustor for the IGCC was developed for middle and small capacity gas turbines. The flat multi-cluster combustor was employed for the test because it was applicable to hydrogen-rich syngas fuels with intermediate hydrogen contents and its structural reliability was ensured by the simple structure of the flat perforated plate.

The syngas fuel burned in the tests was comprised mainly of CO, H_2, and N_2. The syngas fuel contained approximately 50 vol% CO, 20 vol% H_2, and 20 vol% N_2. Distillate oil was also burned for oil fuel operation. The EAGLE pilot plant test was conducted from startup on distillate oil to the maximum load (corresponding to 80% of the gas turbine load) on syngas produced in the test series.

The measuring systems consisted mainly of a gas analyzer, fluctuating-pressure-measuring system, and metal-temperature-measuring system. The gas analyzer measured the concentrations of NOx, CO, O_2, and CO_2 contained in the exhaust gas. The exhaust gas was sampled at multiple points in a cross section located in the exhaust duct downstream from the turbine. The fluctuating-pressure-measuring system measured pressure fluctuations at a point inside the combustion chamber on each can combustor. The metal-temperature-measuring system was equipped with thermocouples to measure metal temperatures of the liner and burner perforated plate [27].

4.3.2. Combustor performance at maximum load

This subsection evaluates combustor performance at a maximum gas turbine load of 80% [19]. The combustor operated with all the syngas-fueled burners in the final mode at the maximum load.

Figure 20 shows the maximum amplitudes of pressure fluctuations in all the cans at the maximum load versus the outer-fuel ratio. The amplitudes were maintained at low values well

http://dx.doi.org/10.5772/66742

below the criterion over the whole test range. This result demonstrated that the multi-cluster combustor achieved stable operation with low levels of pressure fluctuation amplitudes. The stable combustion performance was probably due to the stable lifted flames formed by the cluster burners.

Figure 21 shows the NOx emissions at the maximum load as a function of the outer-fuel ratio. The NOx decreased with increasing outer-fuel ratio until reaching the target ratio (80%); it yielded the minimum value at the target ratio, and then increased again with increasing outer-fuel ratio above the target ratio. The minimum NOx value was 10.9 ppm at the target ratio. The minimum NOx value at the target ratio was achieved by homogeneous lean combustion with a uniform equivalence ratio over the region in the main burners. The higher NOx values at outer-fuel ratios below and above the target ratio were due to the formation of high-temperature flames with a higher equivalence ratio in the inner region and the outer region, respectively.

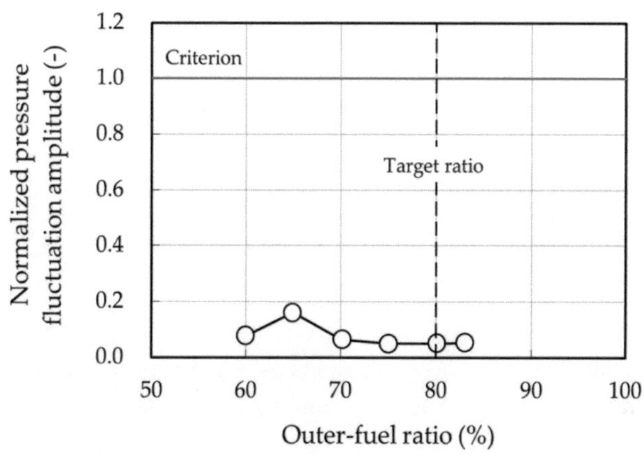

Figure 20. Pressure fluctuation amplitude at maximum gas turbine load of 80%.

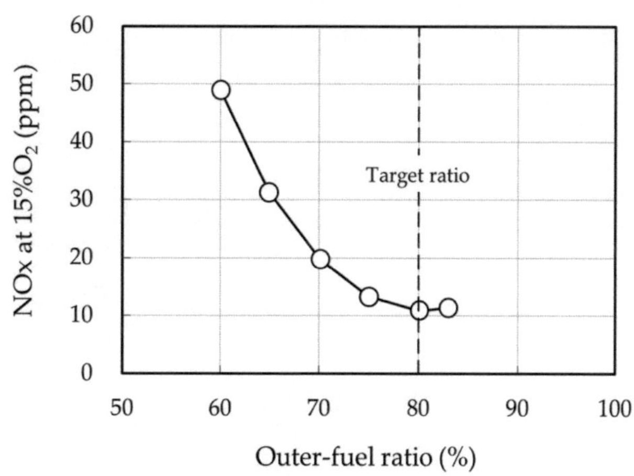

Figure 21. NOx emissions at maximum gas turbine load of 80%.

The multi-cluster combustor features diluent-free (dry), low NOx combustion. In order to demonstrate this feature, the dry low NOx performance of the multi-cluster combustor was compared with the diluent-controlled low NOx performance of the diffusion-flame combustor. **Figure 22** compares the NOx emissions for the multi-cluster combustor and the diffusion-flame combustor at the maximum load plotted against the normalized mass-flow-rate ratio of diluent nitrogen to syngas fuel [19]. The data for the multi-cluster combustor yielded the minimum NOx value of 10.9 ppm as attained at the target ratio. The data for the diffusion-flame combustor were obtained in tests of the combustor in the same plant. This figure plots the data in the operating range. The prediction curve represents predicted NOx values for the diffusion-flame combustor. This curve was predicted on a correlation between NOx and the stoichiometric flame temperature of a fuel/air/diluent mixture, which is representative of the actual flame temperature closely associated with the NOx formation rate in diffusion flames [33, 34]. This figure shows that the multi-cluster combustor achieved low NOx below around 10 ppm at a N_2/fuel ratio of zero (diluent-free (dry)). In contrast, the diffusion-flame combustor yielded much higher NOx around 200 ppm at a N_2/fuel ratio of zero, and needed diluent nitrogen to decrease NOx to the same level as that achieved by the multi-cluster combustor as a diluent-free condition. This comparison demonstrated that the multi-cluster combustor could feasibly achieve dry low NOx combustion of the syngas fuel in the IGCC pilot plant.

4.3.3. Combustor performance at part load

The combustor is required to operate stably from ignition through part load to the maximum load in practical IGCC plants. This subsection evaluates the performance of the combustor at part load in the plant [22].

Figure 23 shows the variations in NOx emissions as a function of the gas turbine load. From 0% load (full speed no load) to 30% load, the combustor operated on distillate oil with diluent

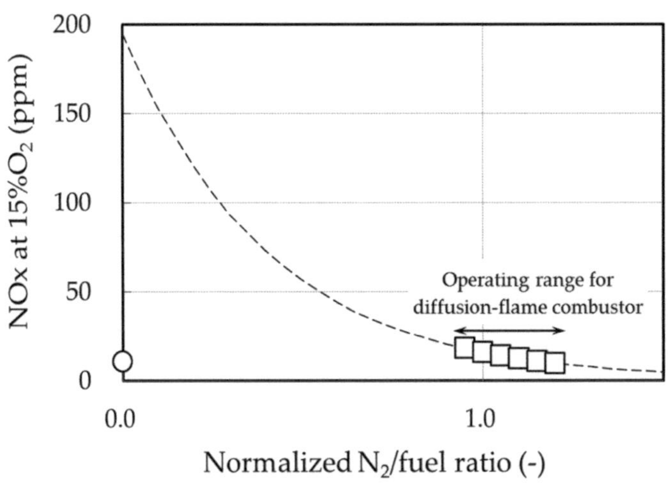

Figure 22. NOx comparison between multi-cluster combustor and diffusion-flame combustor at maximum gas turbine load. Symbols: circles, multi-cluster combustor; squares, diffusion-flame combustor; dotted line, prediction curve for diffusion-flame combustor.

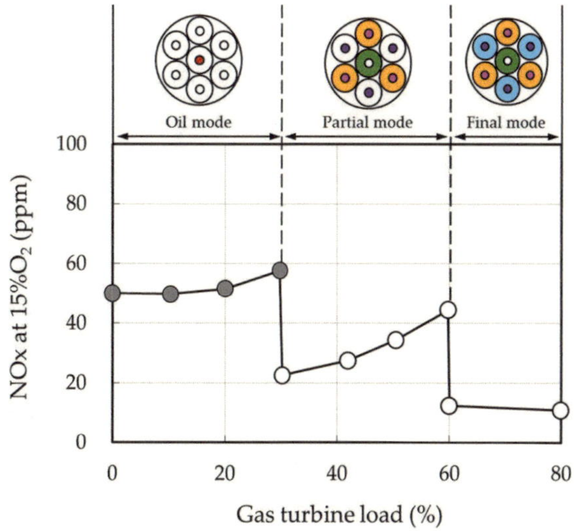

Figure 23. Variations in NOx emissions with gas turbine load.

nitrogen injection to control NOx in the oil mode. NOx increased with the load during oil operation between 0 and 30% load. At 30% load, the combustor switched from distillate oil to syngas. During this switching, the combustor disconnected the supply of diluent nitrogen for NOx control. NOx decreased from 58 to 23 ppm when the combustor switched the diluent-controlled oil operation to diluent-free syngas operation. From 30 to 60% load, the combustor operated on syngas in the partial mode. NOx increased with the load between 30 and 60% load. At 60% load, the combustor switched from the partial mode to the final mode, where the pilot burner and all the main burners were operating. NOx decreased from 44 to 12 ppm at this mode-switching load. This NOx decrease was due to the dispersion of fuel to all the burners. Finally from 60 to 80% load, the combustor operated on syngas in the final mode. NOx abruptly decreased at 60% load, and then gradually decreased with the load between 60 and 80% load. These results demonstrated that the multi-cluster combustor achieved dry low NOx of less than 12 ppm in the final mode above 60% load.

High values of combustion efficiency close to 100% indicate complete combustion, whereas low values of combustion efficiency indicate incomplete combustion, as manifested mainly in the form of CO emissions in the exhaust gas. Combustion efficiency is defined as the ratio of actual heat energy released in combustion to the theoretical heat energy available in fuel. In this study, the actual heat energy was calculated by subtracting the waste heat due to CO emissions in the exhaust gas from the theoretical heat energy. The theoretical heat energy was calculated as the heat liberated when fuel was completely burned [26].

Figure 24 shows combustion efficiency as plotted against the gas turbine load. The combustion efficiency periodically increased and decreased with the load. The combustion efficiency increased with the load in each mode, and decreased at each mode-switching load. The combustion efficiency decreased from 99.8 to 98.7% when switching from the oil mode to the partial mode at 30% load, and decreased from 99.9 to 99.5% when switching from the partial mode to the final mode at 60% load. In summary, the multi-cluster combustor attained combustion efficiencies over 99.1% on oil operation between 0 and 30% load, and attained combustion efficiencies over 98.7% on syngas operation between 30 and 80% load. This result demonstrated that the combustor attained high values of combustion efficiency during oil and syngas operation at part load.

Figure 25 shows the maximum pressure fluctuation amplitudes in all the cans as a function of the gas turbine load. The combustor maintained the amplitudes at values below the criterion

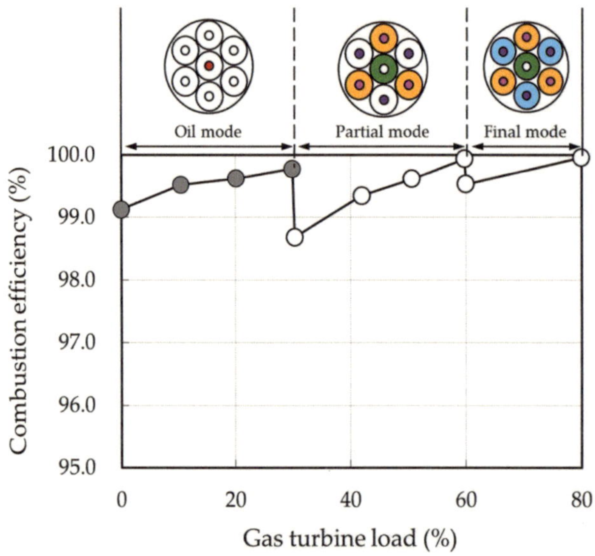

Figure 24. Variations in combustion efficiency with gas turbine load.

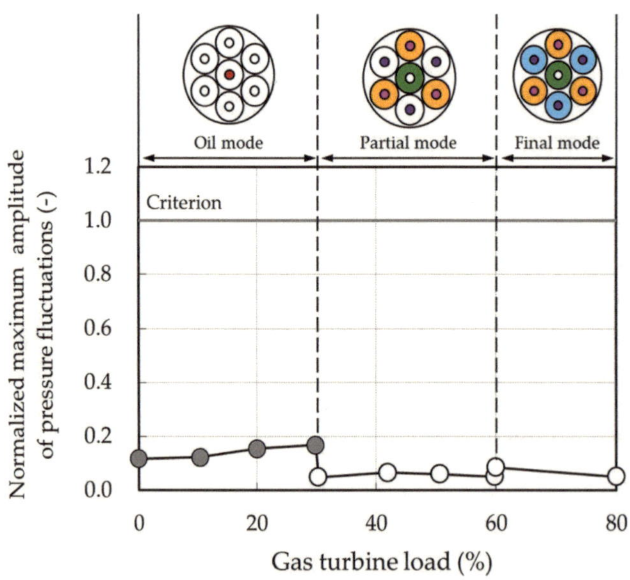

Figure 25. Variations in maximum amplitude of pressure fluctuations with gas turbine load.

over the entire load range. The result demonstrated that the combustor achieved stable operation with low pressure fluctuation amplitudes at part load.

Figure 26 shows the maximum values of all liner metal temperatures in all the cans as a function of the gas turbine load. The combustor maintained the liner metal temperatures at values below the criterion over the load range. The liner metal temperatures yielded the maximum values around the liner end tip.

The multi-cluster combustor achieved flashback-free combustion throughout part load. The test results demonstrated the feasibility of dry low NOx combustion of the practical syngas fuel in the IGCC pilot plant.

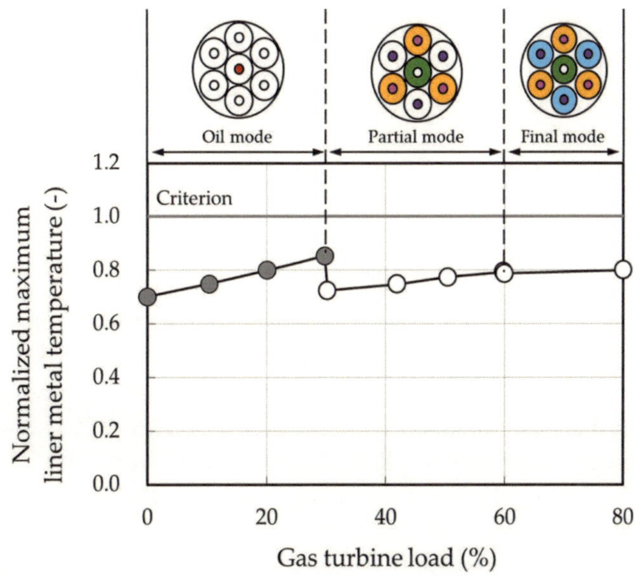

Figure 26. Variations in maximum liner metal temperature with gas turbine load.

5. Conclusions

This chapter described the development of the multi-cluster combustor as a state-of-the-art dry low NOx combustor for hydrogen-rich syngas fuels that can achieve both low NOx and high plant efficiency. The development approach consisted of three steps: burner development; combustor development; and feasibility demonstration for practical plants. This chapter focused mainly on the second and third steps. The main findings from these steps are summarized as follows.

- The single-can combustor test results showed that the convex combustor was effective in suppressing the occurrence of combustion oscillation.
- The test results in the IGCC pilot plant demonstrated the feasibility of the multi-cluster combustor for achieving dry low NOx combustion of the hydrogen-rich syngas fuel in the plant.

6. Next steps

6.1. IGCC demonstration test

On the basis of the experiences in the IGCC pilot plant test, a multi-cluster combustor was developed and installed on a middle capacity gas turbine in an oxygen-blown IGCC demonstration plant of the Osaki CoolGen Corporation, Japan [35–37]. The Osaki CoolGen project has been implemented as an "integrated coal gasification fuel cell combined cycle (IGFC) demonstration project" [37, 38]. This demonstration project is aiming at innovative low-carbon coal-fired thermal power generation by combining IGFC technology with innovative CO_2 capture technologies, thereby dramatically cutting CO_2 emissions from coal-fired thermal power plants. IGFC technology is expected to be an extremely efficient coal-fired thermal power generation technology. This demonstration project consists of three stages. The first stage is an oxygen-blown IGCC demonstration test. The project for the first stage is subsidized by the Ministry of Economy, Trade and Industry (METI) of Japan. The second stage is an oxygen-blown IGCC with CO_2 capture demonstration test. The final stage is a CO_2-capturing IGFC demonstration test. The project for the second and final stages is subsidized by the New Energy and Industrial Technology Development Organization (NEDO) of Japan.

6.2. Applications of multi-cluster combustors

The single-can combustor test results described in Section 4.2 showed that the multi-cluster combustors are capable of achieving dry low NOx combustion of hydrogen-rich fuels with hydrogen content to 65 vol%. Thus, the multi-cluster combustor is also applicable for by-product gases with the same range of hydrogen contents. **Figure 27** shows suitable hydrogen-rich fuels including O_2-blown IGCC/CCS syngas and by-product gases based on the hydrogen content and the volumetric lower heating value. By-product gases include coke oven gas (COG) from ironworks and off-gas from oil refineries. Use of by-product gases as a gas turbine fuel can offer low-cost power generation because it provides fuel cost economy.

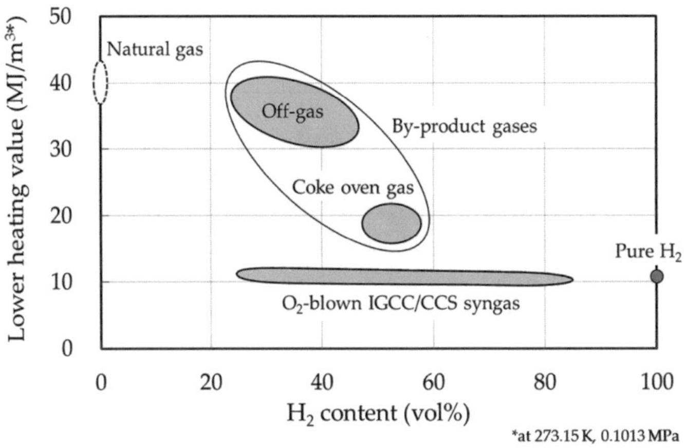

Figure 27. Suitable hydrogen-rich fuels.

The development of the multi-cluster combustors is expected to progress to application to by-product gases and further higher hydrogen content fuels (ultimately pure hydrogen) in order to expand the applicable hydrogen content range.

Acknowledgements

The multi-cluster combustors for the IGCC plant have been developed under the "Innovative Zero-Emission Coal Gasification Power Generation Project: Development of Low NOx Combustion Technology for High-Hydrogen Syngas in IGCC" by the New Energy and Industrial Technology Development Organization (NEDO) of Japan. The tests in the EAGLE plant were performed with the support of the Wakamatsu Research Institute of the Electric Power Development Co., Ltd. (J-POWER). The combustor development has been performed with the support of the Osaki CoolGen Corporation. The authors sincerely appreciate the valuable guidance and support offered by staff of NEDO, J-POWER and Osaki CoolGen Corporation. This chapter includes copyrighted materials of the American Society of Mechanical Engineers (ASME). The authors sincerely appreciate the ASME for granting permission to use them.

Author details

Tomohiro Asai[1]*, Yasuhiro Akiyama[1] and Satoschi Dodo[2]

*Address all correspondence to: tomohiro3_asai@mhps.com

1 Thermal Power Systems Research Department, Research & Development Center, Mitsubishi Hitachi Power Systems, Ltd., Hitachinaka, Japan

2 Hitachi Gas Turbine Engineering Department, Gas Turbine Technology & Products Integration Division, Mitsubishi Hitachi Power Systems, Ltd., Hitachi, Japan

References

[1] International Energy Agency (IEA). Key World Energy Statistics 2015. Paris: IEA; 2015.

[2] Metz B, Davidson O, Coninck H, Loos M, Meyer L, editors. Intergovernmental Panel on Climate Change (IPCC) Special Report on Carbon Dioxide Capture and Storage. Cambridge: Cambridge University Press; 2005. 442 p.

[3] Asai T, Koizumi H, Dodo S, Takahashi H, Yoshida S, Inoue H. Applicability of a Multiple-Injection Burner to Dry Low-NOx Combustion of Hydrogen-Rich Fuels. In: Proceedings of ASME Turbo Expo 2010; 14–18 June 2010; Glasgow, UK. 2010. pp. 183–192. DOI: 10.1115/GT2010-22286

[4] NEDO (New Energy and Industrial Technology Development Organization). Report (FY2004) in Clean Coal Technology Promotion Program: Investigation for Co-Production

System Based on Coal Gasification (in Japanese). Kawasaki, Japan: NEDO; 2005. Report No. 100005208 p.

[5] Lieuwen T, Yang V, Yetter R, editors. Synthesis Gas Combustion: Fundamentals and Applications. Boca Raton: CRC Press; 2009. 400 p.

[6] Lieuwen T, Yang V, editors. Gas Turbine Emissions. New York: Cambridge University Press; 2013. 382 p.

[7] York W D, Ziminsky W S, Yilmaz E. Development and Testing of a Low NOx Hydrogen Combustion System for Heavy-Duty Gas Turbines. Journal of Engineering for Gas Turbines and Power. 2013;**135**(2):022001-022001-8. DOI: 10.1115/1.4007733

[8] Wu J, Brown P, Diakunchak I, Gulati A, Lenze M, Koestlin B. Advanced Gas Turbine Combustion System Development for High Hydrogen Fuels. In: Proceedings of ASME Turbo Expo 2007; 14–17 May 2007; Montréal, Québec, Canada. 2007. pp. 1085–1091. DOI: 10.1115/GT2007-28337

[9] Reiss F, Reyser K, Griffin T. The ALSTOM GT13E2 Medium BTU Gas Turbine. In: Proceedings of ASME Turbo Expo 2002; 3–6 June 2002; Amsterdam, The Netherlands. 2002. pp. 705–712. DOI: 10.1115/GT2002-30108

[10] Littlejohn D, Cheng R K, Noble D R, Lieuwen T. Laboratory Investigations of Low-Swirl Injectors Operating With Syngases. Journal of Engineering for Gas Turbines and Power. 2009;**132**(1):011502-011502-8. DOI: 10.1115/1.3124662

[11] Roediger T, Lammel O, Aigner M, Beck C, Krebs W. Part-Load Operation of a Piloted FLOX® Combustion System. Journal of Engineering for Gas Turbines and Power. 2013;**135**(3):031503-031503-9. DOI: 10.1115/1.4007754

[12] Hollon B, Steinthorsson E, Mansour A, McDonell V, Lee H. Ultra-Low Emission Hydrogen/Syngas Combustion with a 1.3 MW Injector Using a Micro-Mixing Lean-Premix System. In: Proceedings of ASME Turbo Expo 2011; 6–10 June 2011; Vancouver, British Columbia, Canada.2011. pp. 827–834. DOI: 10.1115/GT2011-45929

[13] Funke H, Keinz J, Kusterer K, Haj Ayed A, Kazari M, Kitajima J, Horikawa A, Okada K. Experimental and Numerical Study on Optimizing the DLN Micromix Hydrogen Combustion Principle for Industrial Gas Turbine Applications. In: Proceedings of ASME Turbo Expo 2015; 15–19 June 2015; Montréal, Québec, Canada. 2015. p. V04AT04A008. DOI: 10.1115/GT2015-42043

[14] Horikawa A, Okada K, Kazari M, Funke H, Keinz J, Kusterer K, Haj Ayed A. Application of Low NOx Micro-Mix Hydrogen Combustion to Industrial Gas Turbine Combustor and Conceptual Design. In: Proceedings of International Gas Turbine Congress 2015 Tokyo; 15–20 November 2015; Tokyo, Japan. 2015. pp. 141–146.

[15] Oda T, Sakurazawa T, Kinoshita Y. The Development of Hydrogen Content Gas Combustion Technology for Kawasaki DLE Combustor. In: Proceedings of International Gas Turbine Congress 2015 Tokyo; 15–20 November 2015; Tokyo, Japan. 2015. pp. 292–297.

[16] Reichel T G, Goeckeler K, Paschereit O. Investigation of Lean Premixed Swirl-Stabilized Hydrogen Burner With Axial Air Injection Using OH-PLIF Imaging. Journal of Engineering for Gas Turbines and Power. 2015;**137**(11):111513-111513-10. DOI: 10.1115/1.4031181

[17] Alavandi S K, Etemad S, Baird B. Low Single Digit NOx Emissions Catalytic Combustor for Advanced Hydrogen Turbines for Clean Coal Power Systems. In: Proceedings of ASME Turbo Expo 2012; 11–15 June 2012; Copenhagen, Denmark. 2012. pp. 53–62. DOI: 10.1115/GT2012-68128

[18] Bolaños F, Winkler D, Piringer F, Griffin T, Bombach R, Mantzaras J. Study of a Rich/Lean Staged Combustion Concept for Hydrogen at Gas Turbine Relevant Conditions. In: Proceedings of ASME Turbo Expo 2013; 3–7 June 2013; San Antonio, Texas, USA. 2013. p. V01AT04A031. DOI: 10.1115/GT2013-94420

[19] Asai T, Dodo S, Karishuku M, Yagi N, Akiyama Y, Hayashi A. Performance of Multiple-Injection Dry Low-NOx Combustors on Hydrogen-Rich Syngas Fuel in an IGCC Pilot Plant. Journal of Engineering for Gas Turbines and Power. 2015;**137**(9):091504-091504-11. DOI: 10.1115/1.4029614

[20] Lefebvre A H, Ballal D R. Gas Turbine Combustion: Alternative Fuels and Emissions. 3rd ed. Boca Raton: CRC Press; 2010. 558 p.

[21] Glassman I, Yetter R, Glumac N. Combustion. 5th ed. Orlando: Academic Press; 2014. 774 p.

[22] Asai T, Dodo S, Karishuku M, Yagi N, Akiyama Y, Hayashi A. Part Load Operation of a Multiple-Injection Dry Low NOx Combustor on Hydrogen-Rich Syngas Fuel in an IGCC Pilot Plant. In: Proceedings of ASME Turbo Expo 2015; 15–19 June 2015; Montréal, Québec, Canada. 2015. p. V04AT04A026. DOI: 10.1115/GT2015-42312

[23] Asai T, Dodo S, Koizumi H, Takahashi H, Yoshida S, Inoue H. Effects of Multiple-Injection-Burner Configurations on Combustion Characteristics for Dry Low-NOx Combustion of Hydrogen-Rich Fuels. In: Proceedings of ASME Turbo Expo 2011; 6–10 June 2011; Vancouver, British Columbia, Canada. 2011. pp. 467–476. DOI: 10.1115/GT2011-45459

[24] Dodo S, Asai T, Koizumi H, Takahashi H, Yoshida S, Inoue H. Performance of a Multiple-Injection Dry Low NOx Combustor with Hydrogen-Rich Syngas Fuels. Journal of Engineering for Gas Turbines and Power. 2013;**135**(1):011501-011501-7. DOI: 10.1115/1.4006691

[25] Yunoki K, Murota T, Asai T, Okazaki T. Large Eddy Simulation of a Multiple-Injection Dry Low NOx Combustor for Hydrogen-Rich Syngas Fuel at High Pressure. In: Proceedings of ASME Turbo Expo 2016; 13–17 June 2016; Seoul, South Korea. 2016. p. V04BT04A056. DOI: 10.1115/GT2016-58119

[26] Asai T, Dodo S, Akiyama Y, Hayashi A, Karishuku M, Yoshida S. A Dry Low-NOx Gas-Turbine Combustor with Multiple-Injection Burners for Hydrogen-Rich Syngas Fuel: Testing and Evaluation of its Performance in an IGCC Pilot Plant. In: Proceedings of ASME 2013 Power Conference; 29 July–1 August 2013; Boston, Massachusetts, USA. 2013. p. V001T01A017. DOI: 10.1115/POWER2013-98122

[27] Asai T, Dodo S, Karishuku M, Yagi N, Akiyama Y, Hayashi A. Multiple-Injection Dry Low-NOx Combustor for Hydrogen-Rich Syngas Fuel: Testing and Evaluation of Performance in an IGCC Pilot Plant. Mechanical Engineering Journal. 2014;**1**(5):TEP0044 1–12. DOI: 10.1299/mej.2014tep0044

[28] Kimura N. EAGLE Project— Perspective on Coal Utilization Technology. In: Proceedings of APEC Clean Fossil Energy Technical and Policy Seminar; 26–29 January 2005; Cebu City, Philippine. 2005. pp. 1–12.

[29] NEDO and JCOAL (Japan Coal Energy Center). Clean Coal Technologies in Japan—Technological Innovation in the Coal Industry. Kawasaki, Japan: NEDO; 2006. 116 p.

[30] Nagasaki N, Takeda Y, Akiyama T, Kumagai T. Progress Toward Commercializing New Technologies for Coal Use — Oxygen-Blown IGCC+CCS. Hitachi Review. 2010; **59**(3):77–82.

[31] Omata K. Oxygen-Blown Coal Gasification System. Journal of the Japan Institute of Energy (in Japanese). 2014;**93**(7):624–630.

[32] Yamaguchi K. CO_2 Capture Technology in Coal Gasification Plant by Physical Absorption. In: The Thermal and Nuclear Power Generation Convention Collected Works (in Japanese); 16–18 October 2013; Hiroshima, Japan. Thermal and Nuclear Power Engineering Society; 2014. pp. 95–100. DOI: 10.14942/tenpes.10.95

[33] Chiesa P, Lozza G, Mazzocchi L. Using Hydrogen as Gas Turbine Fuel. Journal of Engineering for Gas Turbines and Power. 2005;**127**(1):73–80. DOI: 10.1115/1.1787513

[34] Gazzani M, Chiesa P, Martelli E, Sigali S, Brunetti I. Using Hydrogen as Gas Turbine Fuel: Premixed Versus Diffusive Flame Combustors. Journal of Engineering for Gas Turbines and Power. 2014;**136**(5):051504-051504-10. DOI: 10.1115/1.4026085

[35] Nagasaki N, Sasaki K, Suzuki T, Dodo S, Nagaremori F. Near-Zero-Emission IGCC Power Plant Technology. Hitachi Review. 2013;**62**(1):39–47.

[36] Dodo S, Karishuku M, Yagi N, Asai T, Akiyama Y. Dry Low-NOx Combustion Technology for Novel Clean Coal Power Generation Aiming at the Realization of a Low Carbon Society. Mitsubishi Heavy Industries Technical Review. 2015;**52**(2):24–31.

[37] Egusa K, Shiiya M. Outline of Osaki Coolgen Oxygen-Blown IGCC Demonstration Project and State of Construction – Working Progress for Demonstration Plant. Journal of the Gas Turbine Society of Japan (in Japanese). 2016;**44**(4):241–246.

[38] Aiso K. Outline of the Osaki Coolgen Project (IGCC Demonstration Project). In: Proceedings of the International Conference on Power Engineering-13 (ICOPE-13); 23–27 October 2013; Wuhan, Hubei, China. 2013. pp. 606–611.

Chapter 2

Modeling and Evaluation of a Coal Power Plant with Biomass Cofiring and CO_2 Capture

Dumitru Cebrucean, Viorica Cebrucean and Ioana Ionel

Additional information is available at the end of the chapter

http://dx.doi.org/10.5772/67188

Abstract

Coal-fired power plants are the largest source of anthropogenic carbon dioxide (CO_2) emissions into the atmosphere, with more than 9.5 billion tonnes of CO_2 emitted annually. In order to mitigate the emissions of CO_2 from coal-fired plants, several measures were proposed, such as increasing the efficiency of the plants, cofiring biomass with coal, and capturing and storing CO_2 deep underground. Among these measures, the use of biomass, which is considered one of the most cost-effective renewables and, in addition, carbon neutral, combined with CO_2 capture and storage will play an important role toward reducing the fossil-based CO_2 emissions. In this study, we investigated in detail the performances of pulverized coal combustion plants with direct cofiring of biomass and integrated with an amine-based postcombustion capture technology. All the systems were modeled and simulated using the process simulation software Aspen Plus. The results indicate that cofiring 10% of biomass in a coal-based power plant only slightly affects the energy performance of the plant, reducing the net efficiency by 0.3% points. The addition of an amine capture system to both the coal-fired and biomass cofiring plants further reduces the efficiency of the plants by more than 10% points. Analyzing the effect of various CO_2 capture process parameters on the heat, solvent and cooling water requirements, and on the overall plant performance, it was found that the concentration of amine in the solution is the most important parameter. The results showed that the net electrical efficiency increases for systems using higher amine concentrations. Further, we investigated the effect of systems with lower heat requirement for solvent regeneration on the plant gross/net power output and also analyzed the plant performances under a flexible CO_2 capture efficiency.

Keywords: pulverized coal combustion, biomass cofiring, postcombustion CO_2 capture, chemical absorption, monoethanolamine, Aspen Plus simulation

1. Introduction

Among fossil fuel power plants, coal-fired ones are the largest source of anthropogenic carbon dioxide (CO_2) emissions, emitting to the atmosphere more than 9.5 billion tonnes of CO_2 each year [1]. The emissions of CO_2 from coal-fired power plants can be reduced by increasing the efficiency of the plants (1% increase in efficiency reduces CO_2 by 2–3%), cofiring carbon neutral fuels (e.g., woody biomass), and/or capturing CO_2 and storing it in geological formations. The capture of CO_2 can be realized by means of three main approaches [2], namely: postcombustion capture, in which the CO_2 is separated from the flue gas after combustion; oxy-fuel combustion, the combustion takes place in nearly pure oxygen resulting in a flue gas stream consisting mainly of CO_2 and water from which the CO_2 can be easily separated; and precombustion, in which the CO_2 is removed from the fuel before combustion. Among these technologies, postcombustion CO_2 capture is the most advanced one that can be relatively easily integrated into existing or new power plants without altering the combustion process. Currently, postcombustion capture with chemical absorption using an aqueous amine (e.g., monoethanolamine, MEA) solution is the most selected process, offering high capture efficiency and selectivity.

Biomass, as one of the most cost-effective renewables and, in addition, carbon neutral, can also contribute toward reducing fossil-based CO_2 emissions. Furthermore, biomass cofiring in existing or new coal-based power plants is an effective means of producing electricity from biomass at higher conversion efficiencies and lower costs [3]. In addition, cofiring biomass with coal can reduce the emissions of sulfur dioxide (SO_2) and, in some cases, nitrogen oxides (NO_x) [4]. Currently, biomass cofiring is used in over 240 power plants (most of which are located in Europe) [5]. The majority of these plants employ direct cofiring in pulverized coal (PC) boilers. Fluidized bed (bubbling and circulating) boilers and grate-firing boilers have also been used. The share of biomass used in the fuel mix is, in most cases, less than 10% on energy basis.

The impact of cofiring different types of biomass on the technical and economic performances of the PC plants, with and without postcombustion capture, was investigated by a number of studies [6–10]. All studies suggest that the use of cofiring in PC combustion systems will have a negative impact on the overall technical and economic performances of the plants. For example, in the work conducted by the US DOE's National Energy Technology Laboratory [6], biomass (hybrid poplar) was directly cofired with bituminous coal (Illinois no. 6) at different ratios, ranging between 15% and 100% on a mass basis. The results showed that the net electrical efficiency decreases with the increase of cofiring ratio. For plants with 15, 30, and 60% biomass share in the fuel mix, the net efficiency decreases by approximately 0.2, 0.4, and 1.1% points, respectively, in comparison with the reference plant without cofiring, while for the case with 100% biomass firing the efficiency penalty is significant, i.e., 2.7% points. Further, it is worth mentioning that, because of the lower overall efficiency of cofiring plants, the total CO_2 emissions, expressed in kg/MWh of power generated, are higher. However, if biomass is considered as a carbon neutral fuel, then the net CO_2 emissions to the atmosphere decrease with the increase of biomass cofiring ratio. For example, a 550 MW_e (net) supercritical coal-fired power plant with a net efficiency of 40.7% (on a lower heating value (LHV) basis) has a carbon intensity of about 800 kg CO_2/MWh. For cases with cofiring, the total CO_2 emissions increase to 813 kg/MWh with 15% biomass cofiring, 826 kg/MWh with 30%, 866 kg/MWh with 60%, and to around 985 kg/MWh with 100% biomass firing. However, if the net CO_2 emissions are calculated, then they decrease to 746 kg/MWh with 15% biomass cofiring, 692 kg/MWh with 30%, 530 kg/MWh with 60%, and 0 kg/MWh for the case with 100% biomass firing.

Integration of amine capture systems to coal-based plants leads to significant energy penalties [11]. The efficiency of coal-fired as well as biomass cofiring plants can be reduced by more than 10% points as a result of CO_2 capture by means of, for example, a MEA-based chemical absorption process [6, 12–14]. The reduction of power output is mainly caused by the extraction of large quantities of steam from the steam cycle of the power plant for amine solvent regeneration (~65% of total energy penalty) and the auxiliary power consumption for the compression of the CO_2 product (~25%) [15]. As mentioned earlier, compared to coal-fired plants, the plants with cofiring depending on the cofiring ratio, have efficiencies up to 1% points lower. In addition to this, the overall energy penalty in cofiring plants with CO_2 capture is higher, and the resulting cost of electricity is also higher than that of coal-fired plants [6, 10].

Although several studies investigated the impact of amine-based postcombustion CO_2 capture on the energy performance of cofiring plants, none of them investigated the effect of the CO_2 capture process parameters on the performances of the cofiring plants. In this study, we investigated in detail the technical performances of PC power plants with direct cofiring of biomass and integrated with an amine-based postcombustion capture technology. Aspen Plus process simulation tool was used for the modeling and simulation. First, we estimated the performances of coal-fired plants with/without cofiring and with/without CO_2 capture. Then we analyzed the effect of various CO_2 capture process parameters on the heat duty of the reboiler, solvent flow rate necessary to capture 90% of the CO_2 from the flue gas, and cooling water requirements. Further, we investigated the effect of absorption processes with lower heat requirement for solvent regeneration on the gross and net power output of the plants, and also analyzed the plant performances under a flexible CO_2 capture efficiency operation.

2. Process description

2.1. Feedstock

The composition and heating values of the fuels used in this study are shown in **Table 1**. Bituminous coal (Illinois no. 6) was selected as the base fuel. It has a lower heating value of ~29.5 MJ/kg (dry basis, db), a low moisture content of 11.1% (as received, ar), and a relatively high ash content of 10.9% (db). It is further characterized by having high sulfur content of 2.8% (db). The biomass selected for cofiring cases is hybrid poplar with a moisture content of 50% (ar) and a lower heating value of about 18.5 MJ/kg (db). Its ash (1.5% db) and sulfur (0.03% db) contents are very low. In this study, we assumed that hybrid poplar prior to be fed into the boiler was dried to 10% using a fluidized-bed drying system [6].

	Bituminous coal Illinois no. 6 [13]	Hybrid poplar [6]
Moisture (% ar)	11.12	50.00
Ash (%)	10.91	1.48
Carbon (%)	71.73	52.36
Hydrogen (%)	5.06	5.60
Oxygen (%)	7.74	40.16
Nitrogen (%)	1.41	0.37
Chlorine (%)	0.33	–
Sulfur (%)	2.82	0.03
Higher heating value (MJ/kg)	30.53	19.63
Lower heating value (MJ/kg)	29.45	18.46

Table 1. Composition and heating values of coal (Illinois no. 6) and biomass (hybrid poplar) used in this study (by weight, db).

2.2. Power plant

The power plant is a supercritical PC-fired power plant designed to operate with main steam conditions at 242 bar/593°C and steam reheating at 42.4 bar/593°C. In this study, the reference plant (coal-fired, without carbon capture) generates 550 MW_e net power at an efficiency of about 40.7% (LHV).

Figure 1 shows a simplified layout of the power plant without CO_2 capture. From the boiler, the flue gas is sent to a gas cleaning system consisting of a selective catalytic reduction (SCR) for NO_x control, a baghouse (BH) for fly ash removal, and a limestone-based flue gas desulfurization (FGD) unit for SO_x removal. An air preheater (APH) is placed between the SCR and BH units. The primary air after exiting the APH is mixed with the fuel prior to entering the boiler while the secondary air was directly sent to the boiler.

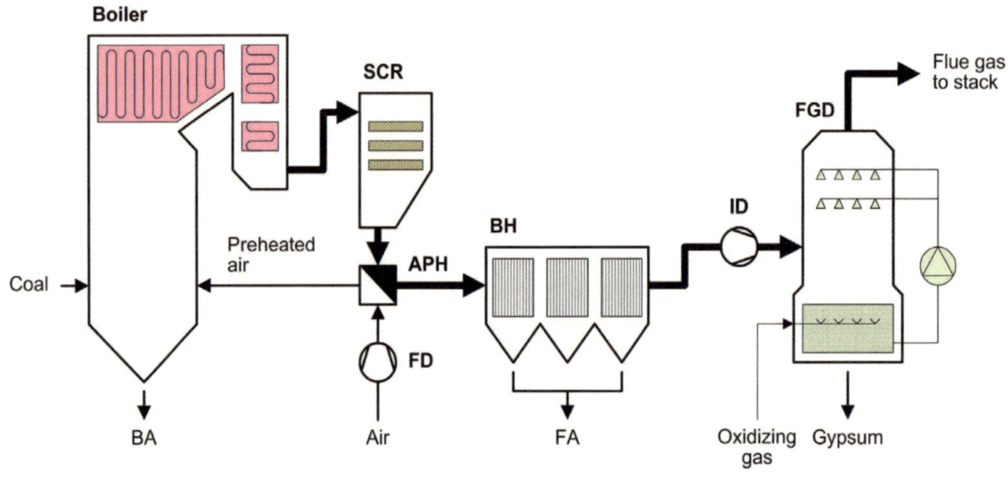

Figure 1. Layout of a pulverized coal-fired power plant (adapted from [16]).

http://dx.doi.org/10.5772/67188

Aspen Plus software was used for the modeling of the whole plant, which included the boiler and the flue gas cleaning section (control of NO_x, ash, and SO_x), the steam cycle, and the capture and compression process.

For the modeling of the boiler and flue gas cleaning section, the PR-BM (Peng-Robinson equation of state with Boston-Mathias alpha function) property method was selected. The following unit operation blocks were used in developing the model: the boiler consisted of an RYIELD block for the fuel decomposition, RGIBBS for the fuel combustion, and several HEATER blocks were used for the steam generation; the flue gas cleaning system was mainly modeled by means of SSPLIT and SEP blocks. The mass flow rate of fuel fed into the boiler, the amount of air required for the combustion, infiltration air, the heat transfer, and several other process parameters were controlled by means of Design Specs.

Figure 2 shows the flow diagram of the power plant steam cycle considered in this study. As shown, the turbine consists of a high pressure (HP), an intermediate pressure (IP), and a low pressure (LP) turbine, all connected to the generator by a common shaft. The main steam from the boiler enters the high pressure steam turbine (HPST) at a pressure of 242 bar and a temperature of 593°C. From the HPST, the steam is reheated in the boiler to 593°C and introduced to the intermediate pressure steam turbine (IPST) at a pressure of 45.2 bar. In the IPST, the steam is expanded to a pressure of 9.3 bar and then sent to a low pressure steam turbine (LPST), where it is further expanded to the condenser pressure of 0.069 bar. For feedwater preheating, five low pressure feedwater heaters (LPFWHs) are used, including the deaerator, and three high pressure feedwater heaters (HPFWHs). The conditions of the feedwater before entering the boiler are 288 bar and 291°C.

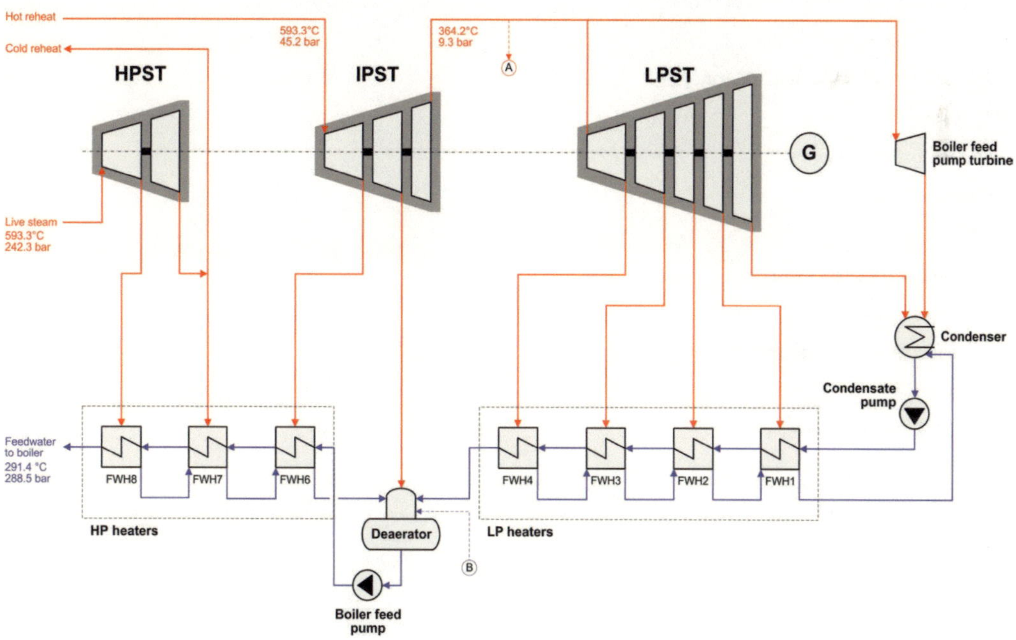

Figure 2. Diagram of the supercritical steam cycle (A: steam extraction point for solvent regeneration; B: reboiler condensate reinjection point in carbon capture cases) (adapted from [16]).

The STEAMNBS (NBS/NRC steam table) property method was selected to model the water-steam cycle in Aspen Plus. The flowsheet was built mainly with the COMPR and HEATER blocks. For example, to model a HPST, two COMPR units were connected in series. The stream exiting the first unit was split into two, one directed to the second unit, and the other one was sent for feedwater preheating. The amount of steam extracted for feedwater preheating, the steam required to drive the boiler feed pump turbine, the steam required for the solvent regeneration and condensate used for desuperheating in carbon capture cases, and the cooling water requirement in the condenser were controlled within the program by means of several Design Spec functions.

The assumptions used for the modeling and simulation of the power plant are presented in **Table 2**.

Parameter	Value
Boiler section:	
Operating pressure (bar)	1.01
Boiler efficiency (% LHV)	91.2
Primary/secondary air (%)	23.5/76.5
Infiltration air (% of FG exiting the APH)	1.6
Air leakage in APH (%)	5.5
Ash distribution, BA/FA (%)	20/80
FG outlet temperature, boiler/APH (°C)	350/170
O_2 content in FG at the APH outlet (mol%)	2.5
PA/FD/ID fans pressure ratio (−)	1.10/1.04/1.08
Fans isentropic/mechanical efficiency (%)	80/95
BH:	
FA removal efficiency (%)	100
Pressure drop (bar)	0.014
FGD:	
SO_2 removal efficiency (%)	98
Limestone purity (wt%)	80.4
Limestone slurry, solid/liquid (%)	30/70
Excess sorbent for SO_2 removal (%)	4
Excess O_2 for oxidation (%)	135
Pressure drop (bar)	0.034
FG outlet temperature (°C)	57.2
Steam cycle:	
Live steam pressure/temperature (bar/°C)	242.3/593.3
Reheated steam pressure/temperature (bar/°C)	45.2/593.3
IP/LP crossover pressure (bar)	9.3
Condenser pressure (bar)	0.069

Parameter	Value
Boiler FW temperature (°C)	291.4
ST isentropic efficiency, HP/IP/LP (%)	83/88/93
ST mechanical efficiency (%)	99
Generator efficiency (%)	98.5
FW and condensate pump hydraulic/mechanical efficiency (%)	85/99.6
Steam extraction pressures for FW preheating (bar)	76.9/49.0[(a)], 21.4/9.5[(b)], 5.01/1.32/0.58/0.24[(c)]
Number of FWHs, HP/LP (including deaerator) (–)	3/5
Temperature difference in FWHs (hot outlet—cold inlet) (°C)	5.56
Pressure drop in FWHs (cold-side) (bar)	0.34

[(a)]Steam from HPST to FWHs 8 and 7.
[(b)]Steam from IPST to FWH 6 and deaerator.
[(c)]Steam from LPST to FWHs 4-1.

Table 2. Main assumptions for the simulation of the reference plant (without CO_2 capture) [6, 12, 13, 16, 17].

The power requirements of various subsystems in the power plant, such as solids handling and processing, emission control and other plant auxiliaries, are given in **Table 3**.

Parameter	Value
Coal handling and milling (kWh/t coal)	17
Biomass handling, processing, and drying (% of biomass heat input)	3.1
Ash handling (kWh/t TA)	30
SO_2 sorbent handling and reagent preparation (kWh/t limestone slurry)	15
BH unit (kWh/t FA removed)	5
FGD unit (kWh/t SO_2 removed)	325
ST auxiliaries (MW_e)	0.4
Condenser auxiliaries (% of heat rejected)	1.2[(a)]
Miscellaneous BOP (MW_e)	2[(b)]
Transformer losses (% of gross power)	0.3

[(a)]Includes power consumption by circulating water pumps, ground water pumps, and cooling tower fans.
[(b)]Includes power consumption by plant control systems, lighting, heating, ventilating, and air conditioning.

Table 3. Power consumption of various subsystems in the power plant [6, 13].

The power plant model developed in Aspen Plus was validated against the data from the NETL studies [6, 13]. The validation of the steam cycle parameters (i.e., mass flow rates, temperatures, and pressures of steam) is shown in **Figure 3**. As can be observed, the results are in good agreement with the reference [13]. The calculated steam turbine power output was

589.21 MW_e, and the gross power output was 580.37 MW_e, which is very close to the reference value of 580.40 MW_e. **Table 4** compares the simulation results on the performance of coal-fired and biomass cofiring plants with the reference data [6, 13]. The deviation between the reference and the present results for the coal-fired case is insignificant. For the biomass cofiring case, with 10% heat input, the results are within the values reported in the reference [6].

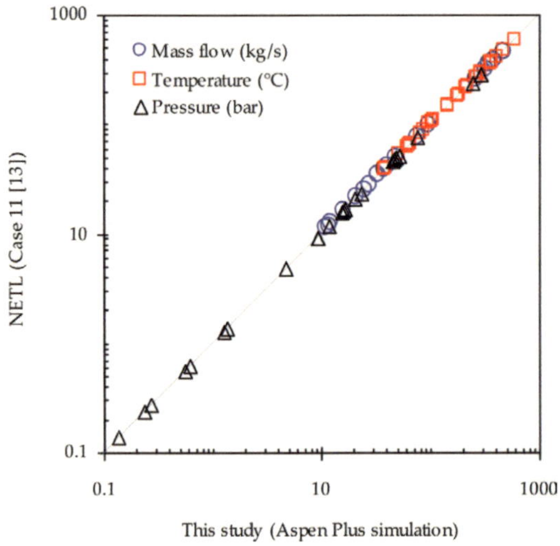

Figure 3. Validation of the water/steam parameters of the supercritical steam cycle.

2.3. CO_2 capture and compression

A simplified process flow diagram of the MEA-based chemical absorption process for CO_2 capture used in this study is shown in **Figure 4**. The flue gas after pretreating in a direct contact cooler (DCC), with reduced temperature and low impurities level, enters the absorber column where it contacts, countercurrently, with the aqueous amine solution (30 wt% MEA and 0.25 mol CO_2/mol MEA loading) introduced from the top of the column. The CO_2 from the flue gas reacts with the absorbent forming a CO_2-rich solution (~0.49 mol CO_2/mol MEA), which is then pumped to the desorber column via a lean/rich heat exchanger (HX). The clean

Case study	Coal-fired		Biomass cofiring		
	NETL[(a)]	This study	NETL[(b)]	This study	NETL[(c)]
Biomass share (% heat input)	0	0	6	10	13
Net plant efficiency (% LHV)	40.73	40.67	40.54	40.35	40.32
Fuel consumption (kg/MWh)	337.7	338.2	375.8	403.5	423.4
CO_2 emissions (kg/MWh)	802.0	803.0	812.6	821.4	825.8

[(a)]Case 11 [13].
[(b)]Case PN4 [6].
[(c)]Case PN3 [6].

Table 4. Comparison of plant performance without CO_2 capture.

flue gas exits the absorber and is further washed in a water washing section in order to remove any amine residues. In the desorber, the CO_2 is released from the liquid absorbent as a result of the heat provided by the LP steam in the reboiler. The CO_2-lean solution is then sent to the absorber column for the next cycle. The CO_2 product stream from the desorber column is further compressed, dehydrated, and transported to a storage site. The main assumptions used to model and simulate the capture and compression process are presented in **Table 5**.

In Aspen, the ELECNRTL (electrolyte nonrandom two-liquid model with Redlich-Kwong equation of state) property method was selected for the simulation of the absorption process. The chemical reactions taking place during the absorption process are as follows:

$$MEA^{+} + H_2O \leftrightarrow MEA + H_3O^{+} \tag{R1}$$

$$CO_2 + 2\,H_2O \leftrightarrow H_3O^{+} + HCO_3^{-} \tag{R2}$$

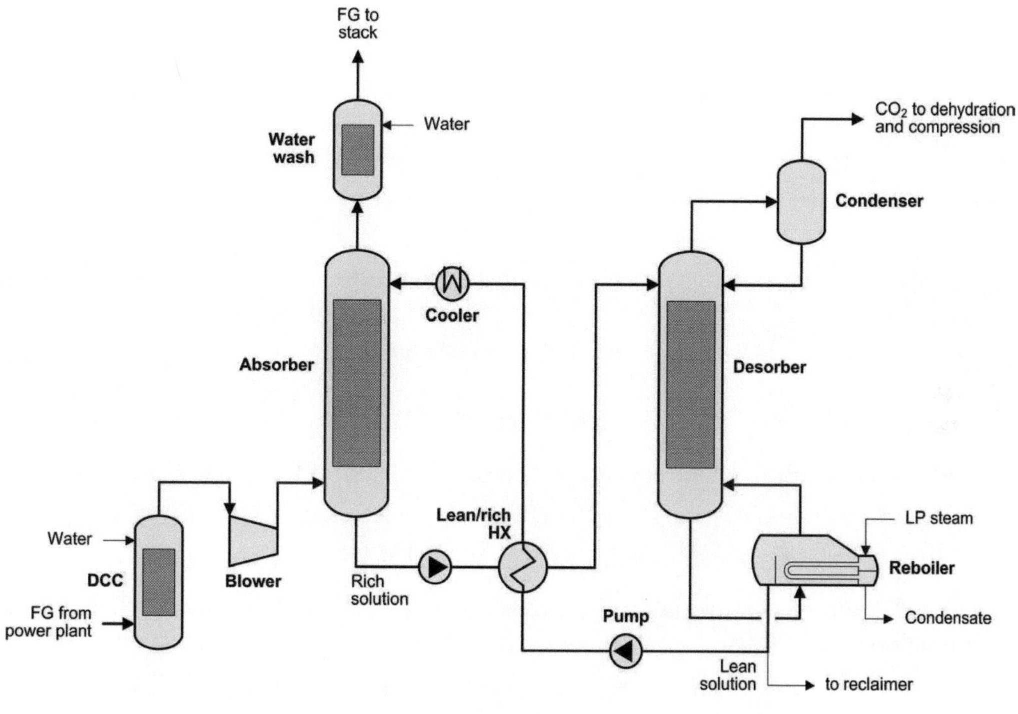

Figure 4. Process flow diagram of the chemical absorption process for CO_2 capture from flue gas.

Parameter	Value
CO_2 capture:	
CO_2 removal rate (%)	90
MEA concentration (wt%)	30
CO_2 loading, lean/rich (mol CO_2/mol MEA)	0.25/~0.49[(a)]

Parameter	Value
FG/lean solvent temperature at the absorber inlet (°C)	~45[(a)]/40
Lean/rich HX temperature difference (hot outlet—cold inlet) (°C)	5
Reboiler temperature difference (°C)	10
Operating pressure, absorber/desorber (bar)	1.0/1.7
Pressure drop, absorber/desorber (bar)	0.14/0.20
Number of equilibrium stages, absorber/desorber (–)	18/12
Booster fan pressure ratio (–)	1.1
Booster fan isentropic/mechanical efficiency (%)	85/95
CO_2 compression:	
Final delivery pressure/temperature (bar/°C)	110/30
Number of compression stages (–)	7
Compressor pressure ratio (–)	1.8
Compressor isentropic/mechanical efficiency (%)	80/95
Intercoolers outlet temperature (°C)	30
Intercoolers pressure drop (% of inlet stream)	2

[(a)]Calculated.

Table 5. Main assumptions for the simulation of the CO_2 capture and compression process [13, 14, 16, 18].

$$HCO_3^- + H_2O \leftrightarrow H_3O^+ + CO_3^{2-} \tag{R3}$$

$$MEACOO^- + H_2O \leftrightarrow MEA + HCO_3^- \tag{R4}$$

$$2\,H_2O \leftrightarrow H_3O^+ + OH^- \tag{R5}$$

For these reactions, the equilibrium constants were calculated with the following equation:

$$\ln(K_{ech}) = A + B/T + C\ln(T) + DT \tag{1}$$

in which, T is the temperature (K) and the coefficients A, B, C, and D are given in **Table 6** for each reaction.

To determine the loading of the solution (mol CO_2/mol MEA), Eq. (2) was applied:

$$\text{Loading} = \frac{[CO_2] + [HCO_3^-] + [CO_3^{2-}] + [MEACOO^-]}{[MEA] + [MEA^+] + [MEACOO^-]} \tag{2}$$

where the components in the numerator represent moles of all CO_2 species in the solution, whereas the components in the denominator represent moles of all MEA species.

The following unit operation blocks were used to develop the process flowsheet: RADFRAC columns were used to model the absorber and desorber columns using 18 and 12 equilibrium

Reaction no.	*A*	*B*	*C*	*D*
1	−3.038325	−7008.357	0	−0.00313489
2	231.465	−12092.1	−36.7816	0
3	216.049	−12431.7	−35.4819	0
4	−0.52135	−2545.53	0	0
5	132.899	−13445.9	−22.4773	0

Table 6. Coefficients from Eq. (1) for the calculation of the equilibrium constants in the CO_2-MEA system.

stages, respectively [16]. For the modeling of the flue gas blower a COMPR block was used. For cooling and heating purposes, several HEATER blocks were employed. The desorber condenser was modeled with a FLASH2 block.

For the development of the compression model, three unit operation blocks were used, namely, COMPR for compression, HEATER for cooling of the product stream, and FLASH2 for excess liquid removal. As specified in **Table 5**, the CO_2 product stream from the capture unit was compressed to 110 bar in a multistage compressor using seven compression stages with intercooling to 30°C.

The modeling results of the capture and compression model are presented in **Table 7**. These are compared with other sources. As can be seen, the results are in good agreement with the values reported in the open literature for conventional absorption/desorption processes with 30 wt% MEA. In this study, the minimum reboiler heat duty of 3.5 MJ/kg CO_2 captured was obtained for the lean loading of 0.25 mol CO_2/mol MEA. The solution leaving the absorber column had a loading of 0.49 mol CO_2/mol MEA. In the simulation, the liquid to gas mass flow rate ratio used was about 3.9, and the lean solvent requirement was about 20 kg/kg CO_2 captured. The total CW needed to cool (i) the FG before entering the capture unit, (ii) the lean solution after exiting the lean/rich HX, and (iii) the CO_2 product stream in the compression train was estimated at about 71 kg/kg CO_2 captured. The specific energy requirement was estimated at about 110 kWh/kg CO_2 captured of which more than 75% were consumed by the compression unit. Furthermore, it was found that the specific steam used for solvent regeneration was 1.45 kg steam/kg CO_2 captured, which is in agreement with the values reported in [13, 21]. For example, in reference [21] about 1.42 kg steam/kg CO_2 captured were used for the case with the steam extracted at 9 bar from the IP/LP crossover pipe and 1.47 kg steam/kg CO_2 for the case with steam extracted at 3 bar from the LPST.

2.4. Integration of CO_2 capture with power plant

The amine capture unit requires significant amounts of energy for solvent regeneration. This energy is usually provided by the steam extracted from the main power plant. It can also be delivered by, for example, an additional boiler in which steam is generated at sufficient quality and quantity necessary for regeneration [9, 22]. However, this measure would be more costly than that of direct extraction from the plant. In this study, the required steam for solvent regeneration is extracted at 9.3 bar from the crossover pipe between the

Study	This study	Abu-Zahra et al. [19] [(a)]	CAESAR [14]	Liu et al. [20]
L/G mass flow rate ratio (–)	3.87–3.92	3.48/4.83	4.05	2.75
Lean loading (mol CO_2/mol MEA)	0.25	0.24/0.32	0.26	0.23
Rich loading (mol CO_2/mol MEA)	0.49	0.48/0.49	0.48	0.54
Reboiler heat duty (MJ/kg CO_2 captured)	3.5	3.89/3.29	3.73	4.6
Lean solvent requirement (kg/kg CO_2 captured)	20	19.3/26.9	21.8	15.7
CW requirement (kg/kg CO_2 captured)	70.1–71.4	106/103	61.7	
Power consumption (kWh/kg CO_2 captured)	109		129.0	84.4[(b)]

[(a)] Values refer to the baseline/optimum case.
[(b)]Only the energy used for CO_2 product compression.

Table 7. Main parameters of the capture and compression process for 90% CO_2 capture with 30 wt% MEA.

intermediate pressure and low pressure turbine. It is first expanded in an auxiliary turbine, desuperheated, and then enters the reboiler (**Figure 5**). Since the MEA solvent is regenerated at ~121°C, and the reboiler temperature approach is assumed to be 10°C, the conditions of the saturated steam before entering the reboiler are 132°C/2.86 bar. As will be further shown in this study, depending on the MEA concentration and other process conditions, approximately half of the steam from the IP/LP crossover will be extracted, and thus significantly reducing the gross power output. From the reboiler, the resulting condensate is pumped to 9.2 bar to the deaerator. The reboiler condensate can also be returned to one of the LPFWHs, provided that the temperature level is close to that of the condensate. In reference [16], with the same steam cycle, it was shown that the most appropriate location for the condensate reinjection is the deaerator.

2.5. Plant performance indicators

The performances of plants with/without cofiring and with/without MEA-based postcombustion CO_2 capture were evaluated using the following plant performance indicators:

Net plant efficiency, η_{net} (%):

$$\eta_{net} = \frac{W_{net}}{\dot{m}_C \cdot LHV_C + \dot{m}_B \cdot LHV_B} \quad (3)$$

Efficiency penalty due to cofiring and/or carbon capture, $\Delta\eta_{net}$ (% points):

$$\Delta\eta_{net} = \eta_{net,ref} - \eta_{net,cofiring\ and/or\ CCS} \quad (4)$$

Specific fuel consumption, SC_{fuel} (kg/MWh):

$$SC_{fuel} = \frac{(\dot{m}_C + \dot{m}_B) \cdot 3600}{W_{net}} \quad (5)$$

Specific CO_2 emissions, SE_{CO_2} (kg/MWh):

$$SE_{CO_2} = \frac{\dot{m}_{CO_2} \cdot 3600}{W_{net}} \quad (6)$$

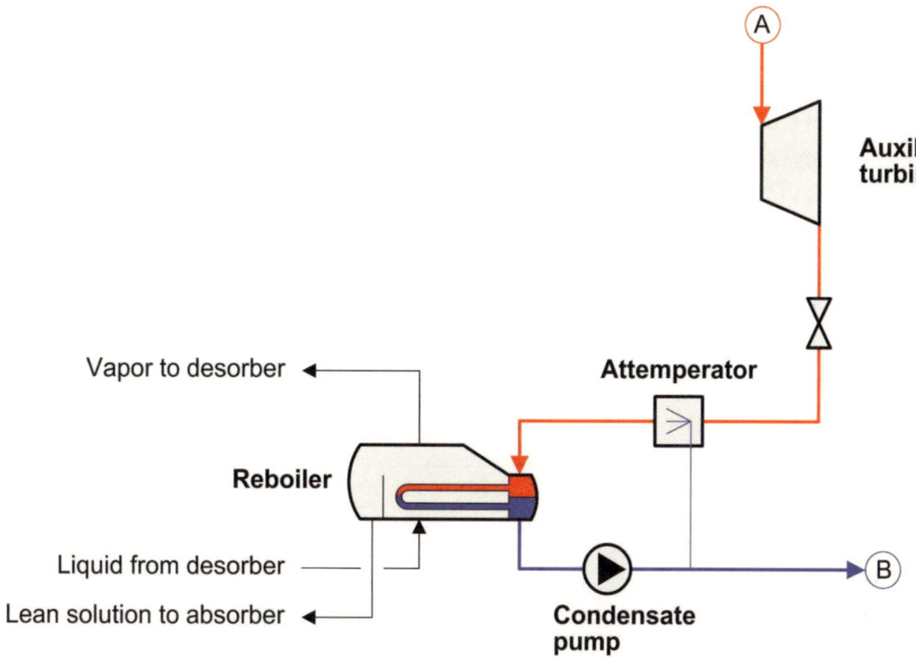

Figure 5. Integration of the steam with the stripper reboiler.

Net CO_2 emissions, NE_{CO_2} (kg/MWh):

$$NE_{CO_2} = \frac{\dot{m}_{CO_2} \cdot 3600}{W_{net}} \tag{7}$$

Here, $\dot{m}_C$ is the flow rate of coal entering the plant (kg/s), $\dot{m}_B$ is the flow rate of raw biomass entering the plant in the case with cofiring (kg/s), LHV_C and LHV_B are the lower heating values of coal and, respectively, biomass (MJ/kg), $\eta_{net,ref}$ is the net efficiency of the reference plant, without cofiring and without CO_2 capture (%), $\eta_{net,cofiring\ and/or\ CCS}$ is the net efficiency of the plant with biomass cofiring and/or with CO_2 capture (%), $\dot{m}_{CO_2}$ is the total flow rate of CO_2 generated (kg/s), $\dot{m}_{CO_2,C}$ is the flow rate of CO_2 generated only from coal combustion (kg/s), and W_{net} is the plant net power output (MW_e), which is obtained after subtracting the plant auxiliary power consumption.

3. Results and discussion

3.1. Performance of coal-fired and biomass cofiring plants with CO_2 capture

Simulation results of the investigated plants with/without biomass cofiring and with/without MEA-based postcombustion carbon capture are summarized in **Table 8**. As can be seen, the reference coal-fired power plant has a net electrical efficiency of 40.67% and releases 803 kg CO_2/MWh. The results further show that the performances of the plant with cofiring are slightly derated in comparison with the coal-fired plant. Cofiring 10% of biomass in a supercritical coal-based plant leads to a reduction in efficiency to 40.35% (i.e., 0.33% points efficiency penalty compared with coal-fired case). This reduction is mainly attributed to the fact that the biomass fuel considered in this study has a lower calorific value and significantly higher moisture content, and the energy needed for its processing and drying is substantial. However, one can also note that cofiring biomass has a positive effect on the ash and SO_2 flow rates reducing the power demand of subsystems associated with their removal. The specific CO_2 emissions from the cofiring plant are estimated at 821.4 kg CO_2/MWh. But if considering only the emissions resulted from coal combustion, then they decrease to around 740 kg CO_2/kWh.

The addition of a MEA-based postcombustion CO_2 capture system significantly reduces the energy performance of both plants. For the coal-fired power plant with 90% CO_2 capture rate, the net efficiency drops to 30.47% (i.e., an efficiency penalty of 10.21% points with respect to

CO_2 capture	**No**		**Yes**	
Cofiring	**No**	**Yes**	**No**	**Yes**
Fuel input:				
Coal (kg/s ar)	51.69	46.60	51.69	46.60
Biomass (kg/s ar)		14.67		14.67
Heat input (MW_{th} LHV)	1352.80	1354.98	1352.80	1354.98
Power generated/consumed:				
ST output (MW_e)	580.37	580.37	485.07	483.26
Coal handling and milling (MW_e)	−3.16	−2.85	−3.16	−2.85
Biomass handling, processing and drying (MW_e)		−4.20		−4.20
PA/FD/ID fans (MW_e)	−9.83	−9.91	−9.83	−9.91
BH and ash handling system (MW_e)	−0.61	−0.57	−0.61	−0.57
FGD and limestone handling/reagent preparation (MW_e)	−3.89	−3.51	−3.89	−3.51
CO_2 capture and compression (MW_e)			−43.03	−43.69
Condensate pumps (MW_e)	−0.80	−0.80	−0.42	−0.41
Condenser auxiliaries (MW_e)	−7.65	−7.65	−8.07	−8.14
Miscellaneous BOP, ST auxiliaries and transformer losses (MW_e)	−4.20	−4.20	−3.90	−3.90
Total auxiliary consumption (MW_e)	−30.14	−33.69	−72.93	−77.19
Overall plant performance:				
Net power output (MW_e)	550.22	546.68	412.14	406.07
Net plant efficiency (% LHV)	40.67	40.35	30.47	29.97
Efficiency penalty (% points)		0.33	10.21	10.70
Specific fuel consumption (kg/MWh)	338.2	403.5	451.5	543.2
Specific CO_2 emissions (kg/MWh)	803.0	821.4	107.2	110.6
Net CO_2 emissions (kg/MWh)	803.0	739.3	107.2	0

Table 8. Performance of PC plants with/without cofiring, with/without CO_2 capture.

the reference plant) while for the cofiring plant the net efficiency decreases to 29.97% (i.e., an efficiency penalty of 10.7% points) after integrating the CO_2 capture and compression process. As can be noted from the table, the capture and compression process consumes more than 55% of the total auxiliary load. The results further show that in order to generate the same amount of energy, the systems with carbon capture should use 35% more fuel than the reference plant without capture. The CO_2 emissions reduce to 107.2 kg CO_2/kWh in case of coal-fired and to 110.6 kg CO_2/kWh in case of cofiring. For the cofiring case, if we assume that all the CO_2 resulted from the combustion of coal is captured from the plant, then the net CO_2 emissions would be zero.

3.2. Effect of operating parameters on CO_2 capture process

One of the main objectives of this study is to investigate the effect of different process parameters on the energy, solvent, and CW requirements in the CO_2 capture process. The effect of the MEA concentration in the solution (20–40 wt%), the FG temperature at the absorber inlet (40–50°C), the lean solvent temperature at the absorber inlet (30–50°C), the temperature difference in the lean/rich HX between hot outlet and cold inlet (5–10°C), and the stripper operating pressure (1.5–1.9 bar) were investigated. **Figure 6** shows the simulation results on the effect of different process variables on the heat, solvent, and CW requirements in the CO_2 capture process with respect to the base case (30 wt% MEA, 45°C FG inlet temperature, 40°C lean solvent temperature, 5°C lean/rich HX temperature difference, and 1.7 bar stripper operating pressure). For all simulation cases, the CO_2 capture rate was fixed at 90%. As shown, the concentration of MEA is the most important parameter with great effect on the heat, solvent, and CW requirements. Operating the capture unit with a lower MEA concentration (20 wt%) leads to a significant increase of the reboiler heat duty (>12% more compared with the base case), solvent requirement (>35%), and CW requirement (>30%). This is because as the MEA concentration decreases more solvent needs to be fed into the absorber column to remove 90% of CO_2 from the FG stream. The increased solvent flow rate then leads to higher cooling requirements. Further, the temperature of the rich solution entering the desorber column is lower, which needs more heat for solvent regeneration. Contrary to this, increasing the MEA concentration from 30 wt% (base case) to 40 wt% results in a decrease of the reboiler heat duty (>9%), solvent flow rate (>17%) and CW requirement (>17%). It should be mentioned, however, that the use of more concentrated solutions can lead to higher corrosion rates and increased amine emission from the system. In addition, the reboiler temperature increases for cases operating with higher MEA concentrations, which can also lead to thermal degradation of the solvent.

From **Figure 6**, it can be further noted that the FG inlet temperature has almost no effect on the heat requirement, solvent flow rate, and CW requirements. The same was also observed when varying the lean solvent temperature and only influencing the CW requirements. The use of solvent at lower temperatures than that of the base case (40°C) increases the CW requirements by more than 20%. This increase is mainly used in the lean solvent cooler. The temperature difference in the lean/rich HX and the operating pressure of the stripper were found to influence only the heat and CW requirements. If the lean/rich HX is operated with a larger temperature difference, then the rich solvent before entering the desorber column is cooler and, in consequence, more heat is required for solvent regeneration, and since the lean solvent

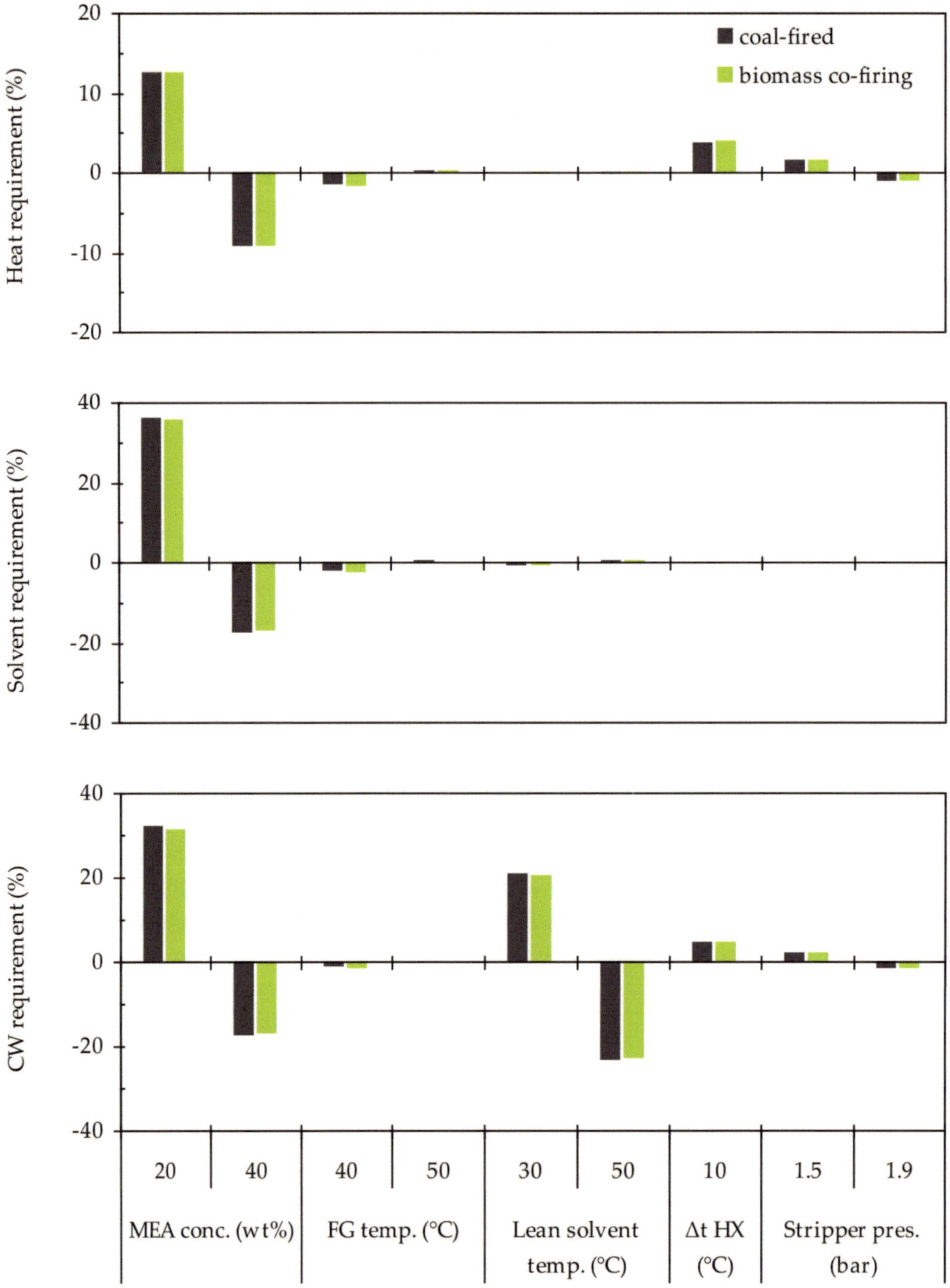

Figure 6. Effect of different parameters on the heat, solvent, and CW requirement in the CO_2 capture process with respect to the base case.

leaving the HX is warmer, then more CW is required to cool the stream to 40°C. Furthermore, operating the stripper at lower pressure leads to higher heat requirement and CW consumption. However, increasing the pressure from 1.7 bar (base case) to 1.9 bar reduces both the reboiler duty and the CW requirement, and in addition, the energy consumption for the compression of CO_2 is also reduced.

3.3. Effect of MEA concentration

As was shown earlier, the concentration of MEA in the solution can significantly influence the CO_2 capture process requirements. Therefore, it is necessary to investigate its effect on the plant performance. **Table 9** presents the simulation results for the coal-fired and cofiring cases using different MEA concentrations in the capture process. As can be seen, increasing the concentration of MEA has a positive effect on the plant energy performance. For coal-fired cases, the net electrical efficiency increases from 29.56% with 20 wt% MEA to 31.13% with 40 wt% MEA, i.e., an improvement of 1.57% points. It can be noted that the power demand of the CO_2 capture and compression process in all three cases is almost the same and only slightly decreases with the increase of amine concentration. This is because the solvent flow rate decreases and leads to lower pumps work. The amount of steam required for solvent regeneration decreases from about 180.5 kg/s, representing 57.1% of the total IP/LP crossover with 20 wt% MEA, to 145.4 kg/s, i.e., 46% of the IP/LP steam with 40 wt% MEA. For the cofiring cases, the electrical efficiency is 0.5% points lower than that of coal-fired cases. As noted, the gross power output is lower because the amount of steam extracted for solvent regeneration is higher in the cofiring cases. For example, the amount of extracted steam in the cofiring case with 20 wt% MEA is about 4 kg/s higher than that of the coal case. Moreover, the auxiliary power consumption in the cofiring cases with capture is higher, by approximately 4.2 MW_e, which is mainly consumed by the biomass processing system. The results further show that the solvent flow rate, the CW requirement, and the heat requirement for solvent regeneration for the cofiring plants with CO_2 capture are slightly higher than the values for coal cases.

CO_2 capture (90%)	**No**	**Yes**		
MEA concentration (wt%)		**20**	**30**	**40**
Coal-fired cases:				
Cross power output (MW_e)	580.37	473.78	485.07	493.74
CO_2 capture and compression (MW_e)		−43.27	−43.03	−42.95
Other auxiliary loads (MW_e)	−30.14	−30.61	−29.89	−29.61
Net power output (MW_e)	550.22	399.90	412.14	421.19
Net plant efficiency (% LHV)	40.67	29.56	30.47	31.13
Solvent requirement (kg/s)		2995.1	2202.9	1819.7
CW requirement (kg/s)		10218.1	7739.0	6400.0
Heat requirement (MW_{th})		434.3	385.8	349.7
Steam requirement (% of total IP/LP)		57.1	50.7	46.0

CO_2 capture (90%)	**No**	**Yes**		
MEA concentration (wt%)		**20**	**30**	**40**
Biomass cofiring cases:				
Gross power output (MW_e)	580.37	471.95	483.26	492.01
CO_2 capture and compression (MW_e)		–43.93	–43.69	–43.59
Other auxiliary loads (MW_e)	–33.69	–34.23	–33.50	–33.20
Net power output (MW_e)	546.68	393.80	406.07	415.22
Net plant efficiency (% LHV)	40.35	29.06	29.97	30.64
Solvent requirement (kg/s)		3051.8	2248.9	1859.3
CW requirement (kg/s)		10526.2	8013.7	6636.3
Heat requirement (MW_{th})		442.3	393.4	356.8
Steam requirement (% of total IP/LP)		58.2	51.7	46.9

Table 9. Effect of MEA concentration on the energy performance of coal-fired/biomass cofiring plants with CO_2 capture.

3.4. Effect of heat requirement

In this section, the effect of the heat requirement for solvent regeneration on the power plant gross output was investigated. The simulation results are presented in **Figure 7**. The heat duty of the stripper reboiler was varied between 3.5 MJ/kg CO_2 captured (base case) and 2 MJ/kg CO_2 captured. The results showed that for the chemical absorption systems with heat requirement of 3.5 MJ/kg CO_2 captured, the gross power output of the plant decreases by more than 16% compared with the reference plants without CO_2 capture and the steam extracted from the IP/LP crossover amounts to ~50% of the total flow rate. In comparison, for systems with reduced heat requirement, for example, of 2 MJ/kg CO_2 captured, the power output decreases by only 9%, and the proportion of steam extracted is reduced to less than 30%. The amount of steam extracted for solvent regeneration is reduced from 1.45 kg steam/kg CO_2 captured (base case) to about 0.85 kg steam/kg CO_2 captured for the case with 2 MJ/kg CO_2 captured.

Figure 8 shows the effect of a capture system with lower heat requirement for solvent regeneration on the net power output and efficiency of the biomass cofiring plant. The simulations were carried out using the Cansolv technology for CO_2 capture with the following specific requirements: 2.48 MJ/kg CO_2 reboiler heat duty and 33.3 kWh/t CO_2 power duty [23]. It should be noted here that this technology is currently used at the SaskPower Boundary Dam power plant in Canada being the first commercial scale postcombustion carbon capture project [11]. The simulation results show that the net power output of the biomass cofiring plant integrated with the Cansolv capture technology would increase to about 428 MW_e, which is 5.3% higher than that of the plant using conventional MEA system. Compared with the reference biomass cofiring plant without carbon capture, the efficiency penalty due to CO_2 capture reduces to 8.79% points in case with Cansolv.

http://dx.doi.org/10.5772/67188

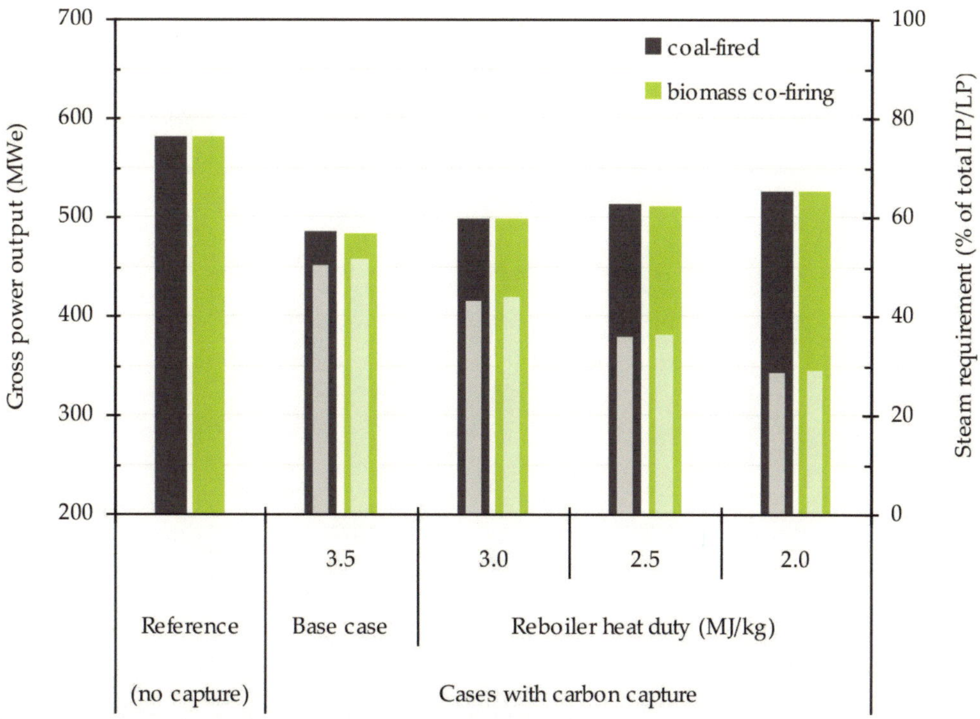

Figure 7. Effect of the reboiler heat duty on the gross power output (bars with a lighter color show the percentage of steam extracted from the IP/LP crossover).

3.5. Effect of CO_2 capture efficiency

To investigate the effect of the capture efficiency on the performances of coal-fired and biomass cofiring plants, the FG exiting the FGD unit was split into two streams, one directed to the capture unit and the other one sent directly to the stack. The amount of FG sent to the absorber column of the capture system varied between 100% and 56% of the total mass flow in order to achieve capture rates of 90–50%. In another configuration (not considered here), all the FG can be sent to the capture unit; however, in this case, the power requirements and cooling duties of the plant would increase.

The simulation results presented in **Table 10** show that the gross power output of both the coal-fired and biomass cofiring plants increases by ~9% as the capture efficiency decreases from 90 to 50%. This is primarily due to the fact that the quantity of steam extracted for solvent regeneration from the steam cycle is significantly lower (by ~45%) and, therefore, more steam is available for power generation. The net power output of the plants increases by more than 15% and the net electrical efficiency is 4.6% points higher than that of the case with 90% CO_2 capture. While the energy performances improve with a decrease in the capture rate, the plants specific CO_2 emissions increase from around 110 kg/kWh to more than 450 kg/MWh.

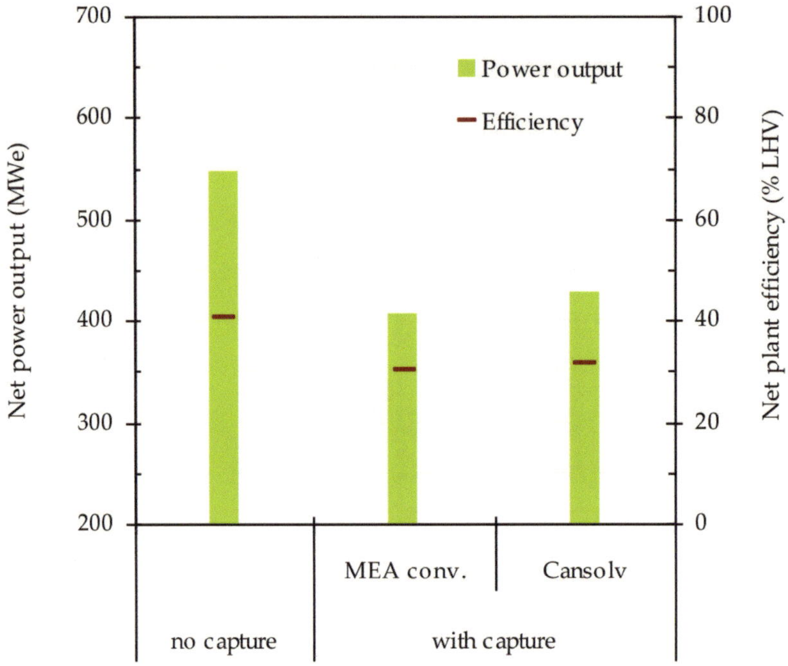

Figure 8. Effect of the capture system with lower reboiler heat duty on the net power output and efficiency of the biomass cofiring plant.

Overall capture efficiency (%)	**90**	**80**	**70**	**60**	**50**
FG to capture unit (% of total FG)	**100**	**89**	**78**	**67**	**56**
Coal-fired cases:					
Gross power output (MW_e)	485.07	495.43	506.11	517.00	528.11
CO_2 capture and compression (MW_e)	−43.03	−38.25	−33.47	−28.70	−23.91
Other auxiliary loads (MW_e)	−29.89	−29.92	−29.95	−29.97	−30.00
Net power output (MW_e)	412.14	427.26	442.69	458.33	474.21
Net plant efficiency (% LHV)	30.47	31.58	32.72	33.88	35.05
Solvent requirement (kg/s)	2203	1958	1713	1469	1224
CW requirement (kg/s)	7739	6878	6020	5160	4299
Heat requirement (MW_{th})	386	343	300	257	214
Steam requirement (% of total IP/LP)	51	45	39	34	28
Specific CO_2 emissions (kg/MWh)	107	207	299	386	466
Biomass cofiring cases:					
Gross power output (MW_e)	483.26	493.75	504.63	515.71	527.00
CO_2 capture and compression (MW_e)	−43.69	−38.83	−33.98	−29.13	−24.27
Other auxiliary loads (MW_e)	−33.50	−33.52	−33.54	−33.56	−33.58

Overall capture efficiency (%)	90	80	70	60	50
FG to capture unit (% of total FG)	**100**	**89**	**78**	**67**	**56**
Net power output (MW_e)	406.07	421.40	437.11	453.02	469.17
Net plant efficiency (% LHV)	29.97	31.10	32.26	33.43	34.63
Solvent requirement (kg/s)	2249	1999	1749	1499	1249
CW requirement (kg/s)	8013	7122	6231	5341	4450
Heat requirement (MW_{th})	393	350	306	262	219
Steam requirement (% of total IP/LP)	52	46	40	34	29
Specific CO_2 emissions (kg/MWh)	111	213	308	397	479

Table 10. Effect of CO_2 capture efficiency on the performances of coal-fired and biomass cofiring plants with CO_2 capture.

4. Conclusions

In this study, we investigated in detail the effect of biomass cofiring and carbon capture on the plant performances. For cofiring cases, 10% of heat input was substituted with hybrid poplar used as the biomass fuel. When carbon capture was considered, the plants were integrated with a MEA-based postcombustion capture technology. The plant's submodels, i.e., the boiler and flue gas cleaning section (deNOx, deDust, deSOx), the steam cycle, and the CO_2 capture and compression process were all modeled and simulated in Aspen Plus software. The simulation showed the following:

(i) The addition of a MEA-based capture system to a supercritical coal-fired plant reduces the net plant efficiency to 30.47% (with 90% CO_2 capture rate), i.e., an efficiency penalty of 10.21% points compared with the reference plant without capture. The plant specific CO_2 emissions were decreased to 107.2 kg/MWh and the avoided emissions were 86.6%. Compared with these figures, the net efficiency of the plant with cofiring, after capturing 90% of the CO_2, decreases to 29.97%, i.e., an efficiency penalty of 10.7% points. Lower efficiency of the biomass cofiring plant with CO_2 capture of 0.5% points compared with the coal-fired plant with CO_2 capture is mainly due to the additional power demand for biomass processing and drying. Because of the lower performances and higher flue gas CO_2 flow rates, it was calculated that the plant specific CO_2 emissions were 110.6 kg/MWh. However, taking the biomass as carbon neutral and assuming that all the fossil-CO_2 is captured than the plant would have ~0 net CO_2 emissions.

(ii) Among the investigated parameters that may affect the CO_2 capture process, it was found that the MEA concentration greatly influences the performances of the capture unit. For capture systems, operating with lower MEA concentrations, a higher reboiler heat duty, solvent flow rate, and CW requirement were achieved. However, the use of higher MEA concentrations, although lowers the consumption of steam, solvent, and CW, can lead to corrosion, solvent degradation, and higher amine emissions. The concentration of MEA also influences the energy performance of the plant. The results showed that the net efficiency increases as the MEA concentration increases. For the case with 40 wt% MEA, the net efficiency of both the coal-fired and biomass cofiring plants improves by ~0.7 and ~1.6% points compared with the cases of 30 and 20 wt% MEA in the solution, respectively. The steam requirement reduces by ~9 and ~20% compared with 30 and 20 wt% MEA cases, respectively.

(iii) The heat requirement for solvent regeneration in conventional MEA-based capture systems significantly affects the gross power output of the plant and, in consequence, the overall plant energy performance. It was found that the gross power output increases with decreasing the heat duty of the reboiler. For both the coal-fired and biomass cofiring plants with 90% CO_2 capture and an assumed heat requirement of 2 MJ/kg CO_2 captured, the gross power output increases by ~8.5% compared with the base case, while the steam requirement decreases by more than 40%. Lower reboiler heat duty of chemical absorption systems can be achieved by, for example, using an improved process configuration (e.g., absorber intercooling, lean vapor compression, split-stream, etc. [24–26]) and/or solvents with better characteristics [18, 23, 26–28].

(iv) In addition, we analyzed the effect of the CO_2 capture efficiency on the overall performances of both the coal-fired and biomass cofiring plants. In this case, only a part of the flue gas stream was treated in the capture unit (with a fixed 90% capture rate), and the rest was sent directly to the stack. The results showed that the gross power output of the plants increases with decreasing the capture efficiency. Capturing less CO_2 from the plant requires less steam to be extracted for the solvent regeneration and, consequently, more steam is available for power generation. Furthermore, for lower capture rates, the net power output improves since the auxiliary power demand of the capture and compression process decreases. However, reducing the capture rates would negatively affect the plants CO_2 emissions, generating significantly more CO_2 into the atmosphere, which in case of biomass cofiring will be lower compared with coal-fired if only the net CO_2 emissions would be considered.

Acknowledgements

Financial support provided by the Ministry of Labor, Family and Social Protection (Romania) and cofinanced by the European Social Fund for the second author through the strategic grant POSDRU107/1.5/S/77265 is gratefully acknowledged.

Nomenclature

APH	Air preheater
BA/FA/TA	Bottom/fly/total ash
BH	Baghouse
BOP	Balance of plant
DCC	Direct contact cooler

FD	Forced draft
FG	Flue gas
FGD	Flue gas desulfurization
FWH	Feedwater heater
HHV/LHV	Higher/lower heating value (MJ/kg)
HP/IP/LP	High/intermediate/low pressure
HX	Heat exchanger
ID	Induced draft
PA	Primary air
PC	Pulverized coal
SCR	Selective catalytic reduction
ST	Steam turbine
ar	As received
db	Dry basis
$\dot{m}$	Mass flow rate (kg/s)
NE	Net emissions (kg/MWh)
SC	Specific consumption (kg/MWh)
SE	Specific emissions (kg/MWh)
W	Power (MW)
$\Delta\eta$	Efficiency penalty (% points)
η	Efficiency (%)

Author details

Dumitru Cebrucean*†, Viorica Cebrucean† and Ioana Ionel

*Address all correspondence to: dumitru_cebrucean@yahoo.com

Department of Mechanical Machines, Equipments and Transportation, Politehnica University of Timisoara, Timisoara, Romania

† These authors contributed equally.

References

[1] IEA. CO_2 emissions from fuel combustion. Highlights. OECD/IEA; 2015.

[2] IPCC. Carbon dioxide capture and storage. Cambridge University Press, Cambridge; 2005.

[3] van Loo S, Koppejan J. The handbook of biomass combustion and co-firing. Earthscan, London; 2008.

[4] Spliethoff H. Power generation from solid fuels. Springer, Heidelberg; 2010.

[5] IEABCC. Cofiring database. IEA Bioenergy Task 32; 2016.

[6] NETL. Greenhouse gas reductions in the power industry using domestic coal and biomass. Volume 2: Pulverized coal plants. DOE/NETL-2012/1547; February 2012.

[7] Schakel W, Meerman H, Talaei A, Ramírez A, Faaij A. Comparative life cycle assessment of biomass co-firing plants with carbon capture and storage. Applied Energy 2014;131:441–467. DOI:10.1016/j.apenergy.2014.06.045

[8] Khorshidi Z, Ho MT, Wiley DE. The impact of biomass quality and quantity on the performance and economics of co-firing plants with and without CO_2 capture. International Journal of Greenhouse Gas Control 2014;21:191–202. DOI:10.1016/j.ijggc.2013.12.011

[9] Khorshidi Z, Ho MT, Wiley DE. Techno-economic evaluation of using biomass-fired auxiliary units for supplying energy requirements of CO_2 capture in coal-fired power plants. International Journal of Greenhouse Gas Control 2015;32:24–36. DOI:10.1016/j.ijggc.2014.10.017

[10] Fogarasi S, Cormos CC. Technico-economic assessment of coal and sawdust co-firing power generation with CO_2 capture. Journal of Cleaner Production 2015;103:140–148. DOI:10.1016/j.jclepro.2014.07.044

[11] Cebrucean D, Cebrucean V, Ionel I. CO_2 capture and storage from fossil fuel power plants. Energy Procedia 2014;63:18–26. DOI:10.1016/j.egypro.2014.11.003

[12] NETL. Cost and performance baseline for fossil energy plants. Volume 3b: Low rank coal to electricity: Combustion cases. DOE/NETL-2011/1463; March 2011.

[13] NETL. Cost and performance baseline for fossil energy plants. Volume 1: Bituminous coal and natural gas to electricity. DOE/NETL-2010/1397, Revision 2a; September 2013.

[14] CAESAR. D 4.9 European best practice guidelines for assessment of CO_2 capture technologies. 2011.

[15] Oexmann J, Kather A, Linnenberg S, Liebenthal U. Post-combustion CO_2 capture: chemical absorption processes in coal-fired steam power plants. Greenhouse Gases: Science and Technology 2012;2(2):80–98. DOI:10.1002/ghg.1273

[16] Cebrucean V. The absorption process of CO_2 from flue gases [PhD Thesis]. Politehnica University of Timisoara, Timisoara; 2015.

[17] NETL. Pulverized coal oxycombustion power plants. Volume 1: Bituminous coal to electricity. DOE/NETL-2007/1291, Revision 2; August 2008.

[18] Sanchez Fernandez E, Goetheer ELV, Manzolini G, Macchi E, Rezvani S, Vlugt TJH. Thermodynamic assessment of amine based CO_2 capture technologies in power plants based on European benchmarking task force methodology. Fuel 2014;129:318–329. DOI:10.1016/j.fuel.2014.03.042

[19] Abu-Zahra MRM, Schneiders LHJ, Niederer JPM, Feron PHM, Versteeg GF. CO_2 capture from power plants: Part I. A parametric study of the technical performance based on monoethanolamine. International Journal of Greenhouse Gas Control 2007;1(1):37–46. DOI:10.1016/s1750-5836(06)00007-7

[20] Liu X, Chen J, Luo X, Wang M, Meng H. Study on heat integration of supercritical coal-fired power plant with post-combustion CO_2 capture process through process simulation. Fuel 2015;158:625–633. DOI:10.1016/j.fuel.2015.06.033

[21] Sanpasertparnich T, Idem R, Bolea I, deMontigny D, Tontiwachwuthikul P. Integration of post-combustion capture and storage into a pulverized coal-fired power plant. International Journal of Greenhouse Gas Control 2010;4(3):499–510. DOI:10.1016/j.ijggc.2009.12.005

[22] Romeo LM, Bolea I, Escosa JM. Integration of power plant and amine scrubbing to reduce CO_2 capture costs. Applied Thermal Engineering 2008;28(8–9):1039–1046. DOI:10.1016/j.applthermaleng.2007.06.036

[23] NETL. Cost and performance baseline for fossil energy plants. Volume 1a: Bituminous coal (PC) and natural gas to electricity. DOE/NETL-2015/1723, Revision 3; July 2015.

[24] de Miguel Mercader F, Magneschi G, Sanchez Fernandez E, Stienstra GJ, Goetheer ELV. Integration between a demo size post-combustion CO_2 capture and full size power plant. An integral approach on energy penalty for different process options. International Journal of Greenhouse Gas Control 2012;11:S102–S113. DOI:10.1016/j.ijggc.2012.09.016

[25] Olaleye AK, Wang M, Kelsall G. Steady state simulation and energy analysis of supercritical coal-fired power plant with CO_2 capture. Fuel 2015;151:57–72. DOI:10.1016/j.fuel.2015.01.013

[26] Kvamsdal HM, Ehlers S, Kather A, Khakharia P, Nienoord M, Fosbøl PL. Optimizing integrated reference cases in the OCTAVIUS project. International Journal of Greenhouse Gas Control 2016;50:23–36. DOI:10.1016/j.ijggc.2016.04.012

[27] Kvamsdal HM, Romano MC, van der Ham L, Bonalumi D, van Os P, Goetheer E. Energetic evaluation of a power plant integrated with a piperazine-based CO_2 capture process. International Journal of Greenhouse Gas Control 2014;28:343–355. DOI:10.1016/j.ijggc.2014.07.004

[28] Van Wagener DH, Liebenthal U, Plaza JM, Kather A, Rochelle GT. Maximizing coal-fired power plant efficiency with integration of amine-based CO_2 capture in greenfield and retrofit scenarios. Energy 2014;72:824–831. DOI:10.1016/j.energy.2014.04.117

Section 2

Carbon Capture Methods

Chapter 3

Membrane Separation Technology in Carbon Capture

Guozhao Ji and Ming Zhao

Additional information is available at the end of the chapter

http://dx.doi.org/10.5772/65723

Abstract

This chapter introduces the basics of membrane technology and the application of membrane separation in carbon capture processes. A number of membranes applicable in pre-combustion, post-combustion or oxy-fuel combustion have been discussed. An economic comparison between conventional amine-based absorption and membrane separation demonstrates the great potential in membrane technology.

Keywords: membrane separation, carbon dioxide capture, pre-combustion, post-combustion, oxy-fuel combustion

1. Introduction

Gas separation by membrane is attractive in low carbon emission technologies, as it can be operated in a continuous system, which is preferred by industry, other than the conventional batch systems such as adsorption and absorption. Feeding of mixed gas and exiting of purified gas can happen at the same time. Membrane selectively permeates the desired components and retains the unwanted, resulting in separation of gas mixtures. In carbon capture and storage (CCS) processes, CO_2 has to be separated from the exhaust gas streams before the subsequent transportation and storage. Membrane separation technology is one of efficient solutions for carbon capture.

There have been a number of books regarding membrane technology. However, most of them are about liquid separation and very few are found for CCS. This chapter aims at introducing and demonstrating the membrane technology in CCS. The application of membrane in carbon capture mainly includes H_2/CO_2 separation for pre-combustion, CO_2/N_2 separation for post-combustion and O_2/N_2 separation (air separation) for oxy-fuel combustion. There is a wide variety of membrane types based on its physical and chemical property. Many of them have showed great potentials to fulfill the need of CCS.

2. Overview of membranes

Membrane performs as a filter. It allows certain molecules to permeate through, while blocks other specific molecules from entering the membrane as demonstrated in **Figure 1**. Membrane has already been widely used in liquid separations such as micro-filtration, ultra-filtration, reverse osmosis, forward osmosis, desalination and medical application. However, gas separation using membrane is still developing. Membrane gas separation has attracted intensive researches in CCS field during recent years.

Gas permeation flux across unit membrane area under unit pressure difference through unit membrane thickness is called permeability (mol s^{-1} m^{-2} Pa^{-1}) and the ratio of permeabilities of different gases through the same membrane is defined as selectivity. The gas separation mechanism varies from membrane to membrane. The selectivity of different gases may result from the difference in molecular size, affinity to membrane material, molecular weight, etc., depending on the gas and membrane of interest.

In order to achieve high permeate flux, the feed gas is pressurized, while the permeate gas is connected to atmosphere or vacuum to obtain a higher driving force. However, since the thickness of a membrane is only several hundred nanometers to several microns, it is impossible to resist this force. So a membrane is normally coated onto a thick, porous substrate to achieve enough mechanical strength. The supporting substrate should offer minimum flow resistance, thus containing large pores, which allows free flow of gas that has permeated the top layer. In case of too large pores and highly rough surface on the substrate, membrane defects such as cracking and peeling may occur. An interlayer with much smaller pore size (than substrate pore size) can enable smoother transition in between. This design is referred to as asymmetric structure as shown in **Figure 2**.

In the current Research & Developments (R&Ds) of membrane, the most popular mechanism is size sieving separation. Therefore, the key parameter of a membrane is its pore size. By

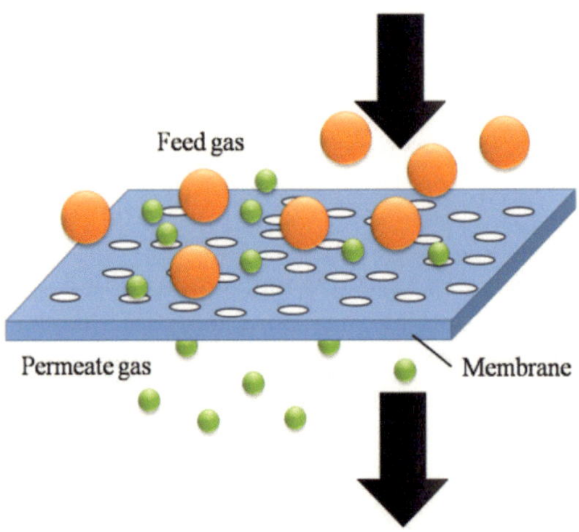

Figure 1. The schematic of membrane separation for binary gas mixtures.

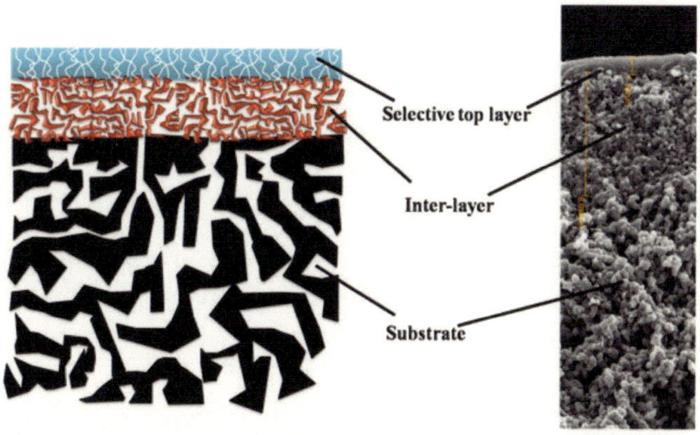

Figure 2. The asymmetric structure of membrane coated on a substrate.

pore size, membranes are classified into three categories that are listed in **Table 1**. In addition, one more type of membrane that is nonporous, therefore called dense membrane, is also discussed in this chapter.

2.1. Advantages of membranes

Compared to conventional CO_2 removal technologies, membrane has shown great potential in CCS owing to its characteristics listed below:

Low capital cost

Membrane requires little material to coat. It does not need additional facilities such as large pretreatment vessel and solvent storage.

Low operating cost

The main operating cost for membrane separation unit is only membrane replacement. Due to the smaller size and weight of membrane, the cost is much lower than the conventional techniques, which replace the large amount of solvent or sorbent.

Simplicity and reliability

Since membrane does not show fast decay in performance that most likely occurs to the traditional solvents or sorbents, it can be running unattended for long periods. Another character of membrane is that gas does not stay and reacts with membrane, so membrane has no saturation and thus avoids frequent shut down and start-up.

Pore classification	Pore size range (nm)
Micropore	<2
Mesopore	2–50
Macropore	>50

Table 1. Membrane classification by pore size.

Adaptability

Membrane system is designed and operated to remove the required percentage of CO_2 instead of the absolute quantity of CO_2 removal. Variations in the feed CO_2 concentration can be adjusted by varying the space velocity to keep constant product quality.

Design efficiency

Membrane system can integrate a number of processes into one unit, such as Hg vapor removal, H_2S removal and dehydration. Traditional CO_2 removal techniques have to operate these steps separately.

Easy for remote area

Multiple membranes could be packed into one module to reduce size and weight, which not only increases membrane area in unit volume but also makes it easier to transport to remote locations. Simple installation is feasible at which spare parts are rare, labors are unskilled and additional facilities (such as solvent storage, water supply and power generation) are short in supply.

2.2. Membrane fabrication

Membrane fabrication involves how to coat the selective layer onto the porous substrate. The fabrication process has significant influence on the membrane property such as membrane uniformity and thickness. The membrane coating technique includes dip-coating, chemical vapor deposition (CVD), spinning and spraying. Among them, the most popular and mature methods are dip-coating and CVD. This section will demonstrate these two technologies.

2.2.1. Dip-coating

Dip-coating involves dipping the macro-porous substrate in a solution and in turn, the solution is coated on the substrate, which is followed by a dehydration process at a lower temperature. It is the oldest and the simplest film deposition method. The dip-coating process can be separated into five stages: immersion, start-up, deposition, drainage and evaporation (**Figure 3**).

Immersion: The substrate is immersed in the solution of the coating material at a constant speed to avoid jitter.

Start-up: The entire substrate has remained inside the solution for a while and is starting to be pulled up.

Deposition: The thin layer of solution deposits itself on the surface of the substrate when it is pulled up. The withdrawing speed is constant to avoid any jitters. The speed determines the thickness of the coating. Faster withdrawal speed gives thicker layer and vice versa.

Drainage: Excess liquid will drain from the surface back to the solution due to the gravity.

Evaporation: The solvent evaporates from the liquid, forming the thin layer. Evaporation normally accompanies the start-up, deposition and drainage stages.

http://dx.doi.org/10.5772/65723

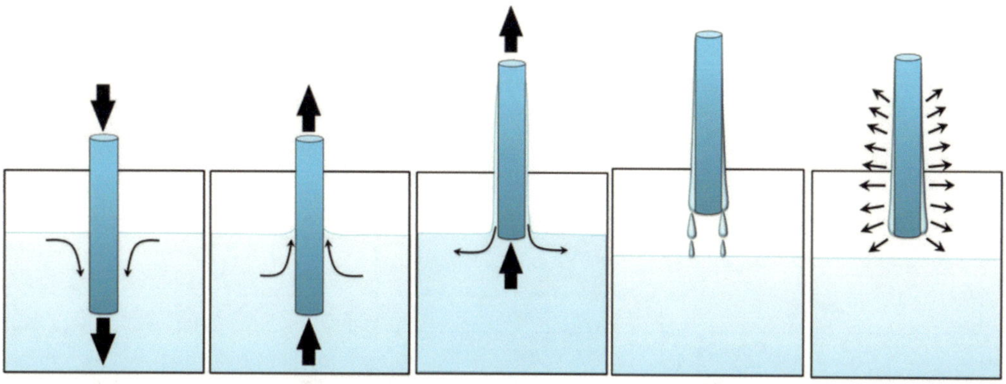

Figure 3. Dip-coating stages: (a) immersion; (b) start-up; (c) deposition; (d) drainage and (e) evaporation. Reproduced from Brinker [1].

2.2.2. *Chemical vapor deposition (CVD)*

Another common membrane coating technique is CVD. CVD modifies the properties of a substrate surface by depositing a thin layer of film via chemical reactions in a gaseous medium surrounding the substrate at elevated temperatures.

The process of CVD includes transporting the reactant gases and/or carrier gas into a reaction chamber, which is followed by a deposition process to form a film. The film coating could be performed by decomposition, oxidation, hydrolysis or compound formation. The reactions normally take place in the gaseous phase and the intermediate gases adsorb on the substrate followed by surface reactions. The detailed steps of CVD process are demonstrated in **Figure 4**.

1. Reactant feeding: Delivering the reactant gaseous species into the reaction chamber.
2. Reaction: Chemical reactions of the reactant gas species under heating condition to form intermediates.
3. Diffusion to substrate: Diffusion of gases through the boundary layer to the substrate surface.
4. Adsorption on the substrate: Adsorption of reactant species or intermediates on substrate surface.
5. Surface migration: Inclusion of coating atoms into the growing surface and formation of by-product species.
6. By-product desorption: Desorption of by-product species of the surface reaction.
7. By-product diffusion: Diffusion of by-product species to the bulk phase.
8. By-product exiting: Transport of by-product gaseous species away from substrate and exit the reaction chamber.

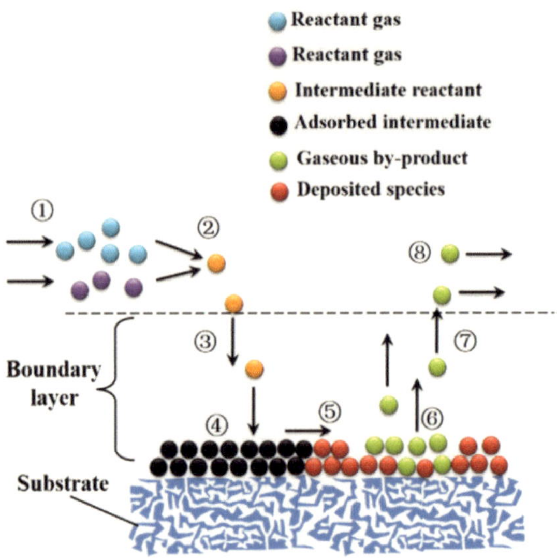

Figure 4. Schematic model of a CVD process. Reproduced from Khatib et al. [2].

As illustrated above, CVD is more complicated technique than dip-coating, thus the manufacture cost of a membrane is relatively higher than that of dip-coating. The advantage of CVD is good reproducibility over dip-coating as the latter may suffer from a lack of reproducibility.

2.3. Membrane separation mechanism

A membrane can separate gas mixture because different gases have different permeability through the membrane. The permeate flux across unit membrane area under unit pressure gradient is called permeability and the ratio between permeability of gas A and that of gas B is defined as selectivity of A to B. In order to achieve separation, a greater difference between gas permeabilities is preferred. This difference comes from their physical and/or chemical properties as well as the interaction with membrane.

2.3.1. Size sieving

The most widely known separation mechanism is size sieving. The membrane pore size is just between the smaller gas molecule and larger gas molecule as depicted in **Figure 5**. The smaller gas molecule A passes the pore channel freely, while the counterpart gas B is not able to enter the pore. As a result, pure component A is obtained in the permeate stream from the gas mixture A–B. This mechanism applies to separating gas mixtures with very different molecular sizes such as H_2 and CO_2, H_2 and hydrocarbons, etc. Some common gas kinetic diameters are given in **Table 2**. Size sieving basically performs in micro-porous membrane.

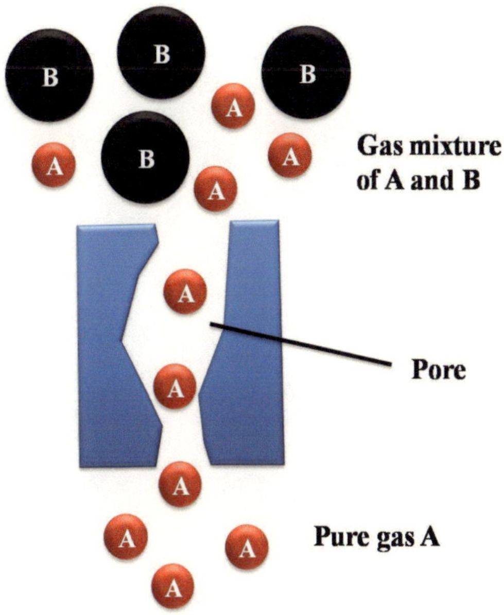

Figure 5. Size sieving separation mechanism.

2.3.2. *Surface diffusion*

When the membrane material has higher affinity to one particular component than the other, this affinitive component is preferentially adsorbed on the membrane surface and then the adsorbed gas molecules move along the pore surface to the permeate side until desorbing to the permeate gas. Since the membrane is occupied by the highly adsorbable component, the less adsorbed component has lower probability to access the pore, which results in a much lower permeability. In such a way, the more adsorbable gas is separated from the gas mixture (**Figure 6**). This type of mechanism is generally used to separate adsorbing gas with non-adsorbing gas such as CO_2 with He, CO_2 with H_2. Surface diffusion generally acts in micro- and meso-porous membranes.

Gas	σ (nm)
He	0.26
H_2	0.289
CO_2	0.33
Ar	0.341
O_2	0.346
N_2	0.364
CH_4	0.38

Table 2. The kinetic diameter of different gases.

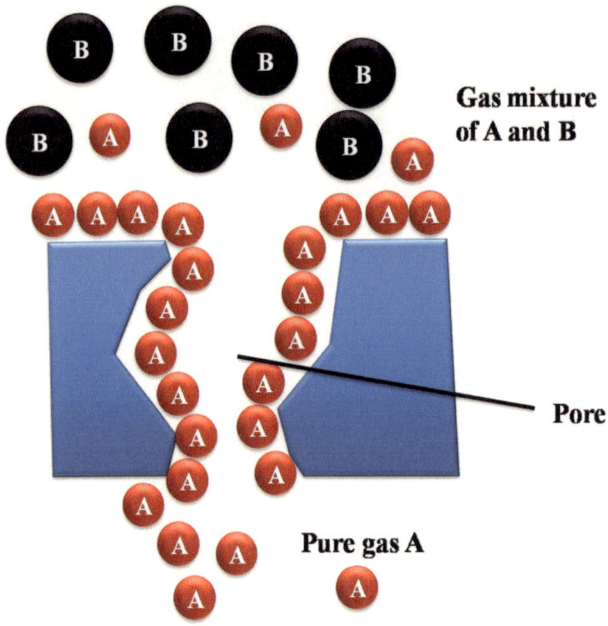

Figure 6. Surface diffusion separation mechanism.

2.3.3. Solution diffusion

Unlike membranes discussed above, dense membrane has no pore channel for gas transportation. However, it follows solution diffusion model. The process of gas separation using dense membranes occurs in a three-step process, which is similar to surface diffusion. The dense membrane has no pore to accommodate gas molecules, however, it can solve specific gas component. As shown in **Figure 7**, due to the difference in solubility or absorbability in the membrane material, gas A solves or absorbs in the membrane after they contact at the feed interface, while gas B still remains as gas phase at the interface. The second step is the solved A component diffusing across the membrane driven by the concentration gradient from feed interface to the permeate interface. Finally, component A desorbs from the permeate interface under a low pressure. This is a common mass transfer mechanism in polymeric membrane.

2.3.4. Facilitated transport

The solution-diffusion process is often constrained by low permeate flux rates due to a combination of low solubility and/or low diffusivity. In contrast, facilitated transport that delivers the target component by a carrier can increase the permeate flow rate. As demonstrated in **Figure 8**, the gas A and carrier C form a temporary product A–C that is from a reversible chemical reaction. The product diffuses across the membrane under the concentration gradient of this product A–C instead of the concentration gradient of A. At the permeate interface, the reverse reaction takes place and A is liberated from this reverse

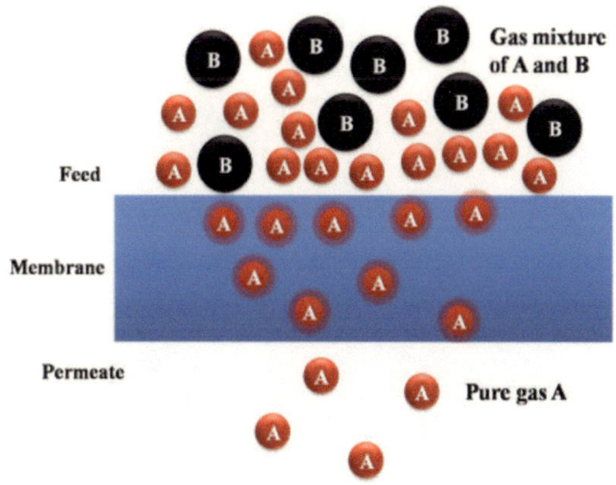

Figure 7. Solution diffusion separation mechanism.

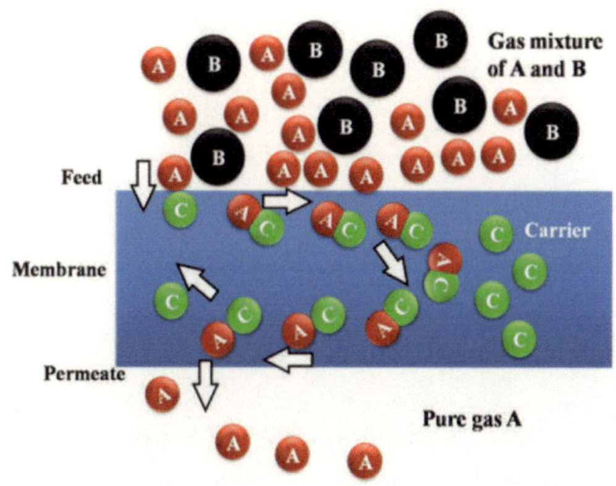

Figure 8. Facilitated transport separation mechanism.

reaction. A is released to the permeate stream and C diffuses back to the feed interface again to attach and deliver a new A. Facilitated transport mechanism normally exists in liquid membrane.

2.3.5. Ion transport

Ion transport is usually applied in air separation (O_2/N_2). As **Figure 9** shows, only oxygen gas molecule (O_2) can be converted into two oxygen ions ($2O^{2-}$) by the surface-exchange reaction on the feed interface. Nitrogen retains in the feed side. Oxygen ions are transported across by jumping between oxygen vacancies in the membrane lattice structure. At the permeate interface, electrons liberated as the oxygen ions recombine into oxygen molecules. To maintain electrical neutrality, there is a simultaneous electrons flux going back to the feed interface neutralizing the charge caused by oxygen flux.

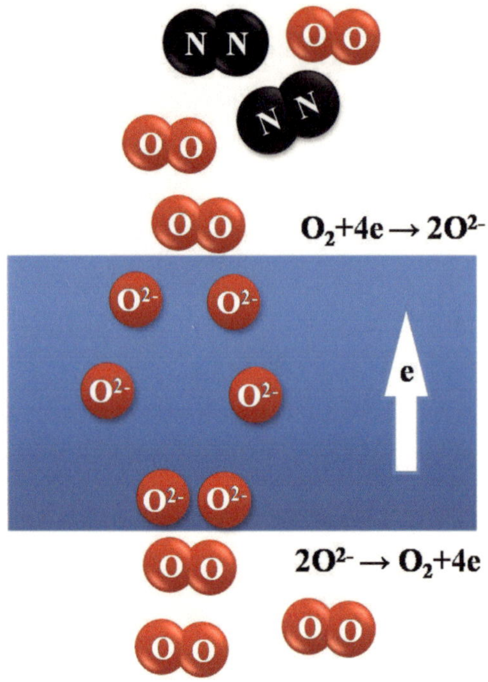

Figure 9. Ion transport separation mechanism.

3. Membranes for pre-combustion capture

Pre-combustion capture is a process that separates CO_2 from the other fuel gases before the gas combustion. First, it involves the processes of converting solid, liquid or gaseous fuels into a mixture of syngas (H_2 and CO) and CO_2 by coal gasification or steam reforming. Afterwards water-gas shift (WGS) reaction is conducted to reduce the content of CO, thus more H_2 and CO_2 are generated. Membrane separation is then applied to separate H_2 and CO_2. Upon compression, the CO_2 rich stream is transported to a storage or utilization site. Meanwhile, the nearly pure H_2 stream enters the combustion chamber for power generation that emits mainly water vapor in the exhaust.

Coming from the upstream gasification, reforming and WGS, the feed gas of pre-combustion CO_2 capture is hot with a temperature between 300 and 700°C. In addition, the pre-combustion separation can happen at high pressures up to 80 bar.

Pre-combustion membranes are basically classified into two categories: H_2-selective membrane and CO_2-selective membrane. The former favors H_2 permeation but retains CO_2 in the feed side, while the latter preferentially permeates CO_2.

In principle, metallic membrane is the ideal candidate for separating H_2/CO_2 due to the infinite selectivity. H_2 molecule dissociates as two H atoms at the membrane surface and then the atomic H diffuses to the permeate side of the membrane driven by the partial pressure drop, which is followed by the association and desorption at the permeate interface. The permeate flux is given by

$$J_{H_2} = \frac{P_{H_2}}{L}\left(\sqrt{p_{\text{feed}}} - \sqrt{p_{\text{permeate}}}\right). \tag{1}$$

This mechanism is similar to solution diffusion and ion transport. The reason for the infinite selectivity of H_2 over CO_2 is that this dissociation-diffusion mechanism only applies to diatomic gases such as H_2 and CO_2 cannot permeate by the same mechanism. For ultrathin membrane, the rate-limiting step is the dissociation of hydrogen on the membrane surface and Pd material performs the best in hydrogen dissociation. Consequently, Pd membrane was intensively investigated in the past several decades. H_2 permeability through palladium membrane varies in the range between 10^{-7} and 10^{-8}mol s^{-1} m^{-1} $Pa^{-0.5}$ (**Table 3**). However, the permeability was not satisfactory for the industrial requirement yet. This is due to the slow permeation of H atom in the lattice of Pd, which is one order of magnitude lower than in other metals. In order to promote the permeability, a number of palladium-based alloys have been examined. A list of reported permeability data are summarized in **Table 4**. The alloy membranes dramatically improve the H_2 permeability by 2–3 orders of magnitude.

Still, a few barriers need to be overcome for the commercialization of palladium-based membrane. First, the cost of palladium is around 18,000 US$/Ounce (in June 2016), which is 150 times more expensive than silica membrane. Second, the H_2 permeation driving force is not from pressure; instead, it is from the root square of pressure (Eq. (1)). Therefore, the effect of compressing feed gas is not as significant as in other permeation mechanisms. In

Membrane	Permeability (mol s^{-1} m^{-1} $Pa^{-0.5}$)	Temperature (°C)	Reference
Pd	9×10^{-9}	227	[3]
Pd on sliver disk	1.47×10^{-7}	407	[4]
Pd disk	1.08×10^{-7}	300	[5]
Pd disk	1.06×10^{-7}	350	[6]
Pd disk from Pd sheet	7.25×10^{-7}	400	[7]
Pd on Vycor support	3.10×10^{-7}	350	[8]
Pd on Nickel	2.00×10^{-12}	200	[9]
Pd on Vycor support	1.18×10^{-7}	500	[10]
Pd on γ alumina	1.47×10^{-7}	480	[11]
Pd on alumina	6.27×10^{-8}	300	[12]
Pd on alumina	3.75×10^{-8}	400	[13]

Table 3. Hydrogen permeability through palladium membrane.

Membrane	Permeability ($mol\ s^{-1}\ m^{-1}\ Pa^{-0.5}$)	Temperature (°C)	Reference
$Pd_{59}Cu_{41}$	1.59×10^{-7}	400	[6]
$Pd_{60}Cu_{40}$	1.57×10^{-7}	350	[5]
$Pd_{60}Cu_{40}$	1.78×10^{-7}	400	[5]
$Pd_{94}Cu_{6}$	3.65×10^{-8}	400	[14]
$Pd_{50}Ni_{50}$	7.00×10^{-6}	450	[15]
$Pd_{69}Ag_{30}Ru_{1}$	1.03×10^{-6}	400	[13]
$Pd_{70}Ag_{30}$	2.35×10^{-7}	400	[13]
$Pd_{77}Ag_{23}$	1.35×10^{-7}	350	[16]
$Pd_{77}Ag_{23}$	5.00×10^{-5}	450	[17]
$Pd_{93}Ag_{7}$	7.25×10^{-8}	400	[14]

Table 4. Hydrogen permeability through palladium-based alloy.

addition, at temperatures lower than 300°C, hydrogen embrittlement causes catastrophic failure. Furthermore, the contaminations such as CO, NH_3 and sulfur compounds inhibit H_2 permeation through palladium membrane. Currently, palladium membrane separation still remains in small laboratory scale.

Besides metal membrane, inorganic membrane also plays an important role in separating H_2/CO_2 at elevated temperatures. The separation by inorganic membrane is generally achieved by the molecular size sieving effect. Carbon molecular sieve membrane has demonstrated in pilot scale to separate H_2 from refiner gas streams in the early 1990s. The disadvantage of carbon membrane is that it is only feasible in non-oxidizing condition. Another type of inorganic membrane is alumina membrane. However, the majority pore size is not in the range of micropore and cannot separate gas by the size sieving mechanism. Due to the large pore size, the selectivity of alumina membrane is fairly low.

Silica membrane shows great commercial potential for separating H_2 and CO_2. It is one of the most abundant materials on the planet, thus the cost is significantly reduced. Also the good thermal and chemical stability makes it possible to work in long term without frequent replacement or maintenance. The pore diameter could be controlled around 0.3 nm by proper coating-calcining process, which is the ideal size for separating H_2 (σ = 0.26 nm) and CO_2 (σ = 0.33 nm). The performance of some reported silica membranes is summarized in **Table 5.** Due to the difficulty in measuring the membrane thickness on porous substrate, permeability of H_2 divided by thickness is lumped together as permeance.

The H_2 permeance of silica membrane could reach up to the order of 10^{-6} mol s^{-1} m^{-2} Pa^{-1}, which strongly suggests that silica membrane is competitive in pre-combustion capture. However, exposure to high concentration water vapor leads to a decline in performance of silica membrane. Such a steady decay over long time can cause the H_2 permeance decrement by an order of magnitude. This still inhibits the commercialization of silica membrane.

Membrane	Permeance (mol s^{-1} m^{-2} Pa^{-1})	H_2/CO_2 selectivity	Temperature (°C)	Reference
Silica (Si400)	2.01×10^{-6}	7	200	[18]
Silica (hydrophobic)	1.51×10^{-6}	6	200	[19]
Silica on zirconia	1.34×10^{-6}	4	300	[20]
Silica	1.34×10^{-6}	8	300	[20]
Silica (Si600)	5.02×10^{-7}		200	[18]
Silica (hydrophilic)	6.70×10^{-9}	11	200	[19]
Silica with Co	5.00×10^{-9}	1000	250	[21]
Silica	1.80×10^{-8}	15–80	150	[22]
Silica with Co&Pd	6.00×10^{-6}	200	500	[23]
Silica (ES40)	1.01×10^{-6}	12	450	[24]
AKP-30 tubular silica	1.8×10^{-6}	3.5	200	[25]
Silica with C6 surfactant	1.5×10^{-6}	6	200	[19]
Silica without C6 surfactant	7.0×10^{-9}	10	200	[19]

Table 5. H_2/CO_2 separation performance by silica-based membrane.

As a nonporous membrane, polymeric membrane permeates gases via the solution-diffusion mechanism. Permeability is a function of gas diffusivity and solubility. The hydrogen molecules diffuse faster than other gases due to the small molecular size. However, the lower solubility of hydrogen within the polymeric membrane reduces its permeability. For H_2-selective polymeric membranes, the permeability is limited by the low solubility of H_2. There is a wide range of polymeric membranes available for H_2 separation from CO_2. The performance of some polymeric membranes is shown in **Table 6**. High permeabilities are observed for polyimides such as 6FDA-durene. Higher selectivities are reported for polybenzimidazole and poly(vinyl chloride), but H_2 permeability is compromised.

Membrane	Permeability (mol s^{-1} m^{-2} Pa^{-1})	H_2/CO_2 selectivity	Temperature (°C)	Reference
6FDA-Durene	1.89×10^{-9}	1	35	[26]
Polybenzimidazole	3.15×10^{-12}	45	35	[27]
Poly(vinyl chloride)	5.36×10^{-12}	11	35	[28]
Poly(vinyl chloride)	6.30×10^{-12}	11	30	[29]
Polybenzimidazole	2.89×10^{-13}	9	20	[30]
	4.10×10^{-11}	20	270	
	3.41×10^{-11}	3	300	

Table 6. H_2/CO_2 separation performance by polymeric membrane.

The only shortcoming of polymeric membranes is the poor thermal stability at operating temperatures more than 100°C. Only polybenzimidazole was examined at the temperature range (300–700°C) for syngas purification. For polybenzimidazole membrane, the greatest performance in H_2 permeability and H_2/CO_2 selectivity is observed between 200 and 270°C. This peak performance can be related to the increasing diffusivity of the smaller H_2 molecule as temperature increases. More importantly, the performance of polymeric membranes depends on its stability in the environment of the real process. For example, exposure to gases such as CO_2, water vapor and H_2S may results in plasticization and mechanical fouling.

Due to the good thermal and hydrothermal stability, zeolite membranes were also viewed as another possible candidate for separation of H_2 and CO_2. Zeolite has ordered pore structure. If the pore channel size is proper, efficient size sieving could be achieved. Despite the relative simple concept, only a few types of zeolite are workable since this molecular sieve mechanism requires perfect membranes. This remains a challenge for zeolite membranes. The performance of a number of reported H_2/CO_2 separation using zeolite membranes is summarized in **Table 7**. In general, neither H_2 permeance nor H_2/CO_2 selectivity can exceed ~10^6 mol s^{-1} m^{-2} Pa^{-1} and ~50 to meet the industrial demands.

Metal organic framework (MOF) membrane has been an emerging candidate for H_2/CO_2 separation. In MOF materials, metal or metal oxide cluster cations are interconnected by organic anions. The coordination polymers form flexible frameworks, therefore such MOFs are called 'soft porous crystals'. **Table 8** summarizes the H_2 permeance and H_2/CO_2 selectivity using

Membrane	Permeance[a] (mol s^{-1} m^{-2} Pa^{-1}) or Permeability[b] (mol s^{-1} m^{-1} Pa^{-1})	H_2/CO_2 selectivity	Temperature (°C)	Reference
MFI	2.82×10^{-7}[a]	42.6	500	[31]
MFI	1.50×10^{-7}[a]	5	200	[32]
MFI template free	1.50×10^{-8}[a]	3	500	[33]
DDR	5.00×10^{-8}[a]	5	500	[34]
DDR by CVD	2.24×10^{-8}[a]	5.9	500	[35]
Zeolite-A	9.45×10^{-10}[a]	10	35	[36]
MFI	1.76×10^{-9}[a]	18	450	[37]
$AIPO_4$-5 Zeolite	3.15×10^{-9}[a]	24	35	[38]
ZSM-5	5.68×10^{-8}[a]		110	[39]
ZIF-69	6.60×10^{-8}[a]	1.8	25	[40]
13X with PI	6.93×10^{-11}[b]	2.8	25	[41]

[a]Permeance.
[b]Permeability.

Table 7. H_2/CO_2 separation performance by zeolite membranes.

Membrane	Permeance (mol s^{-1} m^{-2} Pa^{-1})	H_2/CO_2 selectivity	Temperature (°C)	Reference
MOF5	2.80×10^{-6}	4.3	25	[42]
MOF5	4.40×10^{-7}	4.4	25	[43]
MOF5	8.00×10^{-7}	3.5	25	[44]
Ni-MOF-74	1.27×10^{-5}	9.1	25	[45]
NH_2-MIL-53 (Al)	1.98×10^{-6}	30.9	25	[46]
MIL-53	5.00×10^{-7}	4	25	[47]
ZIF-7	7.40×10^{-7}	6.7	200	[48]
ZIF-7	4.55×10^{-7}	13	220	[49]
ZIF-7	4.57×10^{-6}	9.6	25	[50]
ZIF-7	3.05×10^{-6}	18.3	170	[50]
ZIF-8	5.00×10^{-8}	3.5	25	[51]
ZIF-8	1.80×10^{-7}	3	25	[52]
ZIF-8	2.66×10^{-5}	8.8	100	[53]
ZIF-22	2.00×10^{-7}	7.2	25	[54]
ZIF-90	2.95×10^{-7}	16.9	225	[55]
ZIF-95	1.90×10^{-6}	25.7	52	[56]
JUC-150	1.83×10^{-7}	38.7	25	[57]
HKUST-1	1.10×10^{-6}	5.5	190	[58]
MMOF	2.00×10^{-9}	5	190	[59]

Table 8. H_2/CO_2 separation performance by MOF membranes.

different MOF membranes. Despite relatively moderate permselectivity, attractively high permeances are observed. The operating temperature for MOF membranes is normally lower than the pre-combustion temperatures, owing to organic ligands. The synthesis of MOF membranes is relatively sophisticated so that the cost has to be notably reduced toward commercialization. There is still a long way for MOF membranes to fulfill the demands of industrial applications.

Unlike H_2-selective membranes, CO_2-selective membranes preferentially permeate CO_2 and thus they also enable the separation of CO_2 and H_2. Separating CO_2 from H_2 can only be realized through surface diffusion or solution diffusion driven by the difference in adsorb-ability or solubility between the gases. However, retaining the small molecules of H_2 but permeating the larger CO_2 is really challenging. To maximize the difference of adsorption or solution between the two gases, the temperature is required to be low, however, low temperatures are not favored by pre-combustion processes. From this point of view, CO_2-selective membranes are much less applicable than H_2-selective ones.

4. Membranes for post-combustion capture

Another situation where we need to separate CO_2 is after the fuel combustion. The exhaust gas (flue gas) mainly contains CO_2, H_2O and N_2. H_2O vapor is easy to be removed by condensation. More efforts are required to separate CO_2 and N_2 prior to further treatments such as compression. Unlike pre-combustion capture, post-combustion capture separates CO_2/N_2 at moderate temperatures and ambient atmosphere pressure. Such operating conditions seem less severe than those of pre-combustion processes. As a result, post-combustion capture has encountered much less difficulties and is therefore rather closer to practical application. The major challenge for post-combustion capture is the low CO_2 volumetric fraction in flue gas, that is, ~15%, which results in a low driving force of CO_2 permeation.

The separation of CO_2/N_2 mainly rely on surface diffusion and solution diffusion, which is driven by the difference in adsorb-ability and solubility between the gases. The good thing is that, compared to N_2, $CO_2$is more likely to be favored by majority of the membrane materials via adsorption or absorption. Furthermore, the diameter of CO_2 is slightly smaller than that of N_2, which also enhances the diffusion of CO_2 (see **Table 2**). Therefore, for post-combustion capture, CO_2-selective membranes are generally used.

To capture CO_2 from flue gas, a membrane should satisfy a few requirements such as high CO_2 permeability, high CO_2/N_2 selectivity, high thermal and chemical stability and acceptable costs. So far, polymer-based membranes are the only commercially viable type for CO_2 removal from flue gas. The membrane materials include cellulose acetate, polymides, polysulfone and polycarbonates. **Table 9** shows the performance of several such membranes.

Membrane	Permeance[a] (mol s^{-1} m^{-2} Pa^{-1}) or Permeability[b] (mol s^{-1} m^{-1} Pa^{-1})	CO_2/N_2 selectivity	Temperature (°C)	Reference
Cellulose acetate	2.48×10^{-7}[a]	40.17	Not reported	[60]
Polymides-TMeCat	6.30×10^{-10}[b]	25	30	[61]
Polymides-TMMPD	1.89×10^{-9}[b]	17.1	Not reported	[62]
Polymides-IMDDM	6.17×10^{-10}[b]	18.1	Not reported	[62]
Polysulfone-HFPSF-o-HBTMS	3.31×10^{-10}[b]	18.6	35	[63]
Polysulfone-HFPSF-TMS	3.47×10^{-10}[b]	18	35	[64]
PolysulfoneTMPSF-HBTMS	2.27×10^{-10}[b]	21.4	35	[65]
Polycarbonates-TMHFPC	3.50×10^{-10}[b]	15	35	[66]
Polycarbonates-FBPC	4.76×10^{-11}[b]	25.5	35	[67]

[a]Permeance.
[b]Permeability.

Table 9. CO_2/N_2 separation performance by polymer-based membranes.

Selectivity larger than 20 was observed for all the polymer-based membranes with decent permeability. The high solubility of CO_2 in polymers ensures sufficient CO_2/N_2 selectivity. Furthermore, polymers with a high fractional free volume present excellent gas transport properties.

Mixed-matrix membrane is a new option to enhance the properties of polymeric membranes. The microstructure consists of an inorganic material in the form of micro- or nanoparticles in discrete phase incorporated into a continuous polymeric matrix. The addition of inorganic materials in a polymer matrix offers improved thermal and mechanical properties for aggressive environments and stabilizes the polymer membranes against the changes in chemical and physical environments. Carbon molecular sieves membranes also show interesting performance for CO_2 separation applications. Polyimide is the most used precursor for carbon membranes. Carbon membranes improved gas transport properties for light gases (molecular size smaller than 4.0–4.5Å) with thermal and chemical stability. The major disadvantages of mixed-matrix and carbon membranes that hinder their commercialization include brittleness and the high cost that is 1–3 orders of magnitude greater than polymeric membranes.

5. Membranes for oxy-fuel combustion

In oxy-fuel combustion, oxygen is supplied for combustion instead of air. This avoids the presence of nitrogen in the exhaust gas, the major issue to be solved by post-combustion CO_2 capture technologies. With the use of pure oxygen for the combustion, the major composition of the flue gases is CO_2, water vapor, other impurities such as SO_2. Water vapor can be easily condensed and SO_2 can be removed by conventional desulphurization methods. The remained CO_2-rich gases (80–98 vol.% CO_2 depending on fuel used) can be compressed, transported and stored. This process is technically feasible but consumes large amounts of oxygen coming from an energy intensive air separation (O_2/N_2) unit.

The O_2/N_2separation follows the ion transport mechanism as depicted in **Figure 9** for air separation membrane. Oxygen molecules are converted to oxygen ions at the surface of the membrane and transported through the membrane by an applied electric voltage or oxygen partial pressure difference; these ions are reverted back to oxygen molecules after passing through the membrane. These membranes are O_2-selective in principle. Generally, fluorite-based and perovskite-based membranes are used to deliver oxygen through this mechanism.

Air separation is mostly carried out at atmosphere and meanwhile the permeate side connects to high speed sweep gas or vacuum. So, for convenience, the membrane performance is generally described as permeate flux instead of permeance. **Table 10** shows a list of oxygen permeation flux for the fluorite membranes. The oxygen permeation flux of fluorite-based membranes ranges from 10^{-4} to 10^{-6} mol s^{-1} m^{-2} between 650 and 1527°C. The highest oxygen flux was observed for $Bi_{1.5}Y_{0.3}Sm_{0.2}O_3$ compounds.

Membrane	O_2 flux (mol s^{-1} m^{-2})	Thickness (mm)	Temperature (°C)	Reference
$Bi_{0.75}Y_{0.5}Cu_{0.75}O_3$	2.80×10^{-5}–1.06×10^{-4}	2	650–850	[68]
$Bi_{1.5}Y_{0.3}Sm_{0.2}O_3$	4.40×10^{-3}–6.36×10^{-3}	1.2	825–875	[69]
$Ce_{0.8}Pr_{0.2}O_{2-\delta}$	1.33×10^{-4}–3.35×10^{-4}	1	850–950	[70]
$(ZrO_2)_{0.85}(CaO)_{0.15}$	1.70×10^{-4}	1	870	[71]
$[(ZrO_2)_{0.8}(CeO_2)_{0.2}]_{0.9}(CaO)_{0.1}$	1.36×10^{-6}–9.44×10^{-5}	2	1127–1527	[72]

Table 10. Oxygen permeation flux data for fluorite membranes.

Membrane	O_2 flux (mol s^{-1} m^{-2})	Thickness (mm)	Temperature (°C)	Reference
$BaBi_{0.4}Co_{0.2}Fe_{0.4}O_{3-\delta}$	3.064×10^{-3}–5.985×10^{-3}	1.5	800–925	[73]
$BaCo_{0.4}Fe_{0.5}Zr_{0.1}O_{3-\delta}$	1.908×10^{-3}–6.813×10^{-3}	1	700–950	[74]
$CaTi_{0.8}Fe_{0.2}O_{3-\delta}$	7.976×10^{-5}–2.185×10^{-4}	1	800–1000	[75]
$Gd_{0.6}Sr_{0.4}CoO_{3-\delta}$	1.179×10^{-2}	1.5	820	[76]
$LaCo_{0.8}Fe_{0.2}O_{3-\delta}$	1.786×10^{-4}	1.5	860	[76]
$La_{0.6}Sr_{0.4}Co_{0.8}Cu_{0.2}O_{3-\delta}$	1.417×10^{-2}	1.5	860	[76]
$SrCo_{0.8}Fe_{0.2}O_{3-\delta}$	2.485×10^{-2}	1	870	[77]

Table 11. Oxygen permeation flux data for perovskite membranes.

Performance of perovskite membranes are displayed in **Table 11**. Oxygen permeation flux with the magnitude of 10^{-2}–10^{-5} mol s^{-1} m^{-2} between 700 and 100°C was reported. The overall oxygen flux through perovskite membrane is superior to fluorite membrane. $SrCo_{0.8}Fe_{0.2}O_{3-\delta}$ exhibits the best oxygen flux.

In spite of a great number of works that attempt to efficiently separate air using membrane, the membrane technology for oxy-fuel combustion is still at its early stage of development. Compared to the conventional cryogenic air separation technique, the high temperature requirement and the resulting high costs of air separation membrane are unfavorable for commercialization. Some other issues such as high temperature sealing, chemical and mechanical stability and so on also need to be addressed prior to practical application. At present, there has not been any full scale oxy-fuel membrane project reported.

6. Summary of membranes applied in CCS

The aforementioned membranes are compared in **Table 12**. Their application situation, advantages and disadvantages are summarized accordingly.

Membrane type	Application	Advantages	Disadvantages
Metal membrane	Pre-combustion	Infinite H_2/CO_2 selectivity	High cost; poisoning; low driving force
Carbon membrane	Pre-combustion	Size sieving effect; high H_2/CO_2 selectivity	High cost; susceptible to oxygen; brittleness
Alumina membrane	Pre-combustion	Low cost; chemical and physical stability	Low H_2/CO_2 selectivity
Zeolite membrane	Pre-combustion and post-combustion	Low cost; chemical and physical stability	Low H_2/CO_2 selectivity
MOF membrane	Pre-combustion and post-combustion	Large pore volume and surface area	High cost
Silica membrane	Pre-combustion	Proper pore size; low cost; high thermal stability	Poor hydrothermal stability
Polymeric membrane	Post-combustion	Low cost; high CO_2/N_2 selectivity	Low chemical and physical stability; too thick
Fluorite membrane	Oxy-fuel combustion	High O_2/N_2 selectivity	Energy intensive; hard to seal
Perovskite membrane	Oxy-fuel combustion	High O_2/N_2 selectivity	Energy intensive; hard to seal; poisoning

Table 12. The summarization of membranes in CCS.

7. Membrane mass transfer theory

Membrane separation technique has been intensified with the growing needs for CCS. The major two targets of membrane are chasing high permeability and selectivity. The understanding of gas transport through membrane is of great importance in providing the guidance of membrane material design and synthesis improvement.

For all mass transfer problems, a general form is always expressed as a coefficient multiplied by a driving force as

$$J = C \cdot f. \tag{2}$$

where J is the mass transfer flux, C is the general transfer coefficient and f is the general driving force. The driving force can be the gradient of pressure, concentration, chemical potential or even electrical potential depending on the mass transfer mechanism. The coefficient can be permeability, diffusivity or other term depending on the term of driving force. For membrane mass transfer, the pressure difference and permeate flux are generally determined from experimental measurements, so the most common form in membrane industry is

$$J = \left(\frac{P}{l}\right) \cdot \Delta p. \tag{3}$$

Membrane thickness l is lumped together with permeability P into a term called permeance $\left(\frac{P}{l}\right)$, which is a convenient form of addressing permeation due to the difficulty in measuring the exact thickness of thin films. Generally, membrane films interpenetrate into the pores of the interlayer or substrate (interlayer-free membrane). Hence, the thickness is not homogenous.

7.1. Viscous flow model

When the pore size is large, the gas molecule-molecule collision is relatively dominant than gas molecule-wall collision. That means the mean free path is far less than the pore size

$$\frac{\lambda}{d} \ll 1, \tag{4}$$

where λ is the mean free path and d is the diameter of the pore.

In such situation, viscosity plays an important role in the mass transfer and the permeate flux across the membrane is described by viscous flow model:

$$J = -\frac{\varepsilon_p}{\tau_T}\frac{r_p^2}{8\eta}\frac{p}{RT}\frac{dp}{dz}, \tag{5}$$

where η is the viscosity, R is the gas constant, T is the temperature, p is the pressure, ε_p is the porosity of the pore, τ_T is the tortuosity of the pore and r_p is the pore radius. Viscosity increases with temperature for gases. From Eq. (5), it should be noted that if the transportation is in the viscous regime, the flux is a decreasing function of temperature. Although the viscosity is different from gas to gas, gas mixtures share a singular viscosity value when they are well mixed due to the intensive intermolecular collision. Therefore, there is no selectivity for all gases in the viscous regime even if they have different viscosities.

7.2. Knudsen diffusion model

When the pore size is reduced down to the scale much smaller than mean free path, the molecular-wall collision is more dominating than intermolecular collision. In this situation, the viscosity is not playing a role for the gas transportation. Instead, the pore geometry and gas molecule velocity influence more significantly in the mass transfer. This type of transport is called Knudsen diffusion. If the molecule to wall collisions is dominant over intermolecular collision, the Knudsen number must be much higher than 1.

$$Kn = \frac{\lambda}{d} \gg 1, \tag{6}$$

where Kn is called Knudsen number. The permeate flux is described by the Knudsen diffusion model

$$J = -\frac{2}{3}\frac{\varepsilon_p r_p}{\tau_T}\sqrt{\frac{8}{\pi RTM}}\frac{dp}{dz} = -\frac{2}{3}\frac{\varepsilon_p r_p}{\tau_T}\sqrt{\frac{8}{\pi RTM}}\frac{\Delta p}{l}, \tag{7}$$

where M is the molecular weight.

Based on Eq. (3) the permeance of Knudsen diffusion is

$$\left(\frac{P}{l}\right) = -\frac{2}{3}\frac{\varepsilon_p r_p}{\tau_T l}\sqrt{\frac{8}{\pi RTM}}. \tag{8}$$

For the same pore at a fixed temperature, the permeate flux is determined by the molar weight and in principle, the selectivity is the root square of the reciprocal of molar weights. However, due to the limited selectivity, Knudsen diffusion is rarely used in practice for separating real gas mixtures.

7.3. Surface diffusion model

For ultra-micro-porous (d_p < 5Å) material, the Lennard-Jones (L-J) potential from atoms, which forms the pore wall starts to overlap inside the pore. Consequently, there is a very deep potential well around the wall and the distance from wall to the well is around the scale of gas molecule diameter. In this situation, the gas molecule's motion is significantly affected by the potential fields. Since the intrinsic nature of gas is seeking for lower potential, thus adsorption preferentially takes place around the pore wall due to the existence of the potential well. As such, the model is called surface diffusion. A brief introduction has been given in Section 2.3.2. of this chapter, but here a more analytical and mathematical description of surface diffusion will be provided.

The original expression of mass transfer across the membrane is given by

$$J = -qD\frac{1}{RT}\frac{d\mu}{dz}, \tag{9}$$

where q is the molar concentration of gas in the pore, D is the diffusivity, μ is the chemical potential and z is the space coordinate in the membrane thickness direction.

Assuming equilibrium between the membrane surface concentration and the bulk gas phase, the following relationship for the chemical potential is applicable

$$\mu_0 = \mu + RT\ln p, \tag{10}$$

where p is the absolute pressure.

Using Eq. (10), Eq. (9) is converted to

$$J = -D\frac{d\ln p}{d\ln q}\frac{dq}{dz} = -D\Gamma\frac{dq}{dz}. \tag{11}$$

$\Gamma = \frac{d\ln p}{d\ln q}$ is defined as thermodynamic factor. In micro-porous material, the adsorbed gas concentration generally follows Langmuir isotherm,

$$q = q_{sat}\frac{bp}{1+bp}, \tag{12}$$

where b is Langmuir equilibrium constant. Bring Eq. (12) to Eq. (11) gives

$$J = -q_{sat}D\frac{1}{1-\theta}\frac{d\theta}{dz}, \tag{13}$$

where $\theta = \frac{q}{q_{\text{sat}}}$ is called occupancy. Thermal dynamic factor $\Gamma = \frac{1}{1-\theta}$ is derived from Langmuir isotherm. Surface diffusion is often applied in separating gas mixtures, which has very different adsorption capacity in the same material.

However, with elevated temperature, the adsorption is getting weaker and Langmuir isotherm is approaching to Henry's law.

$$q = Kp, \tag{14}$$

where K is Henry's constant. Bring Eq. (14) to Eq. (11), we get Fick's first law

$$J = -D\frac{\mathrm{d}q}{\mathrm{d}z}. \tag{15}$$

Diffusivity D is a function of temperature. The temperature dependence usually obeys an Arrhenius relation

$$D = D_0 \exp\left(-\frac{E_{\mathrm{d}}}{RT}\right), \tag{16}$$

where D_0 is a pre-exponential coefficient depending on the average distance, the frequency and average velocity of gas jump and E_{d} is diffusion activation energy. Henry's constant is a function of temperature according to a van't Hoff relation:

$$K = K_0 \exp\left(\frac{Q}{RT}\right), \tag{17}$$

where K_0 is a pre-exponential coefficient, Q is the heat of adsorption.

Eqs. (14)–(17) can be combined as

$$J = -D_0 K_0 \exp\left(-\frac{E_{\mathrm{d}} - Q}{RT}\right)\frac{\mathrm{d}p}{\mathrm{d}z} = -D_0 K_0 \exp\left(-\frac{E_{\mathrm{a}}}{RT}\right)\frac{\mathrm{d}p}{\mathrm{d}z}. \tag{18}$$

E_{a} is called apparent activation energy, which is defined as

$$E_{\mathrm{a}} = E_{\mathrm{d}} - Q. \tag{19}$$

Apparent activation energy determines whether the permeate flux is an increasing function to temperature or not, so this type of diffusion is called activated transport.

Assuming a uniform pressure gradient, Eq. (18) is simplified to

$$J = -D_0 K_0 \exp\left(-\frac{E_{\mathrm{a}}}{RT}\right)\frac{\Delta p}{l}. \tag{20}$$

The permeance $\left(\frac{P}{l}\right)$ is the coefficient between flux and pressure drop according to Eq. (3)

$$\left(\frac{P}{l}\right) = \frac{D_0 K_0}{l} \exp\left(-\frac{E_{\mathrm{a}}}{RT}\right). \tag{21}$$

http://dx.doi.org/10.5772/65723

Activated transport is generally used to separate gas mixtures, which has different sign of apparent activation energy and the separation performance will be enhanced at elevated temperatures.

7.4. Gas translation diffusion model

If the pore size is further reduced to the molecular level, there is no potential well inside the pore. Instead, the positive potential overlaps, which forms a potential barrier. Only the gas molecules, which have kinetic energy higher than the potential barrier, are possible to make a successful jump to complete permeation. This model is called gas translation diffusion. The permeate flux of gas translation follows Fick's first law as derived in Eq. (15) with the difference in diffusion coefficient term.

$$D_{\mathrm{GT}} = \frac{\lambda}{Z_{\mathrm{n}}}\sqrt{\frac{8RT}{\pi M}}\exp^{\left(-\frac{E_{\mathrm{GT}}}{RT}\right)}, \tag{22}$$

where λ is the jump length, Z_{n} is the number of available jump directions and E_{GT} is the potential barrier. By considering ideal gas law

$$p = cRT. \tag{23}$$

Gas translation permeance should rewrite as

$$\left(\frac{P}{l}\right) = \frac{\lambda}{Z_{\mathrm{n}}}\sqrt{\frac{8}{\pi \mathrm{MRT}}}\exp^{\left(-\frac{E_{\mathrm{GT}}}{RT}\right)}. \tag{24}$$

7.5. Oscillator model

If we assume the pore is a cylinder, the gas molecules are hopping in the pore cylinder from entrance to the exit. The gas molecule trajectory looks like oscillating on the pore cross section. The gas travels with speed between collisions and loses all the momentum when colliding on the wall. This model is a more recent development in mass transfer theory by Bhatia et al. [78, 79].

From Newton's law,

$$\langle \mathrm{v_z} \rangle = \frac{D}{k_B T} f = \frac{f}{m}\langle \tau \rangle, \tag{25}$$

the gas diffusivity in the pore is derived

$$D = \frac{k_B T}{m}\langle \tau \rangle, \tag{26}$$

where $\langle v_z \rangle$ is the average velocity in the permeation direction, k_B the Boltzmann constant, f the force, m the molecule mass and $\langle \tau \rangle$ the average hopping time. The hopping time of each molecule depends on the pore potential distribution, its radial coordinate and momentum

$$\tau(r, p_r, p_\theta) = 2m\int_{\mathrm{r_{c0}(r,p_r,p_\theta)}}^{\mathrm{r_{c1}(r,p_r,p_\theta)}} \frac{\mathrm{d}\,r'}{p_r(r', r, p_r, p_\theta)}. \tag{27}$$

$p_r(r', r, p_r, p_\theta)$ is the radial momentum at r' of a molecule, which had radial momentum p_r at r. $r_{c1}(r, p_r, p_\theta)$ and $r_{c0}(r, p_r, p_\theta)$ are the r' solution of radial momentum $p_r(r', r, p_r, p_\theta) = 0$. The radial momentum is derived from the conservation of total energy or Hamiltonian

$$E_t(r, p_r, p_\theta) = \phi(r) + \frac{p_r^2}{2m} + \frac{p_\theta^2}{2m\,r^2}, \tag{28}$$

where $\phi(r)$ is the radial L-J potential, which could be derived from pore structure and gas property. The force in radial direction is the partial derivative of total energy with respect to r

$$\frac{\mathrm{d}p_r}{\mathrm{d}t} = -\frac{\partial E_t}{\partial r}. \tag{29}$$

Combining Eqs. (28) and (29) gives the radial momentum

$$p_r(r', r, p_r, p_\theta) = \left\{2m\left[\phi(r) - \phi(r')\right] + p_r^2(r) + \frac{p_\theta}{r^2}\left(1 - \frac{r^2}{r'^2}\right)\right\}^{1/2}. \tag{30}$$

Considering a canonical distribution for p_r and p_θ, we have

$$\psi(r, p_r, p_\theta) = \psi_0 \exp\left[-\frac{1}{RT}\left(\phi(r) + \frac{p_r^2}{2m} + \frac{p_\theta^2}{2m\,r^2}\right)\right]. \tag{31}$$

The diffusion coefficient expression is obtained from Eqs. (26), (30) and (31)

$$D(r_p, T) = \frac{2}{\pi m \int_0^\infty r\, e^{-\frac{\phi(r)}{RT}}\,\mathrm{d}r} \int_0^\infty e^{-\frac{\phi(r)}{RT}}\,\mathrm{d}r \int_0^\infty e^{-\frac{p_r^2}{2mRT}}\,\mathrm{d}p_r \int_0^\infty e^{-\frac{p_\theta^2}{2mr^2RT}}\,\mathrm{d}p_\theta \int_{r_{c0}(r,p_r,p_\theta)}^{r_{c1}(r,p_r,p_\theta)} \frac{\mathrm{d}r'}{p_r(r', r, p_r, p_\theta)}. \tag{32}$$

Oscillator model is a pure theoretical and analytical approach without any empirical or semi-empirical factors. It takes account adsorption effect and applies to all pore sizes, pressure and temperatures.

Besides the mass transfer models introduced above, there are some other methods to study the membrane gas transport from a theoretical perspective. Monte Carlo and molecular dynamics are also major techniques to investigate the micropore mass transfer. Because this chapter focused on membrane CCS technology rather than transport phenomena, other sophisticated theories are not demonstrated here.

8. Current status of membrane application

8.1. Membranes for pre-combustion

The membrane separation for pre-combustion is not a mature technology so far. There has not been industry-scale membrane system. However, a few pilot scale pre-combustion membrane systems have demonstrated the potential of extending the system to enlarged scale.

Eltron Research & Development Inc. developed a pilot-scale pre-combustion membrane with 100 kg day^{-1} H_2 production from 2005. They employed alloy membrane to separate H_2 according to Sieverts' Law. This project successfully improved membrane-based integrated gasification combined cycle (IGCC) flow sheets, achieving carbon capture greater than 95%.

Another pilot-scale pre-combustion membrane set-up was constructed by Worcester Polytechnic Institute's (WPI) in 2010. More than 566 L H_2 was produced per day. Stable H_2 fluxes were achieved in actual syngas atmospheres at 450°C for more than 470 h under 12 bar pressure difference. The implement MembraGuardTM (T3's technology) inhibited surface poisoning by hydrogen sulfide (H_2S) and H_2 permeation showed good stability for more than 250 h.

8.2. Membranes for post-combustion

Membrane separation for post-combustion is a relatively mature technique. In 1995, the largest membrane-based natural gas processing plant in the world was built in Kadanwari, Pakistan. Cellulose acetate membrane was applied in this project to separate CO_2. The Kadanwari system is a two-stage unit designed to treat 25×10^5 m^3 h^{-1} of feed gas at 90 bar. The CO_2 content is reduced from 12% to less than 3%.

After Kadanwari plant, the Qadirpur plant started in the same year and the processing capacity exceeded Kadanwari plant with 31×10^5 m^3 h^{-1} of feed gas at 59 bar. The CO_2 content is reduced from 6.5 to 2%. The Qadirpur plant was upgraded to 64×10^5 m^3 h^{-1} of feed gas in 2003.

8.3. Membranes for oxy-fuel combustion

Air separation membrane is still in its early stage. In view of the high energy requirement of ion transport mechanism, air separation membrane can hardly challenge the traditional cryogenic air separation for large scale product.

Air products, which have been developing ion transport membrane technology since 1988 and the DOE (US Department of Energy) are collecting data from a pilot plant near Baltimore in Maryland, with the capacity of 5 tons of oxygen per day. This facility will lead to the next step of designing and building a larger membrane air separation unit (150 tons oxygen per day).

9. Techno-economic of membrane

The conventional CO_2 capture process is absorption (with ammines). Amine-based absorption is the most common technology. However, the corrosion, degradation and high regeneration energy of amine significantly increase the electricity cost. Substantial technological improvements and alternative technologies are highly needed to lower the CO_2 capture cost.

The economic indicator CO_2 avoided ($/ton) is an established term for measuring and comparing different CO_2 capture strategies such as absorption, adsorption, cryogenic separation and membrane separation. It is the additional cost of establishing and running a CO_2 capture facility for an industrial plant or power plant compared to the respective plant without CO_2 capture. The CO_2 avoided is expressed as:

$$CO_2\text{ avoided} = \frac{\text{LCOE(capture)} - \text{LCOE(ref.)}}{CO_2\text{ emission(ref.)} - CO_2\text{ emission(capture)}}, \quad (33)$$

where ref. and capture mean the reference plant without capture and the respective plant with CO_2 capture facility. LCOE is the levelized cost of electricity which is expressed as:

$$LCOE = \frac{\text{sum of cost over lifetime}}{\text{sum of electrical energy produced over lifetime}}. \quad (34)$$

A brief techno-economic comparison was made between two power plants using conventional amine scrubbers in and a power plant using polymer membrane (**Table 13**). The estimates are subject to uncertainty because we cannot accurately predict all input parameters such as fuel price, operational and maintenance cost. The aim of the comparison is not to give absolute costs, but to illustrate indicatively that the costs per ton CO_2 avoided. The overall comparison indicates that the case employing membrane separation results in slightly lower LCOE and CO_2 avoided than traditional amine-based solvent scrubbing. Although this cannot judge the membrane economical advantage, the comparison at least indicates that membrane separation is competitive to the amine-based solvent scrubbing. However, significant efforts are still required to improve the membrane properties so as to achieve higher stability, permeate purity and recovery.

Organization		**Carnegie Mellon University**	**Electric Power Research Institute**	**Membrane Technology and Research, Inc**
CCS technology		Amine-based	Amine-based	Membrane-based
Location		USA	USA	USA
Coal type		Bitcoal	Bitcoal	Illinois#6
Plant size (MW)		575	600	580
Designed CO_2 capture rate (%)		90	85	90
CO_2 emission (kg/MWh)	Reference	811	836	760
	Capture	107	126	87
Net power output (MW)	Reference	528	600	550
	Capture	493	550	461
Net plant efficiency (LHV, %)	Reference	41.4	40	41.4
	Capture	31.5	29.1	34.4
Efficiency penalty (%)		9.9	10.9	7
Capital costs ($/kW)	Reference	1696	2104	1727
	Capture	2759	3516	2627
LCOE ($/MWh)	Reference	62	77	62
	Capture	104	127	93
CO_2 avoided ($/ton)		58	71	46

Table 13. Techno-economic comparisons between amine-based CO_2 removal and membrane separation.

Author details

Guozhao Ji and Ming Zhao*

*Address all correspondence to: ming.zhao@tsinghua.edu.cn

School of Environment, Tsinghua University, Beijing, China

References

[1] Brinker, C. J.; Scherer, G. W., Sol-gel science. *Material Science* **1989**,*6* (1), 475–495.

[2] Khatib, S. J.; Oyama, S. T.; Souza, K. R. D., et al., Chapter 2–Review of silica membranes for hydrogen separation prepared by chemical vapor deposition. *Journal of Membrane Science & Technology* **2011**,*14* (25), 25–60.

[3] Holleck, G. L., Diffusion and solubility of hydrogen in palladium and palladium–silver alloys. *The Journal of Physical Chemistry* **1970**,*74* (3), 503–511.

[4] Govind, R.; Atnoor, D., Development of a composite palladium membrane for selective hydrogen separation at high temperature. *Industrial & Engineering Chemistry Research* **1991**,*30* (3), 591–594.

[5] Mckinley D L. Method for hydrogen separation and purification: US, US 3439474 A[P]. 1969.

[6] Gade, S. K.; Thoen, P. M.; Way, J. D., Unsupported palladium alloy foil membranes fabricated by electroless plating. *Journal of Membrane Science* **2008**,*316* (1–2), 112–118.

[7] Morreale, B. D.; Ciocco, M. V.; Enick, R. M., et al., The permeability of hydrogen in bulk palladium at elevated temperatures and pressures. *Journal of Membrane Science* **2003**,*212* (1–2), 87–97.

[8] Altinisik, O.; Dogan, M.; Dogu, G., Preparation and characterization of palladium-plated porous glass for hydrogen enrichment. *Catalysis Today* **2005**,*105* (3–4), 641–646.

[9] Zhang, Y.; Gwak, J.; Murakoshi, Y., et al., Hydrogen permeation characteristics of thin Pd membrane prepared by microfabrication technology. *Journal of Membrane Science* **2006**,*277* (1), 203–209.

[10] Uemiya, S.; Sato, N.; Ando, H., et al., The water gas shift reaction assisted by a palladium membrane reactor. *Industrial & Engineering Chemistry Research* **2002**,*30* (3), 585–589.

[11] Zhang, X.; Xiong, G.; Yang, W., A modified electroless plating technique for thin dense palladium composite membranes with enhanced stability. *Journal of Membrane Science* **2008**,*314* (1–2), 226–237.

[12] Itoh, N.; Akiha, T.; Sato, T., Preparation of thin palladium composite membrane tube by a CVD technique and its hydrogen permselectivity. *Catalysis Today* **2005**,*104* (2), 231–237.

[13] Wang, L.; Yoshiie, R.; Uemiya, S., Fabrication of novel Pd–Ag–Ru/Al_2O_3 ternary alloy composite membrane with remarkably enhanced H2 permeability. *Journal of Membrane Science* **2007**,*306* (1), 1–7.

[14] Uemiya, S.; Sato, N.; Ando, H., et al., Separation of hydrogen through palladium thin film supported on a porous glass tube. *Journal of Membrane Science* **1991**,*56* (3), 303–313.

[15] Se, N.; Kh, L., A study on the palladium/nickel composite membrane by vacuum electrodeposition. *Journal of Membrane Science* **2000**,*170* (1), 91–99.

[16] Abate, S.; Genovese, C.; Perathoner, S., et al., Performances and stability of a Pd-based supported thin film membrane prepared by EPD with a novel seeding procedure. Part 1–Behaviour in H_2:N_2 mixtures. *Catalysis Today* **2009**,*145* (1–2), 63–71.

[17] Tong, H. D.; Gielens, F. C.; Gardeniers, J. G. E., et al., Microfabricated palladium-silver alloy membranes and their application in hydrogen separation. *Industrial & Engineering Chemistry Research* **2004**,*43* (15), 4182–4187.

[18] de Vos, R. M.; Verweij, H., Improved performance of silica membranes for gas separation. *Journal of Membrane Science* **1998**,*143* (1–2), 37–51.

[19] Giessler, S.; Jordan, L.; Costa, J. C. D. D., et al., Performance of hydrophobic and hydrophilic silica membrane reactors for the water gas shift reaction. *Separation and Purification Technology* **2003**,*32* (1–3), 255–264.

[20] Yoshioka, T.; Nakanishi, E.; Tsuru, T., et al., Experimental studies of gas permeation through microporous silica membranes. *AIChE Journal* **2001**,*47* (9), 2052–2063.

[21] Battersby, S.; Tasaki, T.; Smart, S., et al., Performance of cobalt silica membranes in gas mixture separation. *Journal of Membrane Science* **2009**,*329* (1–2), 91–98.

[22] Diniz da Costa, J. C.; Lu, G. Q.; Rudolph, V., et al., Novel molecular sieve silica (MSS) membranes: characterisation and permeation of single-step and two-step sol–gel membranes. *Journal of Membrane Science* **2002**,*198* (1), 9–21.

[23] Ballinger, B.; Motuzas, J.; Smart, S., et al., Palladium cobalt binary doping of molecular sieving silica membranes. *Journal of Membrane Science* **2014**,*451*, 185–191.

[24] Miller, C. R.; Wang, D. K.; Smart, S., et al., Reversible redox effect on gas permeation of cobalt doped ethoxy polysiloxane (ES40) membranes. *Scientific Reports* **2013**,*3*, 1–6.

[25] Richard, V.; Favre, E.; Tondeur, D., et al., Experimental study of hydrogen, carbon dioxide and nitrogen permeation through a microporous silica membrane. *Chemical Engineering Journal* **2001**,*84* (3), 593–598.

[26] Chung, T. S.; Lu, S.; Pei, S. T., Surface modification of polyimide membranes by diamines for H_2 and CO_2 separation. *Macromolecular Rapid Communications* **2006**,*27* (13), 998–1003.

[27] Choi, S.; Coronas, J.; Lai, Z., et al., Fabrication and gas separation properties of polybenzimidazole (PBI)/nanoporous silicates hybrid membranes. *Journal of Membrane Science* **2008**,*316* (1–2), 145–152.

[28] Tikhomirov, B. P.; Hopfenberg, H. B.; Stannett, V., et al., Permeation, diffusion and solution of gases and water vapor in unplasticized poly(vinylchloride). *Die Makromolekulare Chemie* **1968**,*118* (1), 177–188.

[29] Robeson, L. M., Polymer membranes for gas separation. *Current Opinion in Solid State & Materials Science* **1999**,*4* (6), 549–552.

[30] Pesiri, D. R.; Jorgensen, B.; Dye, R. C., Thermal optimization of polybenzimidazole meniscus membranes for the separation of hydrogen, methane and carbon dioxide. *Journal of Membrane Science* **2003**,*218* (1–2), 11–18.

[31] Zhang, Y.; Wu, Z.; Zhou, H., et al., Hydrogen-selective zeolite membrane reactor for low temperature water gas shift reaction. *Chemical Engineering Journal* **2012**,*197* (14), 314–321.

[32] Burggraaf, A. J.; Vroon, Z. A. E. P.; Keizer, K., et al., Permeation of single gases in thin zeolite MFI membranes. *Journal of Membrane Science* **1998**,*144* (144), 77–86.

[33] Kanezashi, M.; O'Brien, J.; Lin, Y. S., Template-free synthesis of MFI-type zeolite membranes: permeation characteristics and thermal stability improvement of membrane structure. *Journal of Membrane Science* **2006**,*286* (1–2), 213–222.

[34] Tomita, T.; Nakayama, K.; Sakai, H., Gas separation characteristics of DDR type zeolite membrane. *Microporous and Mesoporous Materials* **2004**,*68* (1), 71–75.

[35] Kanezashi, M.; O'Brien-Abraham, J.; Lin, Y. S., et al., Gas permeation through DDR-type zeolite membranes at high temperatures. *AIChE Journal* **2008**,*54* (6), 1478–1486.

[36] Aoki, K.; Kusakabe, K.; Morooka, S., Gas permeation properties of A-type zeolite membrane formed on porous substrate by hydrothermal synthesis. *Journal of Membrane Science* **1998**,*141* (2), 197–205.

[37] Gu, X.; Tang, Z.; Dong, J., On-stream modification of MFI zeolite membranes for enhancing hydrogen separation at high temperature. *Microporous and Mesoporous Materials* **2008**,*111* (1), 441–448.

[38] Guan, G.; Tanaka, T.; Kusakabe, K., et al., Characterization of AlPO4-type molecular sieving membranes formed on a porous α-alumina tube. *Journal of Membrane Science* **2003**,*214* (2), 191–198.

[39] Richter, H.; Voigt, I.; Fischer, G., et al., Preparation of zeolite membranes on the inner surface of ceramic tubes and capillaries. *Separation and Purification Technology* **2003**,*32* (1), 133–138.

[40] Su, J.; Qian, Y.; Teo, J. F., et al., Cellulose acetate nanofiltration hollow fiber membranes for forward osmosis processes. *Journal of Membrane Science* **2010**,*355* (1–2), 36–44.

[41] Chaidou, C. I.; Pantoleontos, G.; Koutsonikolas, D. E., et al., Gas separation properties of polyimide-zeolite mixed matrix membranes. *Separation Science and Technology* **2012**,*47* (7), 950–962.

[42] Cao, F.; Zhang, C.; Xiao, Y., et al., Helium recovery by a Cu-BTC metal–organic-framework membrane. *Industrial & Engineering Chemistry Research* **2012**,*51* (34), 11274–11278.

[43] Zhao, Z.; Ma, X.; Li, Z., et al., Synthesis, characterization and gas transport properties of MOF-5 membranes. *Journal of Membrane Science* **2011**,*382* (1–2), 82–90.

[44] Yoo, Y.; Lai, Z.; Jeong, H. K., Fabrication of MOF-5 membranes using microwave-induced rapid seeding and solvothermal secondary growth. *Microporous and Mesoporous Materials* **2009**,*123* (1–3), 100–106.

[45] Lee, D. J.; Li, Q.; Kim, H., et al., Preparation of Ni-MOF-74 membrane for CO_2 separation by layer-by-layer seeding technique. *Microporous and Mesoporous Materials* **2012**,*163* (1), 169–177.

[46] Zhang, F.; Zou, X.; Gao, X., et al., Hydrogen selective NH_2-MIL-53(Al) MOF membranes with high permeability. *Advanced Functional Materials* **2012**,*22* (17), 3583–3590.

[47] Hu, Y.; Dong, X.; Nan, J., et al., Metal-organic framework membranes fabricated via reactive seeding. *Chemical Communications* **2010**,*47* (2), 737–739.

[48] Li, Y. S.; Liang, F. Y.; Bux, H., et al., Molecular sieve membrane: supported metal–organic framework with high hydrogen selectivity. *Angewandte Chemie International Edition* **2010**,*49* (3), 548–551.

[49] Li, Y.; Liang, F.; Bux, H., et al., Zeolitic imidazolate framework ZIF-7 based molecular sieve membrane for hydrogen separation. *Journal of Membrane Science* **2010**,*354* (1), 48–54.

[50] Melgar, V. M. A.; Kwon, H. T.; Kim, J., Direct spraying approach for synthesis of ZIF-7 membranes by electrospray deposition. *Journal of Membrane Science* **2014**,*459* (2), 190–196.

[51] Bux, H.; Liang, F.; Li, Y., et al., Zeolitic imidazolate framework membrane with molecular sieving properties by microwave-assisted solvothermal synthesis. *Journal of the American Chemical Society* **2009**,*131* (44), 16000–16001.

[52] Mccarthy, M. C.; Varelaguerrero, V.; Barnett, G. V., et al., Synthesis of zeolitic imidazolate framework films and membranes with controlled microstructures. *Langmuir* **2010**,*26* (26), 14636–14641.

[53] Huang, A.; Liu, Q.; Wang, N., Highly hydrogen permselective ZIF-8 membranes supported on polydopamine functionalized macroporous stainless-steel-nets. *Journal of Materials Chemistry A* **2014**,*2* (22), 8246–8251.

[54] Huang, A.; Bux, H.; Steinbach, F., et al., Molecular-sieve membrane with hydrogen permselectivity: ZIF-22 in LTA topology prepared with 3-aminopropyltriethoxysilane as covalent linker. *Angewandte Chemie International Edition* **2010**,*49* (29), 4958–4961.

[55] Huang, A.; Wang, N.; Kong, C., et al., Organosilica-functionalized zeolitic imidazolate framework ZIF-90 membrane with high gas-separation performance. *Angewandte Chemie* **2012**,*124* (42), 10703–10707.

[56] Huang, A.; Chen, Y.; Wang, N., et al., A highly permeable and selective zeolitic imidazolate framework ZIF-95 membrane for H(2)/CO(2) separation. *Chemical Communications* **2012**,*48* (89), 10981–10983.

[57] Kang, Z.; Xue, M.; Fan, L., et al., Highly selective sieving of small gas molecules by using an ultra-microporous metal-organic framework membrane. *Energy & Environmental Science* **2014**,*7* (12), 4053–4060.

[58] Guerrero, V. V.; Yoo, Y.; Mccarthy, M. C., et al., HKUST-1 membranes on porous supports using secondary growth. *Journal of Materials Chemistry* **2010**,*20* (19), 3938–3943.

[59] Ranjan, R.; Tsapatsis, M., Microporous metal organic framework membrane on porous support using the seeded growth method. *Chemistry of Materials* **2009**,*21* (20), 4920–4924.

[60] Ahmad, A. L.; Jawad, Z. A.; Low, S. C., et al., A cellulose acetate/multi-walled carbon nanotube mixed matrix membrane for CO_2/N_2 separation. *Journal of Membrane Science* **2014**,*451* (1), 55–66.

[61] Majdi Al-Masri; Detlev Fritsch; Kricheldorf Hans R., New polyimides for gas separation. 2. Polyimides derived from substituted catechol bis(etherphthalic anhydride)s. *Macromolecules* **2000**,*33* (19), 7127–7135.

[62] Abajo, J. D.; Campa, J. G. D. L.; Lozano, A. E., Designing aromatic polyamides and polyimides for gas separation membranes. *Macromolecular Symposia* **2003**,*199* (1), 293–306.

[63] Dai, Y.; Guiver M. D.; Robertson G. P., et al., Enhancement in the gas permeabilities of novel polysulfones with pendant 4-trimethylsilyl-α-hydroxylbenzyl substituents. *Macromolecules* **2003**,*36* (18), 6807–6816.

[64] Dai, Y.; Guiver, M. D.; Robertson, G. P., et al., Preparation and characterization of polysulfones containing both hexafluoroisopropylidene and trimethylsilyl groups as gas separation membrane materials. *Macromolecules* **2004**,*37* (4), 1403–1410.

[65] Dai, Y.; Guiver, M. D.; Robertson, G. P., et al., Modified polysulfones 5: synthesis and characterization of tetramethyl polysulfones containing trimethylsilyl groups and their gas transport properties. *Polymer* **2002**,*43* (20), 5369–5378.

[66] Hellums, M. W.; Koros, W. J.; Husk, G. R., et al., Fluorinated polycarbonates for gas separation applications. *Journal of Membrane Science* **1989**,*46* (1), 93–112.

[67] Aguilar-Vega, M.; Paul, D., Gas transport properties of polycarbonates and polysulfones with aromatic substitutions on the bisphenol connector group. *Journal of Polymer Science, Part B: Polymer Physics* **1993**,*31* (11), 1599–1610.

[68] Nigara, Y.; Mizusaki, J.; Ishigame, M., Measurement of oxygen permeability in CeO_2 doped CSZ. *Solid State Ionics* **1995**,*79*, 208–211.

[69] Akin, F. T.; Lin, Y. S., Selective oxidation of ethane to ethylene in a dense tubular membrane reactor. *Journal of Membrane Science* **2002**,*209* (2), 457–467.

[70] Fagg, D.; Marozau, I.; Shaula, A., et al., Oxygen permeability, thermal expansion and mixed conductivity of $GdxCe_{0.8-x}Pr_{0.2}O_{2-\delta}$, $x = 0, 0.15, 0.2$. *Journal of Solid State Chemistry* **2006**,*179* (11), 3347–3356.

[71] Teraoka, Y.; Zhang, H.-M.; Furukawa, S., et al., Oxygen permeation through perovskite-type oxides. *Chemistry Letters* **1985**,*14* (11), 1743–1746.

[72] Nigara, Y.; Mizusaki, J.; Ishigame, M., Proceedings of the 20th commemorative symposium on solid state ionics in Japan measurement of oxygen permeability in CeO_2 doped CSZ. *Solid State Ionics* **1995**,*79*, 208–211.

[73] Shao, Z.; Xiong, G.; Cong, Y., et al., Synthesis and oxygen permeation study of novel perovskite-type $BaBi_xCo_{0.2}Fe_{0.8-x}O_{3-\delta}$ ceramic membranes. *Journal of Membrane Science* **2000**,*164* (1–2), 167–176.

[74] Tong, J.; Yang, W.; Zhu, B., et al., Investigation of ideal zirconium-doped perovskite-type ceramic membrane materials for oxygen separation. *Journal of Membrane Science* **2002**,*203* (1–2), 175–189.

[75] Iwahara, H.; Esaka, T.; Mangahara, T., Mixed conduction and oxygen permeation in the substituted oxides for $CaTiO_3$. *Journal of Applied Electrochemistry* **1988**,*18* (18), 173–177.

[76] Teraoka, Y.; Nobunaga, T.; Yamazoe, N., Effect of cation substitution on the oxygen semi-permeability of perovskite-type oxides. *Chemistry Letters* **1988**,*3* (3), 503–506.

[77] Teraoka, Y.; Zhang, H. M.; Furukawa, S., et al., Oxygen permeation through perovskite-type oxides. *Chemistry Letters* **1985**,*11* (11), 1743–1746.

[78] Jepps, O. G.; Bhatia, S. K.; Searles, D. J., Wall mediated transport in confined spaces: exact theory for low density. *Physical Review Letters* **2003**,*91* (12), 126102.

[79] Bhatia, S. K.; Jepps, O.; Nicholson, D., Tractable molecular theory of transport of Lennard-Jones fluids in nanopores. *The Journal of Chemical Physics* **2004**,*120* (9), 4472–4485.

Chapter 4

Emerging New Types of Absorbents for Postcombustion Carbon Capture

Quan Zhuang, Bruce Clements and Bingyun Li

Additional information is available at the end of the chapter

http://dx.doi.org/10.5772/65739

Abstract

Carbon capture is the most probable technology in combating anthropogenic increase of CO_2 in the atmosphere. Works on developing emerging absorbents for improving carbon capture performance and reducing process energy consumption are actively going on. The most worked-on emerging absorbents, including liquid-liquid biphasic, liquid-solid biphasic, enzymatic, and encapsulated absorbents, already show encouraging results in improved energy efficiency, enhanced CO_2 absorption kinetics, increased cyclic CO_2 loading, or reduced regeneration temperature. In this chapter, the latest research and development progress of these emerging absorbents are reviewed along with the future directions in moving these technologies to higher-technology readiness levels.

Keywords: postcombustion capture, biphasic absorbent, lipophilic amine, ionic liquids, amino acids, enzymes, encapsulated

1. Introduction

Postcombustion carbon capture is considered one of the most promising and feasible technologies for reducing carbon dioxide (CO_2) emissions from energy-intensive industries such as coal-fired power plants. This is because postcombustion carbon capture has relatively higher level of technology readiness, lower energy penalty, and favorable cost compared to other carbon capture technologies (e.g., oxy-fuel, integrated gasification combined cycle or IGCC) [1]. Conventional aqueous alkanolamine-based carbon capture adsorbents were developed over half a century ago for natural gas/CO_2 separation as well as syngas/CO_2 separation, both work at high absorption pressures. Research work has been conducted to extend the conventional absorbents for coal-fired power plant CO_2 capture. These are classified as first-generation absorbents [2]. However, for the application of coal-fired power plant CO_2 capture, the flue gas is at about ambient atmosphere. This difference of the CO_2 absorption operation pressure makes the first-generation absorbents not satisfactory. Among the traditional alkanolamines, 30 wt% monoethanolamine (MEA) with a cyclic CO_2 loading of 4–5 wt% and a regeneration temperature of about 120°C is regarded as a benchmark absorbent [3]. In a continuous operation, a huge volume of the liquid absorbent has to be pumped back and forth between the absorber and the stripper during absorption and regeneration. For regeneration, a significant amount of water in the absorbent (an aqueous solution of MEA) is vaporized in the regenerator to flow upward acting as both a heat transfer agent and a stripping gas. The energy penalty of the regeneration could be as high as 4.2 GJ/tCO_2 [4]. The power generation efficiency would be reduced by about eight percentage points from a range of 28–34% to 20–26%. Therefore, research efforts are continuing in the hope to improve carbon capture performance and to reduce energy penalties. As the third-generation absorbents [2] (second generation: demonstration in 2020–2025 time frame; third generation: at early development stage), biphasic absorbents (liquid-liquid as well as liquid-solid phase change), enzymatic-enhanced, and encapsulated absorbents are attracting ever increasing research interest [5–8].

This chapter presents a review on these emerging absorbents and identifies directions for further research at pilot scale and beyond. We will examine the achievement on the CO_2 absorption energy efficiency, enhanced CO_2 absorption kinetics, and increased cyclic CO_2 loading or lower regeneration temperature.

2. Liquid-liquid biphasic absorbent systems

Liquid-liquid biphasic absorbent systems generally have one liquid phase fed into an absorber while upon CO_2 absorption or increase of temperature, the absorbent turns into two immiscible liquid phases (one CO_2-rich and the other CO_2-lean phases) [9, 10]. Because of the separation of the two liquid phases, during regeneration, only the CO_2-rich phase, a smaller flow than in a conventional alkanolamine case, is sent to the stripper/regenerator. The CO_2-lean phase is mixed with the regenerated stream (now in lean state) and sent back to the absorber to perform another round of CO_2 absorption. By doing so, the regeneration heat consumption can be drastically reduced and therefore, compared to the conventional postcombustion carbon capture absorbents, biphasic absorbent systems may reduce energy consumption and capital cost (requiring relatively smaller strippers).

It is found that, up to 2016, the active developers of the liquid-liquid biphasic absorbents are 3H Company, IFP Energies nouvelles, Korea Institute of Energy Research, Norwegian University of Technology, Tsinghua University, and University of Dortmund. In August 2015, DOE approved and funded 16 transformative carbon capture projects, two of which were on biphasic absorbents [11]. It is likely that more biphasic absorbent work will be published in the next few years.

2.1. Mechanism of liquid-liquid phase separation

The solubility or liquid-liquid phase separation in aqueous amine systems is determined by the relative strength of the molecular interactions among the amine molecules, among the water molecules, and between the amine and the water molecules [12, 13]; nonaqueous liquid-liquid biphasic absorbent would follow the same principle. For instance, in a system of pentane and water, the interaction of pentane with water is weaker than the interaction among water molecules, and therefore pentane does not dissolve well in water. By contrast, in a system of ethanol and water, the interaction of ethanol and water is stronger than the interaction among ethanol molecules and as a result ethanol dissolves well in water. The relative strength of the molecular interactions is known to be influenced by temperature, and change in temperature could turn on a liquid-liquid phase separation from a homogeneous solution, or vice versa, i.e., two liquid phases merging into one homogeneous liquid phase [9]. The possibilities of solubility or phase separation are summarized in **Figure 1** [12, 14]. The real situation of an amine and water could be complicated. Only systems with a lower critical solution temperature (LCST, case C) and both upper critical solution temperature (UCST) and LCST (case B), are potentially suitable for phase separation absorbents. When temperature is increased to above LCST, by breakdown of strong cohesive interactions between the solute and solvent [15], a homogeneous solution changes into two immiscible liquid phases (cases B and C in **Figure 1**). Since postcombustion carbon capture is operated at about 40°C, the LCST of a phase separation absorbent should be higher than the absorption temperature. When CO_2 is absorbed, new chemical species (e.g., carbamate, protonated amine, and carbonate and/or bicarbonate ions) are formed. These new species may lower the LCST of the system and result in phase separation at CO2 absorption temperature.

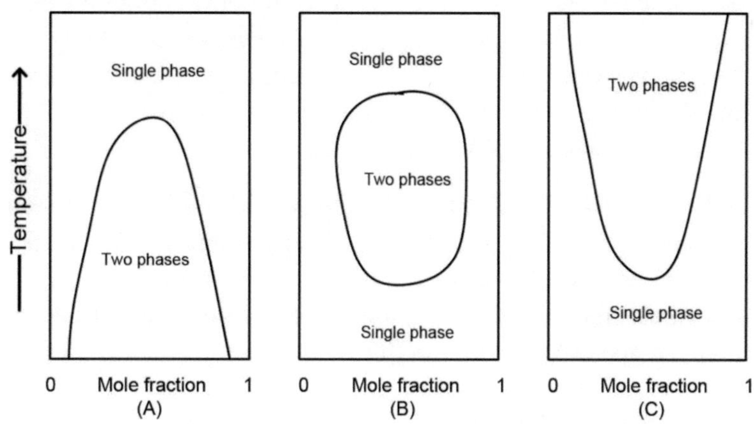

Figure 1. Partial miscibility curves of binary liquid-liquid mixtures. (A) UCST; (B) UCST, and LCST; (C) LCST [15].

2.2. Nonaqueous liquid-liquid biphasic absorbents

3H Company filed seven patents on nonaqueous solution of amine dissolved in an alcohol as self-concentrating absorbents [16]. The amines described in their patents include alamine 336, dibutylamine, diethanolamine (DEA), diisopropylamine, MEA, and piperazine, and the alcohols used as solvents include decylalcohol and isooctanol. Yeo Il Yoon's group at the Korea Institute of Energy Research reported a study on absorbent systems of MEA, DEA (diethanolamine) in 1-heptanol, 1-octanol, and isooctanol [17]. They found that, using a bubbling tube at 40°C with 30% CO_2 in N_2, the absorbent changed into two immiscible liquid phases when CO_2 was absorbed. The CO_2-rich phase was found to be dominant with amine-bonded CO_2 and unreacted amine, while the CO_2-lean phase was mainly alcohol with a small amount of free amine. Proton nuclear magnetic resonance (^1H NMR) characterization further showed that MEA or DEA carbamate and protonated amines existed in the rich phase, possibly in ion pairs such as $MEACOO^-MEAH^+$.

In one of Hu's studies, the CO_2-rich phase had a CO_2 loading of about 27 wt% with a volume only about 30% of the aqueous MEA case [18]. A batch mode of the rich phase stripping of a nonaqueous biphasic absorbent at 115–125°C indicated deeper regenerability down to about 90% of the absorbed CO_2 compared with that of the aqueous MEA absorbent, a regeneration of about 50%. In Hu's study, the regenerated stream, now mainly MEA, was combined with the lean stream (mainly alcohol) and sent back to the absorber. Therefore, the potential net cyclic CO_2 loading (the ratio of the weight of CO_2 released in regeneration to the weight of the absorbent) would double that of the aqueous amine absorbent, and by regenerating the CO_2-rich phase only (a much smaller volume compared to the whole absorbent), less thermal energy consumption and a smaller stripper are expected.

In Hu's study of biphasic absorbents [18], the boiling points of the alcohols were within 176–195°C, which was higher than the regeneration temperature and the alcohols did not evaporate. Improvements could be made by applying, e.g., a stream of CO_2 from the regenerator as a stripping gas, using a much smaller reboiler to raise the temperature of this CO_2 stream 20–30°C higher than the regeneration temperature, and feeding the gas directly into the regenerator. Meanwhile, the energy efficiency of the self-concentrating systems should be examined and compared with the conventional MEA technology.

In addition, in these biphasic absorbent systems, the absorbents are nonaqueous and likely will absorb the moistures from flue gases and may cause potential problems including mutual dissolution amongst amine-alcohol-water and may need for water separation. The relatively high viscosity of the absorbents is another concern; the viscosity of alcohol amine absorbents was found to increase upon CO_2 absorption [10].

2.3. Aqueous liquid-liquid biphasic absorbents

There are more works reported on aqueous liquid-liquid biphasic absorbents. A group from the University of Dortmund theoretically analyzed the solubility of liquid lipophilic amines in water. Different from the hydroxyl group-bearing conventional alkanolamines, lipophilic amines have relatively lower solubility in water. The amines screened are listed in **Table 1** [12, 13], tested either alone or mixed.

Type	Chemical	Abbreviation	Molar mass
Chain amines	Hexylamine	HA	101.19
	Heptylamine	HpA	115.22
	Octylamine	OtA	129.24
	Di-n-propylamine	DPA	101.19
	Diisopropylamine	DIPA	101.19
	Di-n-butylamine	DBA	129.24
	Diisobutylamine	DIBA	129.24
	Di-sec-butylamine	DsBA (B1)	129.24
	N-sec-Butyl-n-propylamine	SBPA	115.22
	N,N-Diisopropylmethylamine	DIMA	115.22
	N,N-Diisopropylethylamine	DIEA	129.24
	N,N-Dimethylbutylamine	DMBA	101.19
	N,N-Dimethyloctylamine	DMOA	157.3
Cycloalkylamines	Cyclohexylamine	CHA	99.17
	Cycloheptylamine	CHpA	113.2
	Cyclooctylamine	COA	127.23
	2-Methylcyclohexylamine	2MCA	113.2
	N-Methylcyclohexylamine	MCA (A1)	113.2
	N-Ethylcyclohexylamine	ECA	127.23
	N-Isopropylcyclohexylamine	IPCA	141.25
	N,N-Dimethylcyclohexylamine	DMCA	127.23
	N,N-Diethylcyclohexylamine	DECA	155.28
	Dicyclohexylamine	DCA	181.32
Aromatic amines	Benzylamine	BzA	107.15
	N-Methylbenzylamine	MBzA	121.18
	N-Ethylbenzylamine	EBzA	135.21
	N-Isopropylbenzylamine	IPBzA	149.23
	Phenylethylamine	PhEA	121.18
	N,N-Dimethylbenzylamine	DMBzA	135.21
Cyclic amines	2,6-Dimethylpiperidine	2,6-DMPD	113.2
	3,5-Dimethylpiperidine	3,5-DMPD	113.2
	2-Methylpiperidine	2MPD	99.17
	2-Ethylpiperidine	2EPD	113.2
	2,2,6,6-Tetramethylpiperidine	TMPD	141.25

Type	Chemical	Abbreviation	Molar mass
	N-Methylpiperidine	MPD	99.17
	N-Ethylpiperidine	EPD	113.2
Other amines	Monoethanolamine	MEA	61.08
	N-Methyldiethanolamine	MDEA	119.16
	2-Amino-2-methyl-1-propanol	AMP	89.14
	2-Amino-2-methyl-1,3-propanediol	AMPD	105.14
	N,N,N′,N′-Tetramethyl-1,6-hexanediami	TMHDA	172.31
	N-Methylmorpholine	MMP	101.15
	N,N-dimethyl-1,3-propanediamine	DMPDA	102.18
	Piperazine	pZ	86.14

Table 1. Lipophilic amines screened in the University of Dortmund's study [9, 12, 13].

Among the lipophilic amines tested, no candidates were found to be suitable as liquid-liquid phase separation absorbents because of their low CO_2 loading, lack of phase separation, or, complicated phase change behaviors or solid precipitation upon CO_2 absorption. By mixing two different amines (the so-called bi-amine systems), however, would enable a CO_2 capture performance that neither of the two components would show alone, such as phase separation, because the physical and chemical properties of the two components may supplement each other. In many cases, one (the so-called activator) of the two amine components has a relatively higher CO_2 absorption capacity and the other one (the so-called promoter) functions to improve phase separation and/or to enhance reaction rates of absorption or regeneration. Two mixtures (i.e., MCA-DSBA and DPA-DMCA) with a ratio of 3:1 have been identified at an optimum total concentration of 3-4 M (**Table 1**). The CO_2 absorption isotherms of these two mixtures were evaluated and compared with those of aqueous MEA [12]. At 40°C (the typical postcombustion capture temperature), both MCA-DSBA and DPA-DMCA were found to have higher CO_2 loadings compared to MEA (**Figure 2**). Due to its favorable CO_2 absorption isotherm, MCA-DSBA performed a little more superior to DPA-DMCA. By contrast, at 65–70°C (could be used as regeneration temperature), the two mixture systems had much lower CO_2 loading than MEA (even at 120°C). This indicated that the cyclic loading of these mixture systems may reach about 10 wt% at a lower regeneration temperature of 65–70°C and could double the cyclic loading of MEA at 120°C. Therefore, the bi-amine systems may have the potential to achieve better regeneration (up to 90%).

Besides the two-component bi-amine mixtures, three-component mixtures can also be developed. Since some aqueous lipophilic amines have LCST lower than 40°C, the absorbent could be in two liquid phases before CO_2 absorption takes place. A solubilizer could be added to increase the LCST. One of the examples is DMCA-MCA-AMP in a ratio of 3:1:1 (see **Table 1**), where AMP was used as a solubilizer to increase the LCST (to > 40°C) of DMCA-MCA [13]. It was found that the CO_2 loading of DMCA-MCA-AMP was 3 mol/L (ca. 13.2 wt %) at 40°C and at a CO_2 partial pressure of 0.15 bar and, at 75°C, over 90% of CO_2 was regenerated. Therefore, it seems that the addition of AMP led to the increase of LCST about 15–20°C but without impairing the CO_2 absorption and desorption performance.

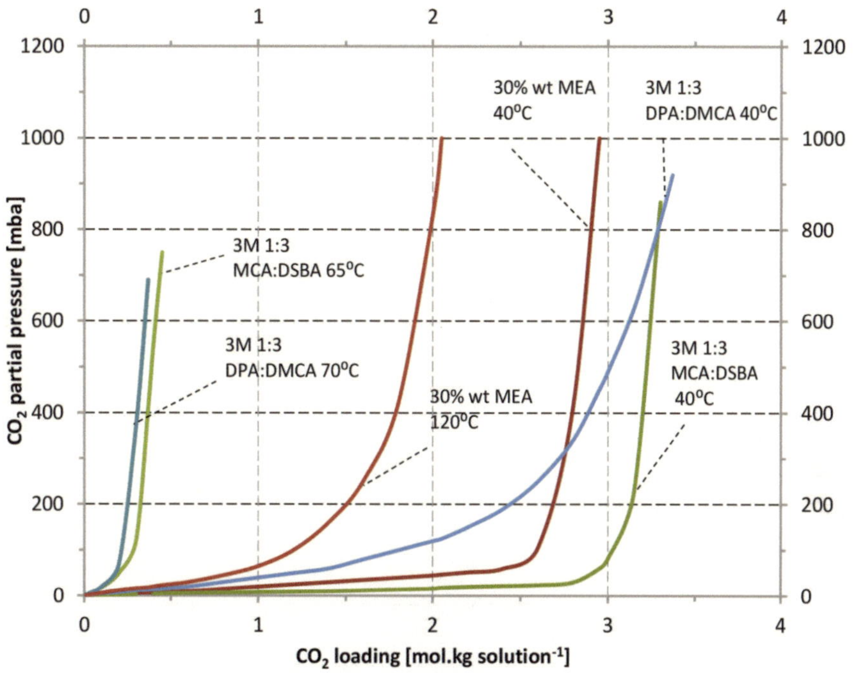

Figure 2. Loading curves (CO_2 absorption isotherms) of 3M 1:3 MCA:DSBA and DPA:DMCA at absorption and regeneration temperatures [12].

The Dortmund group also disclosed some coded proprietary biphasic absorbents [19, 20].

IFP Energies nouvelles has screened a large number of amines and identified DMX-1 [21]. DMX (implying de-mixing, i.e., phase separation) is an aqueous amine solution [22] (US 8,361,424 B2, US 8,500,865 B2, US 8,562,927 B2, US 2011/0185901 A1, WO2007/104856 A1, and US 2007/0286783 A1). After absorbing CO_2 at 40°C, the absorbent is heated to achieve phase separation and form a CO_2-rich phase and a CO_2-lean phase. In the subsequent regeneration process, a decanter is installed between the cross heat exchanger of lean (outlet stream from the stripper) and rich phase (CO_2-loaded stream from the absorber), and the regeneration is operated at 90°C under which two phases are formed. The CO_2-rich phase, up to 75% of the absorbed CO_2, obtained in the decanter is sent to the stripper and the remaining CO_2 is stripped in the stripper. This process could reduce the stripping burden thereby enhancing the regeneration efficiency [23].

In another report, two amine solutions (Amine B and Amine D) of a single tertiary alkanolamine with high dielectric constant have been studied [24]. The high dielectric constant of the alkanolamine is believed to trigger phase separation and to prevent solid precipitation. It has been reported that their CO_2 absorption loadings at 40°C and 0.1 bar are comparable to that of

MEA (**Table 2**), although the CO_2 loading in the CO_2-rich phase is not as concentrated as that with the 3H absorbents [18, 25]. Their CO_2 absorption isotherm is shown to be different from that of MEA (**Figure 3**) with a sharp decrease of the CO_2 loading in the low-pressure region, which is similar to the absorbents developed by the Dortmund's group (**Figure 2**). This feature makes it possible to achieve high cyclic loadings (e.g., 10.56–14.08 wt%). Compared to MEA technology, the use of DMX-1 could lead to 3.8% increase in power plant efficiency and 15.4% reduction in cost [26].

	MEA	Molecule D		Molecule B
Amine wt%	30	30	50	30
mol CO_2 per kg before flash (40°C; 0,1bar CO_2)	2.6	2.4	3.2	2.8
CO_2 wt% in absorbent	11.44	10.56	14.08	12.32
CO_2 % flashed	15	50	65	75
CO_2-rich phase %		89	63	73
Flow reduction in stripper %		11	37	27

Table 2. IFP DMX absorbent performance (data from [24]) [9].

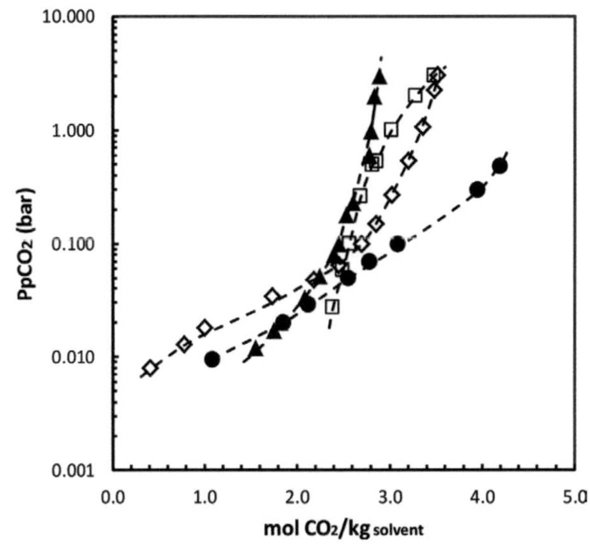

Figure 3. Partial pressure of CO_2 versus loading of a 30% wt MEA (□), a 30% wt molecule B (▲), 30% (◊) and 50% wt (•) molecule D aqueous solutions, at 40°C [24].

In another two studies, approximately 30 aqueous amines (lipophilic amines and alkanolamines) and the combination of them were screened and shown in **Table 3** (The amines appeared in **Table 1** are not included in **Table 3**) [27, 28]. Two promising examples were identified as mixtures of 2M BDA/4M DEEA (2B4D) and 2M DMBA/4M DEEA (2D4D); BDA and DMBA are lipophilic amines, and DEEA is a tertiary alkanolamine. These two mixtures were found to have about 97% of the absorbed CO_2 in the lower phase along with a total loading of 0.51 mol CO_2/mol amine, and had a cyclic loading of 46% higher than MEA (30 wt%). Their CO_2 absorption isotherms were similar to those of the DMX absorbents [24] and the Dortmund's biphasic absorbents [12, 13], and their overall performance was also similar to the DMX absorbents. ^{1}H NMR phase composition analysis and CO_2 absorption kinetics studies showed that the biphasic solvent separation was due to the fast reaction rate of BDA with CO_2 and the limited solubility of DEEA in the reaction products. It was concluded that the phase separation was determined by thermal dynamics of all of the species existing in the CO_2-loaded system, the temperature, and pressure of the CO_2 (**Table 4**).

Chemical	Abbreviation	Molar mass
Diisopropylamine	DIPA	101.19
N-Ethyl-n-butylamine	EBA	101.19
Triethylamine	TEA	101.19
Diallylamine	DAA	97.16
Benzylamine	EMAA	107.15
N,N-Diethylethanolamine	DEEA	117.19
N-Ethylethylenediamine	EEDA	88.15
N,N-Dimethyl-1,3-propanediamine	DMPDA	102.18
1,4-Diaminobutane	DAB or BDA	88.15
N,N-Dimethylbutylamine	DMBA	101.19
Hexylamine	HA	101.19
1,6-Hexanediamine	HAD	116.2
N-Methyl-1,3-Diaminopropane	MAPA	88.15

Table 3. Amines screened (data from [27, 28]) [9].

	Phase split (wt%)	BDA (mol/kg)	DEEA (mol/kg)	CO_2 (wt%)	CO_2 (mol/mol amine)
Upper CO_2 lean	22	0.115	7.265	0.315	0.026
Lower CO_2-rich phase	78	2.233	2.4	3.291	0.43

Table 4. Phase composition of CO_2-loaded 2B4D (data from [27, 28]).

An absorbent with a composition of BDA and DEEA the same as those found in the CO_2-rich phase was further studied. Phase separation was observed in this absorbent and the single liquid phase started to become two liquid phases at CO_2 loadings at 0.099 mol CO_2/mol amine, and from loadings of 0.187 to 0.313 mol CO_2/mol amine, BDA further reacted with CO_2 while DEEA transferred to the upper phase. Between loadings of 0.313 and 0.345 mol CO_2/mol, DEEA reacted with CO_2 with the products transferred to lower phase until the equilibrium loading of 0.505 mol CO_2/mol amine was achieved [29] (**Table 5**).

Amine or mixture of amines	Concentration	Initial CO_2 absorption rate (NL/min)	CO_2 absorption capacity (wt%)	Cyclic capacity (wt%)	Regeneration depth (%)
AMP/PZ	3 M/1 M	0.47	11.44	0	0
AMP/MAPA	3 M/1 M	0.44	11.44	8.36	73.1
DMMEA/PZ	3 M/1 M	0.44	10.12	9.24	91.3
DMMEA/MAPA	3 M/2 M	0.49	13.2	7.48	56.7
DMMEA/PZ	5 M/2 M	0.48	12.76	11.44	89.7
DMMEA/MEA	5 M/2 M	0.44	10.56	9.68	91.7
DMMEA/MAPA	5 M/2 M	0.49	12.76	9.24	72.4
DMMEA/MAPA	5 M/1 M	0.47	14.08		
DEEA/PZ	3 M/1 M	0.5	12.32	0	0.0
DEEA/MEA	5 M/2 M	0.44	6.6	0	0.0
TRIZMA/PZ	3 M/1 M	0.44	7.92	7.04	88.9
TRIZMA/MEA	3 M/1 M	0.33	4.4	3.96	90.0
TRIZMA/MAPA	3 M/1 M	0.43	8.36	5.28	63.2
MEA	5 M	0.48	11.44	4.84	42.3
MAPA	5 M	0.5	19.36	4.84	25.0

Data in the table were read from the graphs in [30]. This operation may introduce an uncertainty of ±5–10%.

Table 5. CO_2 Absorption and regeneration of aqueous amine or binary amine systems.

To identify absorbents with low regeneration energy, researchers from the Norwegian University of Science and Technology screened multiple aqueous amines or mixture of amines [30]. They found that DMMEA/PZ (3M/1M, 5M/2M), especially 3M/1M, and DMMEA/MEA (5M/2M), had high CO_2 absorption rate and characteristics of deep regeneration at low temperature, almost doubling the cyclic loading of the MEA. Further, these researchers developed biphasic absorbents of DEEA/MAPA, tertiary alkanolamine and lipophilic amine [31, 32]. They examined the vapor liquid equilibrium (VLE) of CO_2-DEEA-MAPA-H_2O system at 40, 60, and 80°C, from which the CO_2 absorption loading, absorption phase split ratio, and phase compositions were derived as presented in **Table 6**. The absorbed CO_2 was found to be highly concentrated in the CO_2-rich phase, which alone was shown to have a much deeper regeneration hence a higher cyclic loading compared to aqueous MEA (**Figure 4** (Figure 17 in [31]). However, the CO_2-DEEA-MAPA-H_2O system had a higher viscosity compared to MEA and its CO_2 absorption kinetics was faster but dropped down slower with increasing CO_2 loading. In their pilot plant trials at a scale of 80–90 m^3/h flue gas with CO_2-rich/CO_2-lean phase separation and CO_2-rich phase regeneration, their system was shown to be superior to aqueous MEA, which was tested in the pilot plant as well [33, 34]. This finding was supported by modeling the biphasic absorbent with an energy consumption of 2.2-2.4GJ/tCO_2 and compared to 3.7GL/tCO_2 using 30wt% aqueous MEA [32, 35].

Temperature (°C)	40	60	80
CO_2 pressure (kPa)	13.07	13.07	13.07
CO_2-rich phase fraction	0.68	0.53	0.48
CO_2 loading mol/kg (mol/kg) and in (wt%), total	2.86	2.43	2.11
	12.58	10.69	9.28
CO_2-rich phase loading (mol/L) and in (wt%)	5.55	5.66	5.44
	24.42	24.90	23.94
CO_2 lean phase loading (mol/L)	0.28	0.29	0.1

Table 6. Phase compositions of DEEA/MAPA (5M/2M) [31, 32].

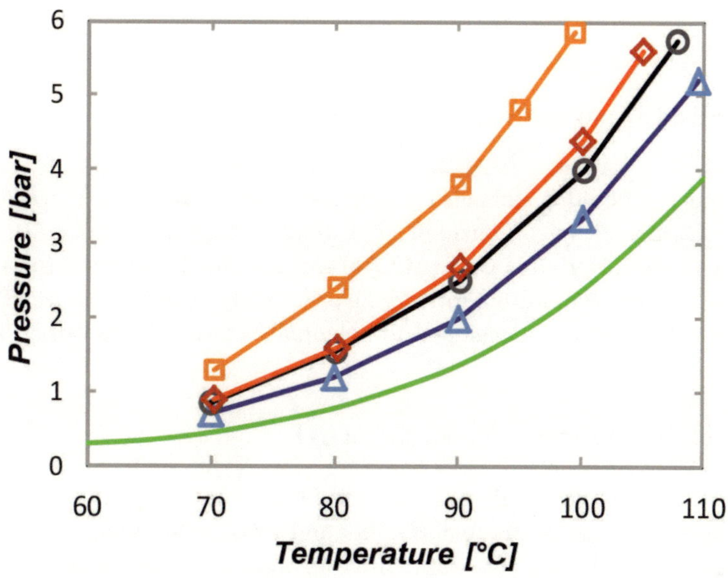

Figure 4. The total pressure from lower phase samples with absorption taken at 40°C from the screening apparatus. P_{CO2}: (Δ) 6 kPa, (O) 8 kPa, (◊) 10 kPa, and (□) 13 kPa; (green line) MEA at loading 0.5 mol CO_2/mol MEA (model from [36]) [31].

In summary, up to now, researchers and developers have achieved encouraging results in the area of liquid-liquid biphasic CO_2 absorbents, and some biphasic absorbents can be regenerated at lower temperatures with deeper regenerability than the bench mark aqueous MEA (**Table 7**).

Developer	Biphasic absorbent	Phase separation temperature	Regeneration temperature	Regeneration depth	Source
3H Company	Nonaqueous, alkanolamine/alcohol	Absorption temperature, 40°C	Up to 125°C	90%	[18, 25]
Korean Institute of Energy Research	Non-aqueous, MEA in 1-heptanol, isooctanol, 1-octanol	Absorption temperature	N/A	N/A	[17]
Dortmund University	MCA/DSBA; DPA/DMCA; DMCA/MCA/AMP	60–70°C	Up to 75°C	90+%	[12, 13]
IFP Energies nouvelles	DMX-1; Amine B; Amine D	90°C	>90°C	90+%	[24]
Tsinghua University	DEEA/BDA	40°C	90°C	90%	[27, 28]
Norwegian University of S&T	DEEM/MAPA	80°C	>80°C	~90%	[31, 34, 35]

Table 7. Summary of the developed biphasic absorbents for CO_2 capture.

3. Liquid-solid biphasic absorbent systems

There is a category of liquid absorbents forming solid precipitates after CO_2 absorption such as carbamate, bicarbonate, or carbonate in solid states. According to Le Chatelier's Principle [37], formation of a solid product during CO_2 absorption and its removal from the solution phase shifts the reaction equilibrium toward the production of more products. This phenomenon could be engineered and developed to potentially more efficient carbon capture technology.

3.1. Emulsion of alkanolamine and ionic liquid (IL)

Research has been going on to use ILs as absorbents for CO_2 capture because ILs have negligible volatility, nonflammability, high thermal stability, and virtually unlimited chemical tunability. However, stand-alone, ILs are not competitive enough when compared to CO_2 capture efficiency of aqueous alkanolamine systems. An idea is to try hybrid system coupling advantages of alkanolamines with those of room-temperature ILs (RTILs) and to achieve potential synergies arising from each of the individual components [38].

A mixture of diethanolamine (DEA) and 1-alkyl-3-methylimidazolium bis(trifluoromethylsulfonyl)imide, which is hydrophobic, was tested for CO_2 absorption (**Figure 5**). This emulsion could capture CO_2 up to the stoichiometric maximum through crystalizing CO_2-bonding product (DEA-carbamate) while avoiding equilibrium limitations and thus making efficient utilization of the absorbent molecules [39]. Similar precipitation of carbamate

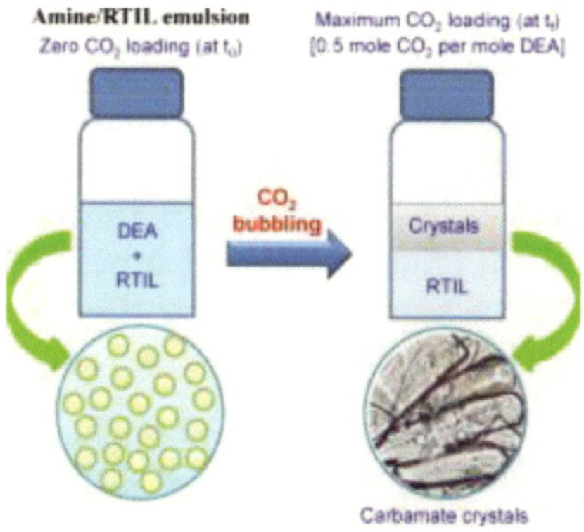

Figure 5. Immiscible alkanolamine/RTIL system for efficient CO_2 captures [39].

upon CO_2 absorption was also observed with ILs such as 1-ethyl-3-methylimidazolium bis(trifluoromethylsulfonyl)imide ([EMIM][Tf_2N]), 1-butyl-3-methylimidazolium bis(trifluoromethylsulfonyl)imide ([BMIM][Tf_2N]), and 1-hexyl-3-methylimidazolium bis(trifluoromethylsulfonyl)imide ([HMIM][Tf_2N]). The density of the solid phase precipitates was lighter thereby quickly rising to the surface and easing the separation for regeneration (**Figure 6a–c**). Hydrophobicity of ILs plays a role in the separation of solid products from

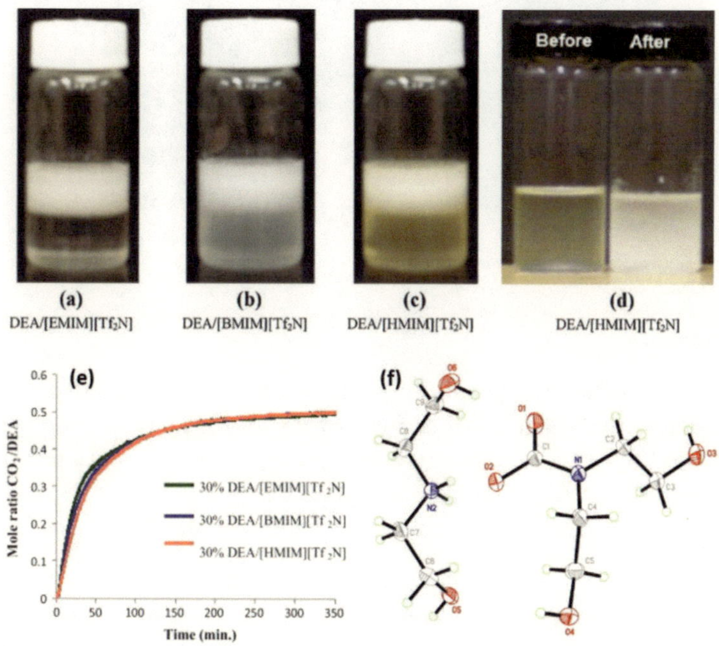

Figure 6. DEA/RTIL system for CO_2 capture: (a–c) (without surfactant) after CO_2 capture; (d) (with surfactant) before and after CO_2 capture; (e) CO_2 capture capacity profiles of the DEA/RTIL system at atmospheric pressure and 25°C; and (f) basic structural unit in DEA-carbamate ($C_9H_{22}N_2O_6$) crystal [39].

the liquid phase. When a surfactant, Triton® X-100, was added to the [HMIM][Tf_2N]-based system, the carbamate product remained dispersed in the suspension (**Figure 6d**). CO_2 loading capacity up to the stoichiometric maximum (0.5 mole of CO_2 per mole of DEA) can be achieved. The three absorbents showed similar CO_2 uptake rates (**Figure 6e**). The crystallization of the carbamate product, which was composed of protonated-DEA cation and DEA-carbamate anion (**Figure 6f**), enabled higher CO_2 uptake, and solid precipitation may have facilitated the separation thereby offering advantages in regenerating a smaller (only solid carbamate) volume with less energy consumption.

Other systems of aqueous solutions of N-methyldiethanolamine (MDEA) and guanidinium tris(pentafluoroethyl) trifluorophosphate $[gua]^+[FAP]^-$ IL showed similar solid formation after the absorption of CO_2 at high pressures. The formed CO_2-bonding solid products could be easily regenerated (**Figure 7**) [40].

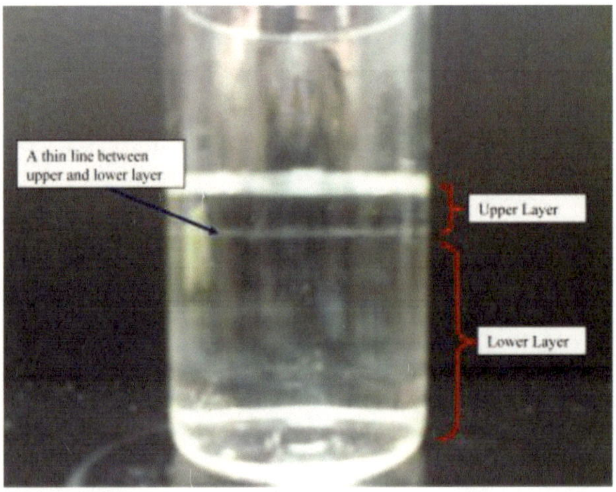

Figure 7. A photo of double layer CO_2-rich mixtures [40].

3.2. Chilled ammonia

Aqueous ammonia can absorb CO_2 to produce solid ammonium carbamate/bicarbonate, which could be separated from the solution thereby allowing an efficient recycling of the unreacted scrubbing solution [31, 41, 42]. Chilled ammonia process could be developed using aqueous ammonia to absorb CO_2 at lower temperature (2–10°C), in which the ammonia slip from the absorber could be reduced and the flue gas volume could be smaller [43].

Precipitation of pheromone in ethanol-water chilled ammonia solution was also observed after CO_2 absorption [44]. In this CO_2 absorption process, solid mixtures of ammonium bicarbonate and ammonium carbamate, or of ammonium carbamate alone were formed. Selective formation or precipitation of solid ammonium carbamate could be obtained by reacting gaseous CO_2 and NH_3 in anhydrous ethanol, 1-propanol, and N,N-dimethylformamide (DMF) in a flow reactor that can operate continuously. After filtering the solid precipitates, the unreacted ammonia solution could be reclaimed into the absorber. Such a chilled ammonia process may be applied to capture CO_2 from flue gas of coal-fired boilers, natural gas combined cycle systems, and other energy heavy industrial applications [45].

3.3. Triethylenetetramine (TETA)/ethanol solution as absorbent

When CO_2 is absorbed into a solution of triethylenetetramine (TETA) dissolved in ethanol, solid precipitates formed (**Figure 8a** and **b**) in contrast to TETA/water solution [46]. Moreover, TETA/ethanol solution showed improvement in CO_2 absorption rate, absorption capacity, and absorbent regenerability. Ethanol not only promoted the solubility of CO_2 in the liquid phase but also facilitated the chemical reaction between TETA and CO_2. The CO_2 capacity of the solid phase as TETA-carbamate accounted for about 81.8% of the total CO_2 absorbed (**Figure 8c**). The TETA/ethanol solution was found to be relatively stable throughout multiabsorption-desorption cycles (**Figure 8d**). One hurdle of applying the TETA/ethanol solution for CO_2 removal is that ethanol has a high vapor pressure and this must be taken into consideration for further development of this absorbent system and later designing for possible commercial applications.

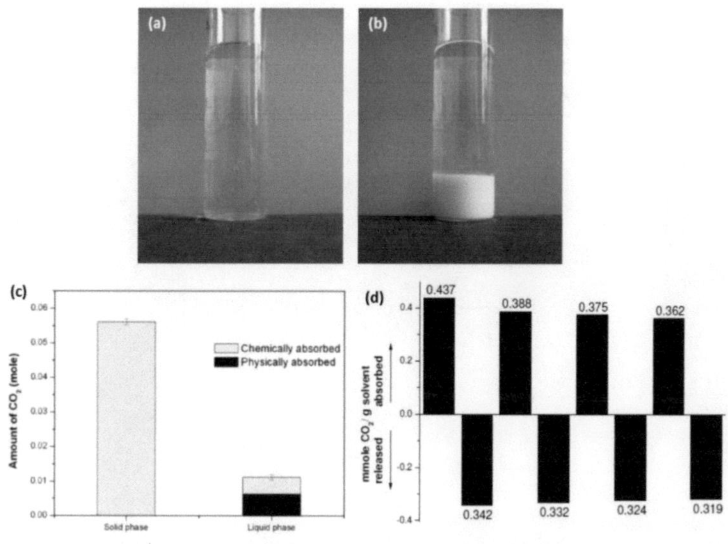

Figure 8. TETA/ethanol solution (a) before CO_2 absorption and (b) after CO_2 absorption. (c) Partition of carbon dioxide in the solid phase and liquid phase. (d) Cycling absorption/regeneration runs of TETA/ethanol solution for CO_2 absorption [46].

3.4. Amino acid salt as liquid-solid phase change absorbent

Being environmental friendly, ionic nature and low volatile, amino acid salts are of great interest as potential solvents for CO_2 capture [47, 48, 49]. Moreover, amino acid salt solutions have good resistance to an oxygen-rich flue gas stream. The reactivity of amino acid salts with CO_2 is similar to those of alkanolamines due to the presence of identical amino functional groups in their molecules. Some of them such as the potassium salts of glycine, sarcosine, and proline, react faster with CO_2 than MEA thereby kinetically favorable [50, 51].

Multiple amino-acid salts were found to precipitate after reacting with CO_2 to a certain degree [52]. During the absorption of CO_2 in aqueous potassium salts of N-methylalanine, DL-alanine, and α-aminoisobutyric acid, solid precipitates were observed [53]. Various types of solid precipitates could be achieved by varying the amino acid structure and solubility. Amino acids with a primary amino group may form only zwitterion species precipitates [54], while amino acids with a hindered amino group and with relatively high zwitterion solubility (e.g., proline) may form potassium bicarbonate precipitates [55].

By Le Chatelier's Principle, the driving force for CO_2 absorption can be maintained at a high level even at high loadings. Thus the absorber performance could be significantly improved. This effect was indicated in **Figure 9** (enhanced absorption) where the possible precipitates were highlighted. In **Figure 9**, besides the heat input necessary to regenerate the solvent, in the case of precipitating amino acids two more effects are possible: Enhanced absorption (purple) due to the precipitation of reaction products during absorption and enhanced

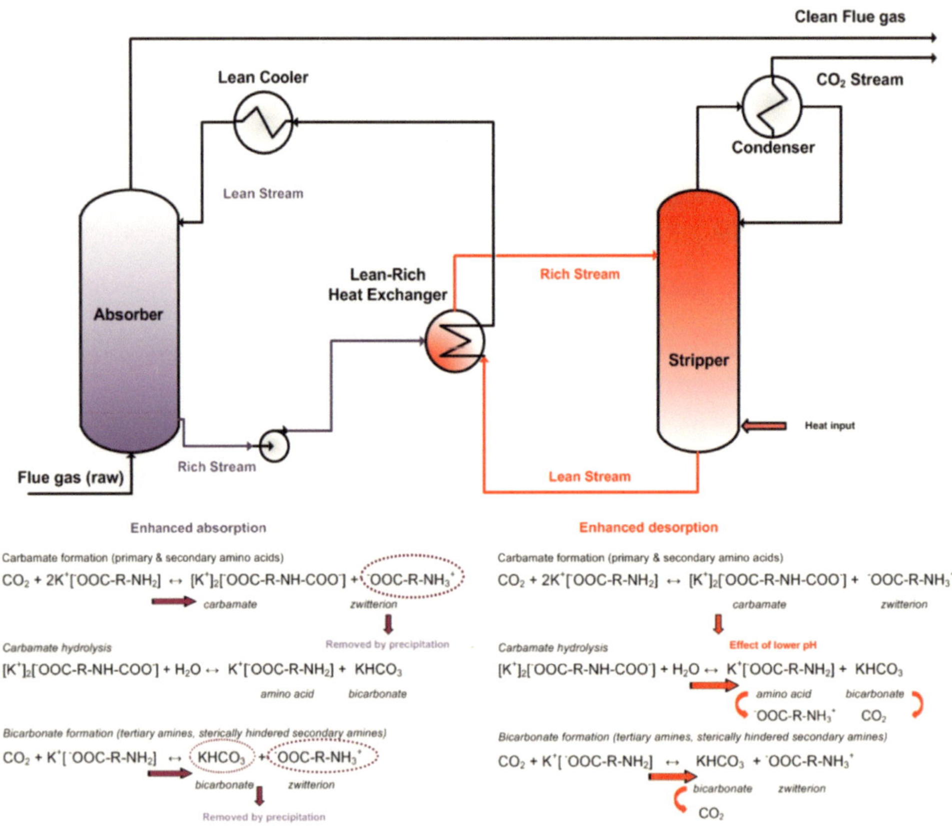

Figure 9. Conventional amine-based process for CO_2 capture where the reactions specific to amino acid salts have been added at the bottom of the absorber and the stripper [56].

desorption (red) due to a lower pH that results from increasing the amino acid to K^+ ratio in solution [56]. Because of the high loadings, the regeneration energy consumption was reduced [57]. As can be seen from **Figure 10**, at a given CO_2 partial pressure, a precipitating-based process would have higher loading than a conventional absorption process without precipitation, and a combined process (simultaneous absorption and precipitating process) is expected to result in increased capacity.

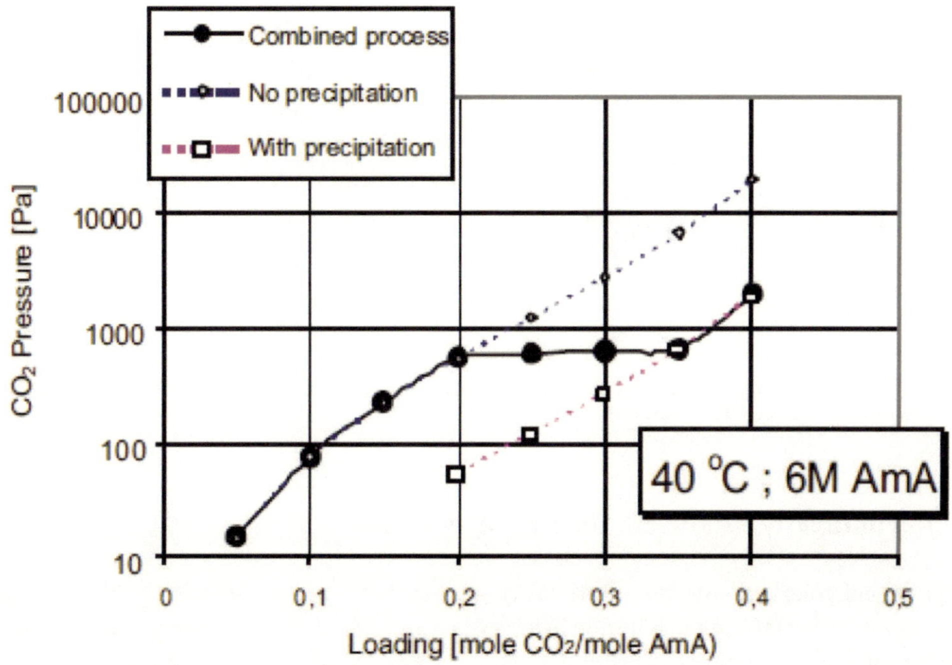

Figure 10. Schematic picture to depict the difference between a precipitating and a nonprecipitating system in terms of CO_2 pressure as a function of loading [58].

Figure 11 shows a schematic representation of the DECAB process for liquid-solid phase change amino acid salt absorbent [56]. The flue gas (at 40°C) is contacted with CO_2 preloaded absorbent in a spray-tower, resulting in that the CO_2 undergoes a chemical reaction with the absorbent that leads to the formation of carbamate and carbonate ions, as shown in **Figure 11**. As absorption goes on, the pH of the absorbent solution as well as solubility of the amino acid decreases. Finally, the CO_2-bonding amino-acid precipitates as an amino acid zwitterion. In the process, the solid precipitates are collected at the bottom of the tower. The remaining CO_2 in the flue gas is captured in the absorption column, where the depleted flue gas is contacted with lean absorbent. The absorption column is a conventional packed absorption column filled with structured packing. There, the CO_2 partial pressure is reduced to the desired value for 90% CO_2 removal. The rich stream containing the solids, is further processed in the stripper, via the lean-rich heat exchanger, to release the CO_2. The lean-rich heat exchanger also needs to be able to handle solids (e.g., spiral heat exchanger).

The CO_2 absorption depth needs to be controlled so that only in the spray-tower the solid products are formed [56].

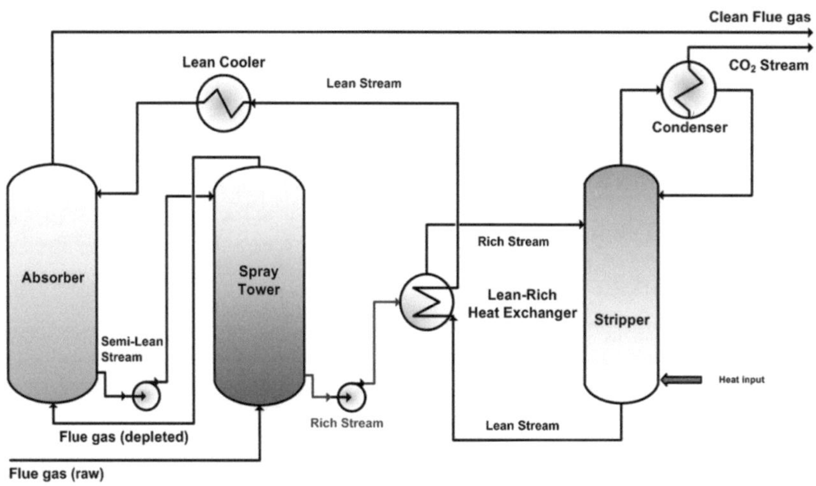

Figure 11. DECAB process concept for CO_2 capture. Enhanced absorption due to the precipitation of reaction products during absorption is highlighted in purple [56].

4. Enzymatically catalyzed absorbent systems

A special enzyme, carbonic anhydrase (CA), works in vertebrates' lungs to facilitate oxygen and CO_2 exchange through respiration in a very fast and effective way. Attempts to incorporate this type of enzyme to carbon capture absorbent systems have shown encouraging results [59]. When a small quantity of the enzyme is used as catalyst in a CO_2 capture absorbent system (usually an aqueous amine absorbent), it enhances the reaction rate and enables rapid approach to equilibrium between dissolved CO_2 and HCO_3^- in aqueous solutions. The idea is a natural extension of the experience that some enzymes have been successfully deployed to increase the efficiency of other industrial processes [60, 61]. The enhancement effect of the enzyme is so significant that the size of an absorber could be reduced up to 90% smaller than the conventional amine case [61]. However, enzymes are bio active compounds. How to maintain its long period activity is an issue. The Canadian company, *CO_2 Solutions*, reported that their developed enzymatic catalytic amine system could work for 15 days for CO_2 capture in temperature ranges of 40–70°C [61].

The way to implement enzyme more effectively in CO_2 capture process is to support or immobilize it onto some sort of carrier to provide sustained stability under working conditions. It may need to be able to stand for the high regeneration temperature if the enzyme is not confined in the absorber only. Report shows that some developed enzyme systems could stand for moderate regeneration temperatures (e.g., around 70–80°C) [59]. The immobilized enzyme should also be strong toward contaminants encountered in the common flue gas.

A porous organosilica coating containing CA to separate CO_2 from a flowing gas stream has been tested [62]. This coating was applied to ceramic random packing and placed in a countercurrent absorber column, where it demonstrated a high rate enhancement for 400 days at 45°C and a total turnover number of ~48 million moles CO_2/mole enzyme. In another development, the same coating formulation was deposited onto stainless steel structured packing. This coating technique was used to produce 275 liters of packing for pilot testing at the National Carbon Capture Center in Wilsonville, AL, on coal-fired flue gas. This unit operated for nearly 5 months at 40°C in the same carbonate solution and exhibited a steady 80% CO_2 capture.

A type of magnetic polymer microspheres functionalized with epoxy group was prepared, and CA enzyme was immobilized on the carriers by selectively covalent binding [63]. The parameters affecting CA immobilization, such as pH, temperature, and enzyme dose were investigated. The kinetic parameters of the immobilized and the free CA were also evaluated. The value of the Michaelis–Menten constant (K_m) and the maximum velocity (V_{max}) of the immobilized CA were 8.077 mmol/L (mM) and 0.027 μmol/(min mL), respectively, while those of the free CA were 6.091 mM and 0.091 μmol/(min mL), respectively. Moreover, the performance of the thermal stability, storage stability, and reusability of the immobilized CA confirmed that CA immobilized on the epoxy-functionalized magnetic polymer microspheres possessed a stable and efficient catalytic ability on CO_2 hydration, which seemed to be a suitable candidate for CO_2 capture [63].

In another development, new materials of Fe_3O_4 magnetic microspheres were functionalized with carboxyl groups and prepared for CA immobilization to capture CO_2. The optimum conditions for immobilization, such as carrier dose, enzyme dose, pH, shaking speed, temperature, and contact time, were examined. The pH and thermal stability of the free and the immobilized CA were compared. The results showed that the immobilized CA had a better enzyme activity, a higher pH and thermal stability than those of free CA. Meanwhile, CO_2 capture was significantly enhanced by the free and immobilized CA in tris(hydroxymethyl) aminomethane (Tris) buffer solution. Moreover, the immobilized CA maintained 58.5% of its initial catalytic activity after 10 recovery cycles due to the protection effect of the magnetic microspheres. All the results confirmed the potential merits of using the carboxyl-functionalized Fe_3O_4 magnetic microspheres immobilized CA to remove CO_2 from air or flue gas [64].

5. Encapsulated absorbents

Many attractive options for carbon capture solvents suffer from high viscosity, making it difficult to generate large surface areas for fast absorption, and amine-based aqueous liquids suffer from potential environmental impacts from ammonia (product of amine decomposition), and amine vapor release. Microencapsulated carbon sorbents (MECSs) are a new class of carbon capture materials consisting of a CO_2-absorbing liquid absorbent contained or confined within solid, CO_2-permeable polymer shells. As part of a US-DOE ARPA-E program, a team from the University of Illinois Urbana-Champaign, Babcock and Wilcox, and the Lawrence Livermore National Laboratory has created this new type of encapsulated form of carbon capture absorbent in which the operating fluid, amines, or carbonates in the tests are enclosed in a thin polymer shell forming 200–400 μm beads [65]. While mass transport across the polymer shell is reduced compared to the bulk liquid, the large surface area of the beads improves overall mass transfer more than off-setting this disadvantage. The liquid, as well as any degradation products or precipitates, remain encapsulated within the beads, which can be thermally regenerated repeatedly. Encapsulated absorbents have the capacity of the liquid absorbents as well as the physical behavior of solid sorbents. It could be imagined for them to be useful in fairly conventional-style capture applications, as well as unprecedentedly new approaches facilitated by their high surface area. The developed beads appear to be both chemically and mechanically stable under typical industrial conditions. There are engineering constraints that the beads must satisfy for several application strategies, including their use in fluidized beds and these should be further studied. The US group has encapsulated MEA, piperazine, and a variety of other carbonate solutions, which appear to be optimal for this application, demonstrating rapid CO_2 uptake and desorption using colorimetric methods, which permit rapid spectroscopic determination of the extent of CO_2 uptake and release. Carbonate capsules are created using a silicone polymer shell which is both rugged and permeable to CO_2. Results of mechanical/thermal cycling tests demonstrate long-term stability of silicone-encapsulated carbonate [65].

Especially, MECS enhances the rate of CO_2 absorption for solvents with slow kinetics and prevent solid precipitates from scaling and fouling equipment, two factors that have previously limited the use of sodium carbonate solution for carbon capture. Researchers have examined the thermodynamics of sodium carbonate slurries for carbon capture [66]. Modeling work has been carried out on the vapor-liquid-solid equilibria of sodium carbonate and several features that can contribute to an energy-efficient capture process have been derived: very high CO_2 pressures in stripping conditions, relatively low water vapor pressures in stripping conditions, and good swing capacity. These would make a more effective and efficient CO_2 absorption and desorption cycle. The high potential energy savings have been indicated compared with an MEA system [66].

6. Direction for further development

6.1. Liquid-liquid biphasic absorbent

The following aspects are recommended for future studies:

- As to liquid-liquid biphasic absorbent systems that have already been identified, detailed, and comprehensive CO_2 absorption and kinetics studies are much needed. Continuous mass transfer from CO_2-lean phase to CO_2-rich phase, diffusion of multispecies within the absorbent, and viscosity should be considered as CO_2 absorption continues. Based on knowledge obtained from these current amines studied, screening, designing, and synthesizing of new amines with improved properties are possible.

- For regeneration, CO_2 taken from the exit of the regenerator may be used as heat-transferring and stripping gas, and a heat exchanging unit may be installed in the stripper/regenerator. It is also possible to increase the regeneration pressure so that the size of the regenerator can be reduced hence reducing capital cost and energy consumption; some biphasic absorbents have high equilibrium CO_2 pressures.
- Phase separation temperature should be optimized to reduce energy consumption during regeneration; phase separation temperatures close to absorption temperature may be preferred. Feasibility studies including modeling and economic evaluation should be conducted to obtain overall performance, absorber/regenerator sizing, energy penalty, and cost of a biphasic absorbent carbon capture plant. In addition, pilot plant demonstrations of promising biphasic absorbents need to be considered.

6.2. Other emerging absorbents

For liquid-solid phase change absorbents, one of the biggest challenges related to their applications is to have a process design that could handle solids/precipitates transportation, heat exchanging, and regeneration. Process designs like the ones (i.e., DECAB process, DECAB Plus process) proposed for liquid-solid phase change amino acid salt absorbent could be adapted for other liquid-solid phase change systems as well [58].

Enzymes are macromolecular proteins and their current research has been focusing on their effects on CO_2 capture performance. The aspects of their active life span, the process of mass production or extraction from natural sources, and the related costs should be further explored.

For the encapsulated absorbents, within capsule kinetics and overall mass transfer studies should be pursued. The technology of mass production and the costs are also key factors in determining its commercial viability for CO_2 capture application. Techno-economic feasibility studies are keenly welcome.

In summary, most of these technologies are still at laboratory research stage, and there remain challenges associated with the scale-up of these technologies to meet the needs of CO_2 capture from power generation as well as other energy heavy industries. Future efforts should be focused on developing basic theoretical and mechanistic understandings of phase change, mass transfer, and CO_2 absorption and desorption phenomena, to perform pilot plant testing to generate design parameters and process requirements, and to create in parallel techno-economic plant design fundamentals and packages for proving the feasibilities of these emerging carbon capture absorbents.

Acknowledgements

Financial supports from EcoEII program, OERD (Office of Energy Research & Development), NRCan (Natural Resources Canada), and WV NASA EPSCoR are greatly acknowledged.

Author details

Quan Zhuang[1*], Bruce Clements[1] and Bingyun Li[2,3]

*Address all correspondence to: quan.zhuang@canada.ca

1 Natural Resources Canada, CanmetENERGY, Ottawa, Canada

2 National Energy Technology Laboratory-Regional University Alliance, Morgantown, West Virginia, USA

3 Department of Orthopaedics, School of Medicine, West Virginia University, Morgantown, West Virginia, USA

References

[1] Bruce, G. M. Clean Coal Engineering Technology. Elsevier; Oxford, UK, 2011.

[2] CSLF. Supporting development of 2nd and 3rd generation carbon capture technologies: Mapping technologies and relevant test facilities. Carbon Sequestration Leadership Forum; 16 December 2015.

[3] Simona, L. L., Eliasa, Y., Graeme, P., Yuli, A., Konrad, H. Rate based modeling and validation of a carbon-dioxide pilot plant absorbtion column operating on monoethanolamine. Chem. Eng. Res. Des, 89(9) (September 2011) 1684–1692.

[4] Yongping, Y., Rongrong, Z. "MEA-Based CO_2 Capture Technology and Its Application in Power Plants," in Paths to Sustainable Energy, Ed. Artie Ng, InTech, Rijeka, Croatia, November 2010.

[5] Li, B., et al. "Phase change solvents for CO_2 capture," in Novel Materials for Carbon Dioxide Mitigation Technology, Elsevier, Amsterdam, 2015.

[6] Wang, X., Li, B. "Phase-change solvents for CO_2 capture," in Novel materials for carbon dioxide mitigation technology (Shi F., Morreale B. Eds.), pp. 3–22. Amsterdam, Elsevier. 2015.

[7] Liang, Z., Fu, K., Idem, R., Tontiwachwuthikul, P. Review on current advances, future challenges and consideration issues for post-combustion CO2 capture using amine-based absorbents. Chin. J. Chem. Eng. 24 (2016) 278–288.

[8] Budzianowski, W. M. Single solvents, solvent blends, and advanced solvent systems in CO_2 capture by absorption: a review. Int. J. Global Warm. 7 (2015) 184–225.

[9] Zhuang, Q., et al., Ten years of research on phase separation absorbents for carbon capture: achievements and next steps. Int. J. Greenhouse Gas Control. 52 (2016) 449–460.

[10] Zhuang, Q., Clements, B. "CO_2 Capture by Biphasic Absorbent—Absorption Performance and VLE Characteristics," in preparation, 2016.

[11] DOE Selects 16 Transformational Carbon Capture Technologies Projects for Funding, August 2015. http://netl.doe.gov/newsroom/news-releases/news-details?id=a3df8c77-500f-4880-8f04-cb19bc1488b0

[12] Tan, Y. H. PhD, "Study of CO_2-Absorption into Thermomorphic Lipophilic Amine Solvents," Universität Dortmund, Germany, 2010.

[13] Zhang, J. Study on CO_2 Capture Using Thermomorphic Biphasic Solvents with Energy-Efficient Regeneration," Ph.D. Thesis, Universität Dortmund, Germany, 2013.

[14] Yalkowsky, S. H. "Solubility and Solubilization in Aqueous Media," American Chemical Society: Washington, DC, 1999.

[15] Counsell, J. F., Everett, D. H., Munn R. J. Recent Redeterminations of the Phase Diagram of the System: Triethylamine + Water, 1961. [online] http://iupac.org/publications/pac/pdf/1961/pdf/0201x0335.pdf (accessed 31 August 2013)

[16] Patents filed by 3H Company: US 2011/0229393 A1, US 2012/0321538 A1, US 6,969,418 B1, US 7,541,011 B2, US 7,718,151 B1, US 7,846,407 B2, US 8,318,116 B2.

[17] Kim, Y. E., Jung, H. P., Soung, H. Y., Sung, C. N., Soon, K. J., Yeo Il Y. Carbon dioxide absorption using a phase transitional alkanolamine–alcohol mixture. J. Ind. Eng. Chem. 20 (2014) 1486–1492.

[18] Hu, L., "Post-Combustion CO_2 Capture for Existing PC Boilers by Self-Concentrating Absorbent," NETL CO_2 Capture Technology Meeting, July 2012.

[19] Zhang, J., Qiao, Y., Agar, D. W. Improvement of lipophilic-amine-based thermomorphic biphasic solvent for energy-efficient carbon capture. Energy Procedia. 23 (2012) 92–101.

[20] Zhang, J., Qiao, Y., Agar, D. W. Intensification of low temperature thermomorphic biphasic amine solvent regeneration for CO_2 capture. Chem. Eng. Res. Des. 90 (2012) 743–749.

[21] Raynal, L., Alix, P., Bouillon, P-A., Gomez, A., Nailly, M. F., Jacquin, M., Kittel, J., Lella, A., Mougin, P., Trapy, J. The DMX^{TM} process: an original solution for lowering the cost of post-combustion carbon capture. Energy Procedia. 4 (2011) 779–786.

[22] Patents filled by IFP: US 8,361,424 B2, US 8,500,865 B2, US 8,562,927 B2, US 2011/0185901 A1, WO2007/104856 A1, US 2007/0286783 A1.

[23] Raynal, L., Bouillon, P-A., Gomez, A., Broutin, P. From MEA to demixing solvents and future steps, a roadmap for lowering the cost of post-combustion carbon capture. Chem. Eng. J. 171 (2011) 742–752.

[24] Aleixo, M., Prigent, M., Gibert, A., Porcheron, F., Mokbel, I., Jose, J., Jacquin, M. Physical and chemical properties of DMX^{TM} solvents. Energy Procedia. 4 (2011) 148–155.

[25] Hu, L. "CO2 Capture from Flue Gasby Phase Transitional Absorption," DOE project report, 2009, DE-FG26-05NT42488.

[26] Gomez, A., Briot, P., Raynal, L., Broutin, P., Gimenez, M., Soazic, M., Cessat, P., Saysset, S. ACACIA Project–development of a post-combustion CO_2 capture process. Case of the DMX™ process. Oil Gas Sci. Technol. 69 (2014) 1121–1129.

[27] Xu, Z., Wang, S., Liu, J., Chen, C. Solvents with low critical solution temperature for CO_2 capture. Energy Procedia. 23 (2012) 64–71.

[28] Xu, Z., Wang, S., Zhao, B., Chen, C. Study on potential biphasic solvents: absorption capacity, CO_2 loading and reaction rate. Energy Procedia. 37 (2013) 494–498.

[29] Xu, Z., Shujuan W., Changhe C. CO_2 absorption by biphasic solvents: mixtures of 1,4-butanediamine and 2-(diethylamino)-ethanol. Int. J. Greenhouse Gas Control. 2013.

[30] Brúder, P., Svendsen, H. F. Solvent comparison for postcombustion CO2 capture. Presented in 1st Post-Combustion Capture Conference, May 2011, Abu Dhabi.

[31] Pinto, D. D. D., Zaidy, S. A. H., Hartono, A., Svendsen, H. F. Evaluation of a phase change solvent for CO_2 capture: absorption and desorption tests. Int. J. Greenhouse Gas Control. 28 (2014) 318–327.

[32] Arshad, M. W., Svendsen, H. F., Fosbol, P. L., Solms, N., Thomsen, K. Equilibrium total pressure and CO_2 solubility in binary and ternary aqueous solutions of 2-(diethylamine) ethanol (DEEA) and 2-(methyalamino) propylamine (MAPA). J. Chem. Eng. Data. 59 (2014) 764–774.

[33] Brúder, P., Owrang, D., Svendsen, H. F. Pilot study-CO_2 capture into aqueous solutions of 3-methylaminopropylamine (MAPA) activated dimethyl-monoethanolamine (DMMEA). Int. J. Greenhouse Gas Control. 11 (2012) 98–109.

[34] Pinto, D. D. D., Knuutila, H., Pytianos, G., Haugen, G., Mejdell, T., Svendsen, H. F. CO_2 post combustion capture with a phase change solvent. Pilot plant campaign. Int. J. Greenhouse Gas Control. 31 (2014b) 153–164.

[35] Liebenthal, U., Pinto, D. D. D., Monteiro, J. G. M.-S., Svendsen, H. F., Kather, A. Overall process analysis and optimisation for CO_2 capture from coal fired power plants based on phase change solvents forming two liquid phases. Energy Procedia. 37 (2013) 1844–1854.

[36] Hessen, E. T., Haug-W., T., Svendsen, H. F. The refined e-NRTL model applied to CO_2-H_2O-alkanolamine systems. Chem. Eng. Sci. 65 (2010) 3638–3648.

[37] Zhuang, Q., Bruce C., Ying L. From ammonium bicarbonate fertilizer production process to power plant CO_2 capture . Int. J. Greenhouse Gas Control. 10 (2012) 56–63.

[38] Camper, D., Bara, J. E., Gin, D. L., Noble, R. D. Room-temperature ionic liquid-amine solutions: tunable solvents for efficient and reversible capture of CO_2. Ind. Eng. Chem. Res. 47 (2008) 8496–8498.

[39] Hasib-Ur-Rahman, M., Siaj, M. Larachi, F. CO_2 capture in alkanolamine/room-temperature ionic liquid emulsions: a viable approach with carbamate crystallization and curbed corrosion behavior. Int. J. Greenhouse Gas Control. 6 (2012) 246–252.

[40] Aziz, N., Yusoff, R., Aroua, M. K. Absorption of CO_2 in aqueous mixtures of N-methyldiethanolamine and guanidinium tris(pentafluoroethyl)trifluorophosphate ionic liquid at high-pressure. Fluid Phase Equilib. 322–323 (2012) 120–125.

[41] Li, X. N., Hagaman, E., Tsouris, C., Lee, J. W. Removal of carbon dioxide from flue gas by ammonia carbonation in the gas phase. Energy Fuels. 17 (2003) 69–74.

[42] Zhuang, Q. Richard P., Ligang Z., Bruce C. Ammonia-based carbon dioxide capture technology: issues and solutions. Energy Procedia. 4 (2010) 1459–1470.

[43] Darde, V., Thomsen, K., Van W., W. J. M. Stenby, E. H. Chilled ammonia process for CO_2 capture. Int. J. Greenhouse Gas Control. 4 (2010) 131–136.

[44] Barzagli, F., Mani, F., Peruzzini, M. From greenhouse gas to feedstock: formation of ammonium carbamate from CO_2 and NH_3 in organic solvents and its catalytic conversion into urea under mild conditions. Green Chem. 13 (2011) 1267–1274.

[45] Mathias, P. M., Reddy, S., O'connell, J. P. Quantitative evaluation of the chilled-ammonia process for CO_2 capture using thermodynamic analysis and process simulation. Int. J. Greenhouse Gas Control, 4 (2010) 174–179.

[46] Zheng, S., Tao, M., Liu, Q., Ning, L., He, Y., Shi, Y. Capturing CO_2 into the precipitate of a phase-changing solvent after absorption. Envir. Sci. Technol. 48(15), 2014, pp 8905–8910. DOI: 10.1021/es501554h

[47] Knuutila, H., Aronu, U. E., Kvamsdal, H. M., Chikukwa, A. Post combustion CO_2 capture with an amino acid salt. Energy Procedia. 4 (2011) 1550–1557.

[48] Liu, A. H., Ma, R., Song, C., Yang, Z. Z., Yu, A., Cai, Y., He, L. N., Zhao, Y. N., Yu, B., Song, Q. W. Equimolar CO_2 capture by N-substituted amino acid salts and subsequent conversion. Angew. Chem Int. Ed. 51 (2012) 11306–11310.

[49] Portugal, A. F., Sousa, J. M., Magalhães, F. D., Mendes, A. Solubility of carbon dioxide in aqueous solutions of amino acid salts. Chem. Eng. Sci. 64 (2009) 1993–2002.

[50] Holst, J. V., Versteeg, G., Brilman, D., Hogendoorn, J. Kinetic study of CO_2 with various amino acid salts in aqueous solution. Chem. Eng. Sci. 64 (2009) 59–68.

[51] Ma'mun, S., Kim, I. Selection and characterization of phase-change solvent for carbon dioxide capture: precipitating system. Energy Procedia. 37 (2013) 331–339.

[52] Majchrowicz, M., Niederer, J. P. M., Velders, A. H., Versteeg, G. F. "Precipitation in Amino Acid Salt CO_2 Absorption Systems," GHGT-8 NTNU VIDERE, Springer Berlin/Heidelberg, Trondheim, Norway, 2006.

[53] Hook, R. J. An investigation of some sterically hindered amines as potential carbon dioxide scrubbing compounds. Ind. Eng. Chem. Res. 36 (1997) 1779–1790.

[54] Kumar, P. S., Hogendoorn, J. A., Timmer, S. J., Feron, P. H. M., Versteeg, G. F. Equilibrium solubility of CO_2 in aqueous potassium taurate solutions: Part 2. Experimental VLE data and model. Ind. Eng. Chem. Res. 42 (2003) 2841–2852.

[55] Majchrowicz, M. E., Brilman, D. W. F., Groeneveld, M. J. Precipitation regime for selected amino acid salts for CO_2 capture from flue gases. Energy Procedia. 1 (2009) 979–984.

[56] Sanchez-Fernandez, E., Mercader, F. D. M., Misiak, K., Van D. H, L., Linders, M., Goetheer, E. New process concepts for CO_2 capture based on precipitating amino acids. Energy Procedia. 37 (2013) 1160–1171.

[57] Brouwer, J., Feron, P., Ten A., N. Amino-acid salts for CO_2 capture from flue gases. Fourth Annual Conference on Carbon Capture and Sequestration, Alexandria, Virginia, USA, May 2005.

[58] Fernandez, E. S., Goetheer, E. L. DECAB: Process development of a phase change absorption process. Energy Procedia. 4 (2011) 868–875.

[59] Sonja, S., House, L. "Chapter 2–Enzyme-Catalyzed Solvents for CO_2 Separation, Novel Materials for Carbon Dioxide Mitigation Technology," Elsevier, Amsterdam, 2015.

[60] Nguyen, L. Low-Cost Enzyme-Based Technology for Carbon Capture, 2012 NETL CO_2 Capture Technology Meeting, Pittsburgh, PA, July 11, 2012.

[61] Carley, J. A. "Enzyme-Enabled Carbon Capture, Lowering the CCS Cost Barrier," PTAC, Calgary, Alberta, Canada, 2011.

[62] Bucholz, T., et al., "Chapter 4–Development of an Organosilica Coating Containing Carbonic Anhydrase for Applications in CO_2Capture," Novel Materials for Carbon Dioxide Mitigation Technology, pp. 117–147. Elsevier 2015.

[63] Jing, G., et al., Immobilization of carbonic anhydrase on epoxy-functionalized magnetic polymer microspheres for CO_2 capture. Process Biochem. 50(12) (December 2015) 2234–224.

[64] Bihong, L., et al., Immobilization of carbonic anhydrase on carboxyl-functionalized ferroferric oxide for CO_2 capture, Int. J. Biol. Macromol. 79 (August 2015) 719–725.

[65] Aines, R. D., et al. Encapsulated solvents for carbon dioxide capture. Energy Procedia. 37 (2013) 219–224.

[66] Stolaroff, J. K., Thermodynamic assessment of microencapsulated sodium carbonate slurry for carbon capture, Energy Procedia. 63 (2014) 2331–2335.

Chapter 5

Bio-inspired Systems for Carbon Dioxide Capture, Sequestration and Utilization

Gonçalo V. S. M. Carrera, Luís C. Branco and Manuel Nunes da Ponte

Additional information is available at the end of the chapter

http://dx.doi.org/10.5772/65861

Abstract

This chapter reviews the study and development of biological, enzymatic and bio-molecular systems for carbon dioxide capture and further sequestration or even utilization. Regardless of the interest on the use of the captured CO_2 as C1 synthon on the manufacture of added-value compounds, there is a tremendous unbalance between the requirements of the contemporary society (leading to a massive production of carbon dioxide) and the framework of commercialization of the products from CO_2 utilization. In this context, viable options are storage as a solid in the form of calcium or magnesium carbonate and conversion into other energetic frameworks. In addition, it is important to highlight that the conventional energy resources are progressively being replaced by renewable resources. While the change in energetic paradigm is not accomplished, systems that capture and convert carbon dioxide are highly sought. To this end, bio-inspired systems will be presented, starting from the use of compounds from the chiral pool, such as amino acids, saccharides and related bio-polymers, involved in the physical and chemical capture, sequestration and/or utilization of CO_2. Additionally, enzymatic systems are presented in the context of sequestration of CO_2 in the form of solid carbonates or even utilization of this C1 synthon in the preparation of fuels and commodity chemicals. Carbonic anhydrase is by far the most studied enzyme, as it catalyses the inter-conversion between CO_2 and hydrogencarbonate in an effective mode. The biological option comprises the utilization of methanogens, acetogens and other organisms leading to the formation of added-value compounds. Most of the described systems are based on microbial electro-synthesis model and microbial carbon-capture cell prototypes.

Keywords: carbon dioxide, amino acids, saccharides, bio-polymers, enzymes, carbonic anhydrase

1. Introduction

In the present context, our civilization's standards of life are grounded on enormous emissions of Green House Gases (GHGs) to the atmosphere, in concrete carbon dioxide. Simultaneously, biological systems available in nature have restricted capacity on the fixation of CO_2, and accumulation of this GHG is creating impact on our environment. It is essential to develop and implement technologies that simultaneously avoid further accumulation and increase the rate of CO_2 incorporation in added value products. The use of renewable energies, such as solar, hydroelectric, wind, geothermic, hydrogen, tides and biofuels are progressively being implemented depending on the specific resources of each country and commercial adjustment of the energetic paradigm that can be valid and affordable for the near 7.5 billion human habitants in our planet (2016) [1]. Regardless of the progresses on the implementation of renewable energy resources, conventional fuels continue to be the main source of energy worldwide, leading to accumulation of CO_2 in the atmosphere.

The most recent data from IPCC is clear [2]. The total cumulative anthropogenic emissions linked to CO_2 (1750–2011) are 2040 ± 310 $GtCO_2$. Nearly 50% of the cumulative emissions took place in the last 40 years (1970–2011), consistent with a steady rise in CO_2 emissions during that period. It is important to highlight that 40% of the anthropogenic emissions (1750–2011) persisted in the atmosphere (880 ± 35 $GtCO_2$). While the anticipated change of the energetic paradigm is not assimilated, systems that capture CO_2 in an effective mode, and incorporate it in safe and useful products, are highly desirable.

An additional point, that should be presented, is illustrated in **Figure 1**, corresponding to the pattern of GHG emissions by economic sector, being useful in the definition of target sectors more able to be optimized in respect to the CO_2 footprint and development of innovative strategies adjusted to a specific challenge. The most representative sector is electricity and heat production applied to the other segments as an indirect source of CO_2 emissions (except other energy). The first sector is followed by agriculture, forestry and other land use (AFOLU). Industry, transport, other energy and buildings are the other sectors, with lower percentage on direct GHG emissions. The data here presented (**Figure 1**) is associated with the year 2010 (the most recent data available from IPCC).

From the more representative GHGs emitted to the atmosphere, CO_2 presents by far, the highest percentage of associated emissions (**Figure 2**, 2010) which demonstrates the importance of the commercially available systems for CO_2 capture and fixation, urgency in the development and implementation of straightforward and sustainable alternative systems complementary to the change in the energetic paradigm already on course.

In order to accomplish an effective CO_2 uptake exist diverse prototypes and mature technology: (A) absorption; (B) adsorption; (C) cryogenic; (D) membrane [3]. All the pointed instances, except (C), incorporate bio-inspired systems, as represented in the literature, and will be described briefly.

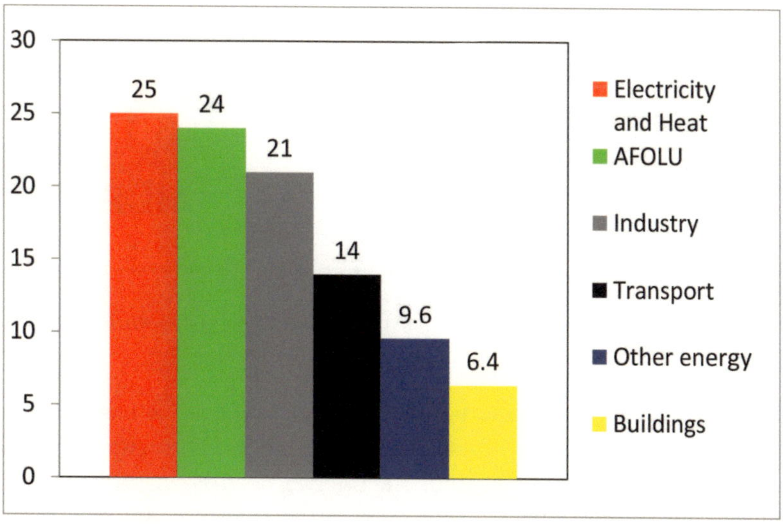

Figure 1. Percentages of direct GHG emissions by economic sector from a total of 49 Gt CO_2-equivalent during 2010. AFOLU: Agriculture, Forestry and Other Land Use [2].

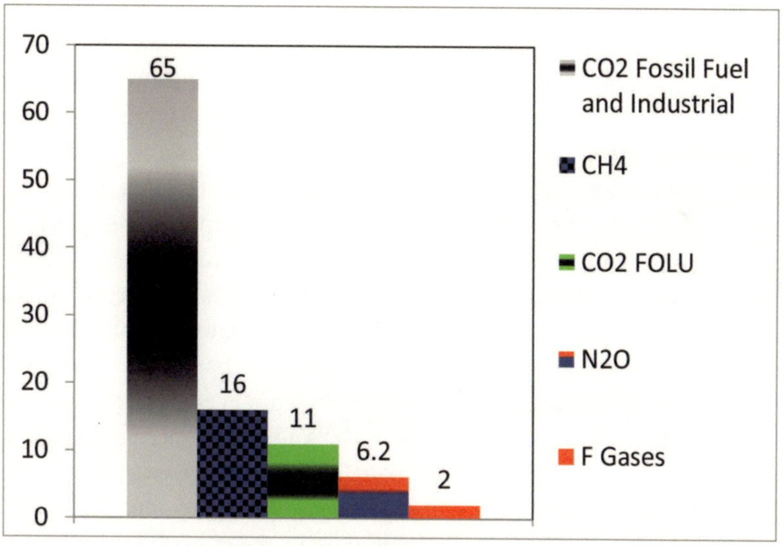

Figure 2. Percentages of GHGs from the 49 Gt CO_2-equivalent emissions during 2010. FOLU: Forestry and Other Land Use, F Gases: fluorinated gases covered under the Kyoto Protocol [2].

(A) Absorption: this topic includes physical and chemical absorption. In both situations, CO_2 is captured in the volume of a solution. The first framework comprises the physical interaction between high-pressure CO_2 and a solution by intermolecular interactions. In this context, molecular solvents and ionic liquids as well carry out this action. Chemical absorption is conventionally performed with solutions of alkanolamines (**Figure 3**):

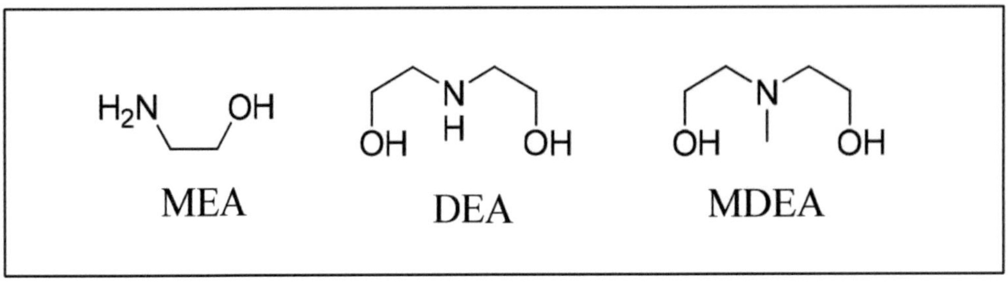

Figure 3. Conventional amines used in chemical absorption of CO_2. MEA: Monoethanolamine; DEA: Diethanolamine; MDEA: Methyldiethanolamine.

This established technology presents various drawbacks, such as the compulsory dilution of the alkanolamine in the aqueous environment to avoid deterioration of materials and excessive release of heat when reaction is performed. The utilization of these systems leads to mitigated CO_2 uptake (7 wt% using a 30 wt% aqueous solution of MEA). Moreover, there is a high-energy penalty incorporated into the system due to a high-heat capacity of the aqueous environment acting as a sink in the process of CO_2 release [4, 5]. Finally, the solvent is volatilized during the operations precluding, an effective regeneration of the CO_2 capture system. These strategies are used when the concentration of CO_2 is low.

The mechanism associated with these systems encompasses the formation of carbamates, in a 0.5:1 stoichiometry (half of the converted alkanolamine is in the form of carbamate and the other is presented as ammonium). Two different paths lead to this same end: The two-step Zwitterion mechanism is started by a nucleophilic attack of the amine group on CO_2 leading to the presence in the same moiety of positive and negative charges. In a further step, a proton is transferred to another alkanolamine, and a salt is formed. In the single-step termolecular mechanism, the nucleophilic attack and proton transfer occurs simultaneously. The other product of reaction is hydrogencarbonate corresponding to the product of reaction between CO_2 and water. This reaction is slower; nevertheless, the stoichiometry of CO_2 incorporation is one, which is a factor of two superior to the formation of carbamate by the same quantity of aminoalcohol. Giving the product, the conventional capture agents can act as generic bases or nucleophiles in the specific instance of CO_2 as well. Another property associated with these systems is that, at given pH, hydrogencarbonate might coexist with carbonic acid and carbonate. The hydroxide anion has a lower expression, though the reaction with CO_2 is faster than with water. Another mechanism associated with these systems is the hydrolysis of carbamate to generate hydrogencarbonate. Given the degree of substitution, the amine functionality can be more CO_2-philic or more alkaline, with a concomitant contribution on the definition of the reaction profile.

(B) Adsorption is obtained by the capture of CO_2 when interacting with a solid surface. Contrarily to chemical absorption, the interaction between CO_2 and the surface is moderated (intermolecular forces). Activated carbons and molecular sieves are conventionally used. Three different modes of action characterize this type of framework, related with the 'switch' used between adsorption/desorption: pressure, temperature and electric power control the behaviour of these systems.

(D) Membrane is a partially permeable structure that separates CO_2 from gas mixtures. With this recent technology, CO_2 is separated from diverse sources, such as post-combustion flue gas, syngas and natural gas. Two operational methods are available: (a) Gas separation membrane: based on the preferential permeation of a specific component of a mixture. (b) Gas absorption membrane: centred on the specific affinity of the previously referred chemical absorber solutions towards CO_2.

The bio-inspired systems described in the following topics represent an effective alternative to the available frameworks, with potential to be integrated in the previously mentioned capture systems as well as in sequestration and utilization frameworks.

Aqueous solutions of amino acids are straightforward alternatives to alkanolamine-based systems. The amino acid frameworks are presented as stand-alone salts, as zwitterionic structures activated by bases, or as salts doped with superbases. Different mechanisms of reaction/association are presented according the structure and composition of the system. A different case study is based on the use of highly abundant saccharides and related biopolymers, which may constitute invaluable systems for CO_2 capture, sequestration and/or utilization, presented as liquid solutions, gels, confined hydrated foams or solid adsorbents.

Enzymes, especially carbonic anhydrase (CA), are useful catalysts for CO_2 sequestration and utilization. In the case of CA, the high rates of reaction and the mild reaction conditions constitute clear advantages of these systems in the laboratory environment. But on a pilot or even industrial scale, harsh conditions of temperature and the presence of contaminants in CO_2 streams hinder the utilization of this enzyme. Possible solutions already on praxis are expression of the genetic code associated with this enzyme from thermophiles on readily available organisms, immobilization on diverse supports or generation of catalysts inspired on the mode of action of carbonic anhydrase.

Finally, the use of microbes is addressed in this chapter on the production of added value products from CO_2, with special focus on the use of methanogens and acetogens in microbial electro-synthesis and microbial capture cell frameworks.

2. Bio-inspired systems

The systems proposed here (biological, enzymatic and bio-molecular) constitute solid alternatives to the conventional platforms available in the market. These bio-inspired frameworks present diverse advantages such as low corrosion, easy disposal and biodegradability, naturally produced and possibility to tune capacity of CO_2-incorporation according to the configuration of the system. Nevertheless, there are practical issues that should be addressed in order to make the current technology available thrive in the various challenges assigned.

2.1. Bio-molecular

2.1.1. *Amino acids*

The amino acid systems highlighted include the use of these frameworks as anions, in the presence of an inorganic cation (sodium or potassium), as aprotic ionic liquids or as protic mixtures, using either organic or inorganic bases. The main advantage of amino acid-based systems over conventional alkanolamines relies on high stability to oxidative degradation, high chemical reactivity with CO_2, low vapour pressures (compatible with temperature of flue gases), and high surface tension (fundamental on the design of membranes). Various studies exist when an inorganic cation is used [6–18]. Here a selection will be emphasized to establish the structure of amino acid/CO_2 absorption-desorption properties relationship (**Figure 4**, **Table 1**). In this comparative study, the conventional MEA aqueous solution is used as a reference. It presents considerably high initial rates of CO_2 absorption and desorption, and high CO_2 uptake is obtained as well (**Table 1**).

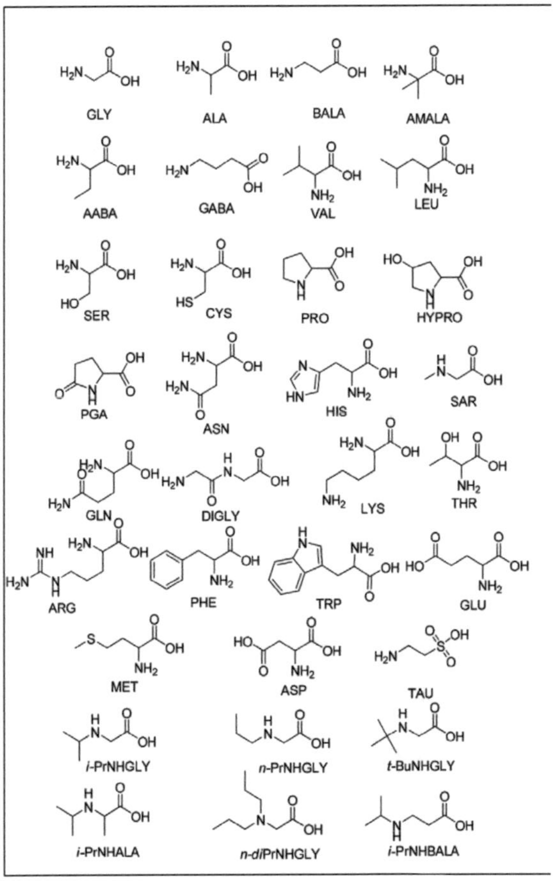

Figure 4. Amino acids used in CO_2-capture studies, and presented along this chapter.

Amino acid system	Solvent / Concentration of the capture agent	P, T_{abs}, T_{desorp} (kPa, K, K)	Absorpt method	Initial rate absorpt (mol CO_2 mol amine^{-1} min^{-1})	Initial rate desorpt (mol CO_2 mol amine^{-1} min^{-1})	CO_2 uptake (mol CO_2 mol amine^{-1})	pKa, parent amine / amino-acid
Reference MEA [6]	1 M aqueous	15, 313, 353	Bulk sol.	3.84×10^{-2}	2.00×10^{-2}	0.736	9.5
[K]GLY [6]	1 M aqueous	15, 313, 353	Bulk sol.	3.93×10^{-2}	1.84×10^{-2}	0.738	2.34, 9.6
[K]ALA [6]	1 M aqueous	15, 313, 353	Bulk sol.	3.16×10^{-2}	1.98×10^{-2}	0.670	2.35, 9.69
[K]BALA [6]	1 M aqueous	15, 313, 353	Bulk sol.	3.65×10^{-2}	1.80×10^{-2}	0.721	3.60, 10.19
[K]AABA [6]	1 M aqueous	15, 313, 353	Bulk sol.	3.76×10^{-2}	1.87×10^{-2}	0.728	2.65, 9.6
[K]GABA [6]	1 M aqueous	15, 313, 353	Bulk sol.	3.87×10^{-2}	1.54×10^{-2}	0.749	4.23, 10.43
[K]AMALA [6]	1 M aqueous	15, 313, 353	Bulk sol.	2.43×10^{-2}	1.20×10^{-2}	0.750	2.36, 10.21
[K]SER [6]	1 M aqueous	15, 313, 353	Bulk sol.	3.16×10^{-2}	2.26×10^{-2}	0.619	2.21, 9.15
[K]CYS [6]	1 M aqueous	15, 313, 353	Bulk sol.	3.18×10^{-2}	2.46×10^{-2}	0.485	1.92, 8.37, 10.70
[K]PRO [6]	1 M aqueous	15, 313, 353	Bulk sol.	4.26×10^{-2}	1.48×10^{-2}	0.746	1.99, 10.96
[K]HYPRO [6]	1 M aqueous	15, 313, 353	Bulk sol.	4.03×10^{-2}	1.55×10^{-2}	0.655	1.82, 9.65
[K]PGA [6]	1 M aqueous	15, 313, 353	Bulk sol.	2.41×10^{-2}	0.38×10^{-2}	0.224	−1.76, 3.48, 12.76
[K]ASN [6]	1 M aqueous	15, 313, 353	Bulk sol.	2.79×10^{-2}	2.34×10^{-2}	0.573	2.02, 8.80
[K]GLN [6]	1 M aqueous	15, 313, 353	Bulk sol.	3.06×10^{-2}	2.08×10^{-2}	0.600	2.2, 9.1
[K]DIGLY [6]	1 M aqueous	15, 313, 353	Bulk sol.	2.99×10^{-2}	2.18×10^{-2}	0.510	-
[K]ARG [6]	1 M aqueous	15, 313, 353	Bulk sol.	3.56×10^{-2}	2.19×10^{-2}	1.107	1.82, 8.99, 12.48
[K]TAU [6]	1 M aqueous	15, 313, 353	Bulk sol.	3.17×10^{-2}	2.26×10^{-2}	0.573	1.5, 9.06
[Na]GLY [7]	PEG_{150} 3 mmol/ 15 mmol sol.	100, 298, 313	Bulk sol.	-	-	0.43 (20 min)	2.34, 9.6
[Na] *i*-PrNHGLY [7]	PEG_{150} 3 mmol/ 15 mmol sol.	100, 298, 313	Bulk sol.	-	-	0.91 (25 min)	-
[Na]*n*-PrNHGLY [7]	PEG_{150} 3 mmol/ 15 mmol sol.	100, 298, 313	Bulk sol.	-	-	0.59 (30 min)	-
[Na]*t*-BuNHGLY [7]	PEG_{150} 3 mmol/ 15 mmol sol.	100, 298, 313	Bulk sol.	-	-	0.85 (25 min)	-
[Na]*i*-PrNHAla [7]	PEG_{150} 3 mmol/ 15 mmol sol.	100, 298, 313	Bulk sol.	-	-	0.73 (30 min)	-
[Na] *n*-*Di*PrNHGLY [7]	PEG_{150} 3 mmol/ 15 mmol sol.	100, 298, 313	Bulk sol.	-	-	0.48 (30 min)	-
[Na] *i*-PrNHBALA [7]	PEG_{150} 3 mmol/ 15 mmol sol.	100, 298, 313	Bulk sol.	-	-	0.65 (15 min)	-

Table 1. CO_2 absorption and desorption properties of amino acid-based systems.

Diverse potassium salts of amino acids from chiral pool dissolved in aqueous media were tested [6] (**Table 1**). Glycine (GLY) salt presents similar performances as conventional MEA solutions. Differently, alanine (ALA) is identified by a different reaction profile with CO_2, the level of CO_2 incorporation and the associated initial rate of absorption are lower. Nevertheless, the rates of CO_2 release at moderate temperatures are higher. The reason for such behaviour are the bulkiness of the associated substituent group and the moderate pKa of the amine functionality, contributing to the mitigated propensity of CO_2 to be chemically captured, either in the form of carbamate (amino acid as nucleophile) or of bicarbonate (alanine as base).β-Alanine (BALA) is an extended amino acid with concomitant bulkiness and high pKa of its amine functionality, contributing to lower CO_2 absorption and desorption performances with respect to GLY, nevertheless, BALA presents improved CO_2 absorption and poorer desorption performances when compared with Alanine. With α-aminobutyric acid (AABA), the amine functionality should act preferably as base, due to probable 5-element-ring-hydrogen bond interaction between carboxyl and amine functionalities owing to the size of the linear substituent group, leading to a more favourable incorporation of CO_2 in the form of hydrogencarbonate after reaction with water. Improved incorporation with a factor of two, with respect to carbamate leads to improved absorption performances when compared with Alanine.

The performance of γ-aminobutyric acid (GABA) in terms of CO_2 uptake is higher than the previously mentioned amino acids, the high pKa leads to the formation of hydrogencarbonate, which is a slower reaction than the formation of carbamate. This information is in line with the lower initial ratio of CO_2 absorption obtained when compared with GLY. α-Methyl alanine (AMALA) presents two methyl groups in the α-position and a high pKa of the amine functionality which, similar to AABA, acts as enhanced base by hydrogen-bond stabilization of the protonated base by the carboxyl group of the amino acid. AMALA is characterized by the highest value of CO_2 uptake among the presented amino acids; nevertheless, it presents the lowest initial rate of CO_2 absorption. These results are compatible with the preferential formation of hydrogencarbonate instead of carbamate. Serine (SER) and cysteine (CYS) are hydroxyl and thio-functionalized amino acids, respectively, that may create destabilization as hydrogen donors when the protonated amine functionality is interacting with the carboxyl group of the amino acid. Considering this aspect, the formation of carbamate should be predominant with respect to the hydrogencarbonate counterpart, and the CO_2 uptake is low. The initial rate of desorption associated with these two amino acids is high due to the steric repulsion between the formed carbamate and the bulky hydroxyl or thiol functionalities.

Proline (PRO), 4-hydroxy proline (HYPRO) and pyroglutamic acid (PGA) are cyclic, secondary amino acids, with PRO presenting high pKa and, possibly due to stabilization of protonated amine by the carboxyl group, may lead to enhanced CO_2 uptake, the highest pKa among the presented amino acids, leading to a higher hydroxide/water ratio and an enhanced kinetics for CO_2 uptake. The amine functionality of HYPRO presents lower pKa than PRO-equivalent, leading, possibly, to a less favourable hydrogencarbonate/carbamate ratio and low CO_2 uptake. PGA presents an electro-tractor carbonyl group conjugated with the amine functionality hampering its performance either as a nucleophile or base, leading to a low level of CO_2 incorporation.

http://dx.doi.org/10.5772/65861

Asparagine (ASN) and Glutamine (GLN) are linear amide-functionalized amino acids, presenting low pKa associated with the corresponding amine functionalities, leading to low levels of CO_2 uptake. Due to steric repulsion, the ratio of the kinetics of CO_2 uptake/desorption is mitigated. Diglycine (DIGLY) leads to a low level of carbon dioxide incorporation due to inherent bulkiness associated to the chemical structure. Arginine (ARG) presents highly basic guanidine functionality and a not so basic amine group. The combination of both functionalities leads to the highest level of CO_2 incorporation, despite the presence of both functionalities. Taurine (TAU) amine group presents low pKa, considering this; the ratio carbamate/hydrogencarbonate should be high.

In another study, [7] mostly secondary but also tertiary bulky amines in PEG_{150} solvent are tested. With [Na] *i*-PrNHGLY (**Figure 4**, **Table 1**), the level of CO_2 incorporation was near 100% with respect to the amine unit. The main reason for such result is the formation of carbamic acid instead of carbamate leading to a nearly full use of the amine functionality to capture CO_2 instead of half (in the carbamate/ammonium salt). To this result contributes the stabilization of the carbamic acid functionality by the carboxyl group, the steric repulsion associated with the amine substituents, avoiding deprotonation of the carbamic functionality, and the type of chemical environment [7]. The carbamic acid reaction profile and steric hindrance lead to low energy requirements for CO_2 desorption (313 K). An interesting work was carried out by Wang et al. [17], where, from a diverse set of amino acid sodium salts (aqueous solutions) was possible to observe with Alanine, a phase-split between a CO_2-rich and a CO_2-lean phase. The CO_2-rich phase was composed of carbamate and hydrogencarbonate. The decrease on the volume of solution leads to a significant decrease in energy input for the CO_2 strip.

In a different study [19], (**Figure 4**, **Table 2**) a diverse set of amino acid ionic liquids (neat) for CO_2 capture was tested. The best performance was obtained with $[N_{66614}]$ LYS, leading to nearly two equivalents of CO_2 reacting with the two amine functionalities existing in lysine. The reason behind this outstanding result relies on the formation of carbamic acid (as in Ref. [7]). One additional point is addressed to the catalytic effect of the carboxyl group, promoting the proton transfer, starting from zwitterionic structure (after nucleophilic attack to CO_2) to the carbamic acid end product. In this work the effect of the cation in the CO_2 uptake was checked. $[N_{66614}]$ and $[P_{66614}]$ were tested in combination with the LYS anion. It was observed that the initial rate of absorption was higher with $[P_{66614}]$, nevertheless $[N_{66614}]$ lead to high CO_2 uptake (**Table 2**). $[N_{66614}]$LYS is associated with stronger hydrogen-bond interactions after chemisorption of CO_2 leading to an increment of viscosity and poorer kinetics. Due to the different stabilized-arrangement of this ionic liquid, after incorporation of CO_2, nearly all the amine groups are converted to carbamic acid functionalities. $[P_{66614}]$LYS presents another configuration after CO_2 uptake with concomitant half of the amine groups presented as carbamic acid functionalities, ¼ as carbamate and the remaining ¼ as ammonium.

A conceptual study was carried out in our laboratories [20] concerning the preparation of reversible ionic liquids using GLY, ALA, valine (VAL), leucine (LEU), phenylalanine (PHE) and tryptophan (TRP) that activated by an organic superbase, 1,8-diazabicyclo[5.4.0]undec-7-ene (DBU) or 1,1,3,3-tetramethylguanidine (TMG), react with CO_2 to obtain carbamate-based ionic liquids. The system can revert back to the early configuration upon heating at an appropriate temperature. It was possible to observe for the DBU series a decrease in the temperature associated with CO_2 release when the bulkiness of the substituent group of the amino acid increases. A similar order was observed with the TMG set, however not completely defined.

Amino acid system	Solvent / Concentration of the capture agent	P, T_{abs}, T_{desorp} (kPa, K, K)	Absorpt method	Initial rate absorpt (mol CO_2 mol amine^{-1} min^{-1})	Initial rate desorpt (mol CO_2 mol amine^{-1} min^{-1})	CO_2 uptake (mol CO_2 mol amine^{-1})	pKa, parent amine /amino – acid
[N_{66614}]LYS [19]	neat	100, r.t, 353	-	-	-	2.1 (24 h)	2.18, 8.95, 10.53
[N_{66614}]ASN [19]	neat	100, r.t, 353	-	-	-	2.0 (48 h)	2.02, 8.80
[N_{66614}]GLN [19]	neat	100, r.t, 353	-	-	-	1.9 (48 h)	2.2, 9.1
[N_{66614}]HIS [19]	neat	100, r.t, 353	-	-	-	1.9 (48 h)	1.80, 9.33, 6.04
[N_{66614}]ARG [19]	neat	100, r.t, 353	-	-	-	1.3 (48 h)	1.82, 8.99, 12.48
[N_{66614}]MET [19]	neat	100, r.t, 353	-	-	-	1.2 (24 h)	2.28, 9.21
[P_{66614}]LYS [19]	neat	100, r.t, 353	-	-	-	1.6 (48 h)	2.18, 8.95, 10.53

Table 2. CO_2 absorption and desorption characteristics of amino acid-based systems.

Other studies concerning the application of amino acid-based ionic liquids in CO_2 capture are presented here. Lv et al. [21] tested 1-aminopropyl-3-methylimidazolium glycinate aqueous solution and could achieve 1.23 mol of CO_2 loading per mol of ionic liquid. The reaction was followed by ^{13}C-NMR and it comprised two steps: (1) initial formation of carbamate, followed by (2) hydrolysis of this functionality to obtain hydrogencarbonate. The same author [22], reported the use of [C_2OHMIM] GLY aqueous solution to 0.575 mol of CO_2 uptake/mol ionic liquid. In that study the effect of O_2 on the absorption performances was determined, which were better than in the case of MEA solution. In another study, Li et al. [23], tested [P_{4444}] salts of GLY, ALA and PRO, in combination with PEG solvents, on the design of membranes for CO_2/H_2 separation. In a different framework, amino acids are combined with inorganic bases (mainly carbonate salts) in aqueous solutions to promote kinetics and capacity of CO_2 uptake. GLY, sarcosine (SAR), PRO [24] and ARG, as well, were used in that context [25].

Finally, as an example for the promoting effect of adding the conventional alkanolamines to amino acid systems, Gao et al. [26] combined MDEA aqueous solution with [N_{1111}] GLY. The

http://dx.doi.org/10.5772/65861

reason behind this option relies on the fact that MDEA, a tertiary amine, acts as base in the considerable slow reaction between water and CO_2 to obtain hydrogencarbonate, leading to high CO_2 uptake. The GLY-based amino acid presents good kinetic performances, however low CO_2 incorporation in the form of carbamate. Considering the reaction of carbamate hydrolysis and the complementary reactive profiles of MDEA and $[N_{1111}]$ GLY, such systems are presented here as an alternative to conventional frameworks.

2.1.2. Saccharides and related bio-polymers

The use of saccharides and related bio-polymers (**Figure 5**) constitute natural and abundant alternatives for CO_2 fixation, chemisorption and adsorption. Concerning fixation, Sun et al. [27] developed a superbase/cellulose catalytic system to obtain cyclic carbonates from epoxides and CO_2. Cellulose acts as a hydrogen bond donor and the superbase as the nucleophile in the activation of the epoxide. High conversions and selectivities are associated with DBU/cellulose. In a different study, Tamboli et al. [28], reported the use of chitosan/DBU dissolved in 1-mesyl-3-methylimidazolium (mesylMIM)-based ionic liquids for preparation of dimethyl carbonate (DMC) from methanol and CO_2. From the chemisorption perspective, it was tested in our laboratories [29] the use of monosaccharides, oligosaccharides or a polysaccharide-activated, by combination with adjustable proportion of liquid DBU or TMG as organic superbases, for CO_2 capture. Of course, it is necessary to consider that low ratios of superbase lead to highly viscous solutions with hampered capacity for CO_2 mass transfer and concomitant poor performances in the capture of this GHG. In a different perspective, an excess of the superbase would lead to high dilution of the capture agent with associated limitation on the wt% of CO_2 uptake. It was necessary to carry out an optimization for maximal performances (D-Mannose:DBU −0.625/1 in equivalents leading to 13.9 wt% of CO_2 uptake and 3.3/5 alcohol functionalities converted to carbonates). It is also important to consider an effective stirring to overcome the increase of viscosity with the progress of reaction, which limits CO_2 uptake.

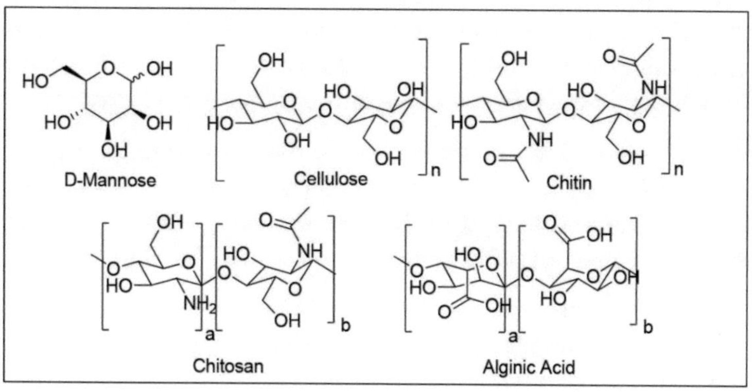

Figure 5. Examples of mono-saccharide and related biopolymers.

In a different study, Eftaiha et al. [30] used chitin acetate in DMSO for CO_2 capture. The mechanism involves activation of the alcohol groups by DMSO followed by conversion into carbonates which are stabilized by the ammonium groups available in chitin acetate. A very interesting concept was reported by Sehaqui et al. [31] and used in the preparation of cellulose-polyethyleneimine foams and study of its properties on CO_2 capture from air. To this end, it is essential a high relative humidity for optimal CO_2 uptake. The confined/activated water might have a crucial role in the mode of action of this system. There are other studies that deal with carbon dioxide chemisorption by the use of chitin or chitosan dissolved in ionic liquids such as the one reported by Xie et al. [32]. The use of saccharide-based polymers is also associated with adsorption systems. In this context, carbon spheres were prepared from alginate and chitosan, after thermal treatment between 673 and 1073 K, leading to an excellent capacity for CO_2 adsorption. The high conductivity presented by alginate-based spheres was crucial for the development of an adsorption/desorption system based on the use of electric power as a "switch" with low energetics associated with CO_2 capture and release.

2.2. Enzymatic

CO_2 fixation and conversion catalysed by enzymatic-cellular systems is essential in the recycling processes inherent to life. Six major routes [33] are available (**Figure 6**):

1. Calvin cycle: it is one of the most important processes in nature, being the major route for carbon dioxide conversion. It is present in a great majority of photosynthetic organisms and besides the CO_2 fixation (thermodynamically favourable), catalysed by 1,5-ribulose bisphosphate carboxylase (RuBisCO), in each cycle are produced saccharides, fatty acids and amino acids.

2. Reductive citric acid cycle: this cycle, comprising four carboxylation steps exists, basically, in the conversion of CO_2 and water into carbon compounds. It is found mainly in some thermophilic bacteria that grow on H_2, and bacteria that reduce sulphate. The first carboxylation step, the conversion of succinyl CoA into 2-oxoglutarate is thermodynamically unfavourable ($\Delta G^{\circ\prime}$ = 19 kJ/mol), such as the conversion of 2-oxoglycerate into isocitrate ($\Delta G^{\circ\prime}$ = 8 kJ/mol) and the production of pyruvate from acetyl CoA ($\Delta G^{\circ\prime}$ = 19 kJ/mol). The fourth carboxylation step, the conversion of phosphoenolpyruvate into oxaloacetate is thermodynamically favourable ($\Delta G^{\circ\prime}$ = −24 kJ/mol) and due to its incorporation into this cycle the promotion of the three previous carboxylation steps is assured.

3. Reductive acetyl CoA route: it is a non-cyclic route, existing in some acetogenic and methanogenic micro-organisms. This process includes the conversion of CO_2 into formic acid ($\Delta G^{\circ\prime}$ = 22 kJ/mol) catalysed by formate dehydrogenase. The other fixation step comprises the conversion of carbon dioxide into CO ($\Delta G^{\circ\prime}$ = 0 kJ/mol).

4. The 3-hydroxypropionate cycle was found in *Chloroflexaceae*, a phototrophic bacterium and comprises two favourable CO_2 conversion steps, from the thermodynamic point of view. Malonyl CoA from acetyl CoA ($\Delta G^{\circ\prime}$ = −14 kJ/mol) and conversion of propionyl CoA into methylmalonyl CoA ($\Delta G^{\circ\prime}$ = −11 kJ/mol) are the two CO_2 conversion steps.

5. A recently found cycle is the 3-hydroxypropionate/4-hydroxybutyrate cycle existing in *Metallosphaera* and includes three CO_2 conversion steps, existing in the (2) reductive citric acid cycle and (4) 3-hydroxypropionate cycle (**Figure 6**).

6. Finally the dicarboxylate/4-hydroxybutyrate cycle is found in *Thermoproteales* and *Desulfurococcales* and comprises two carbon dioxide conversion steps, (2a) and (2b), existing in the reductive citric acid cycle (**Figure 6**).

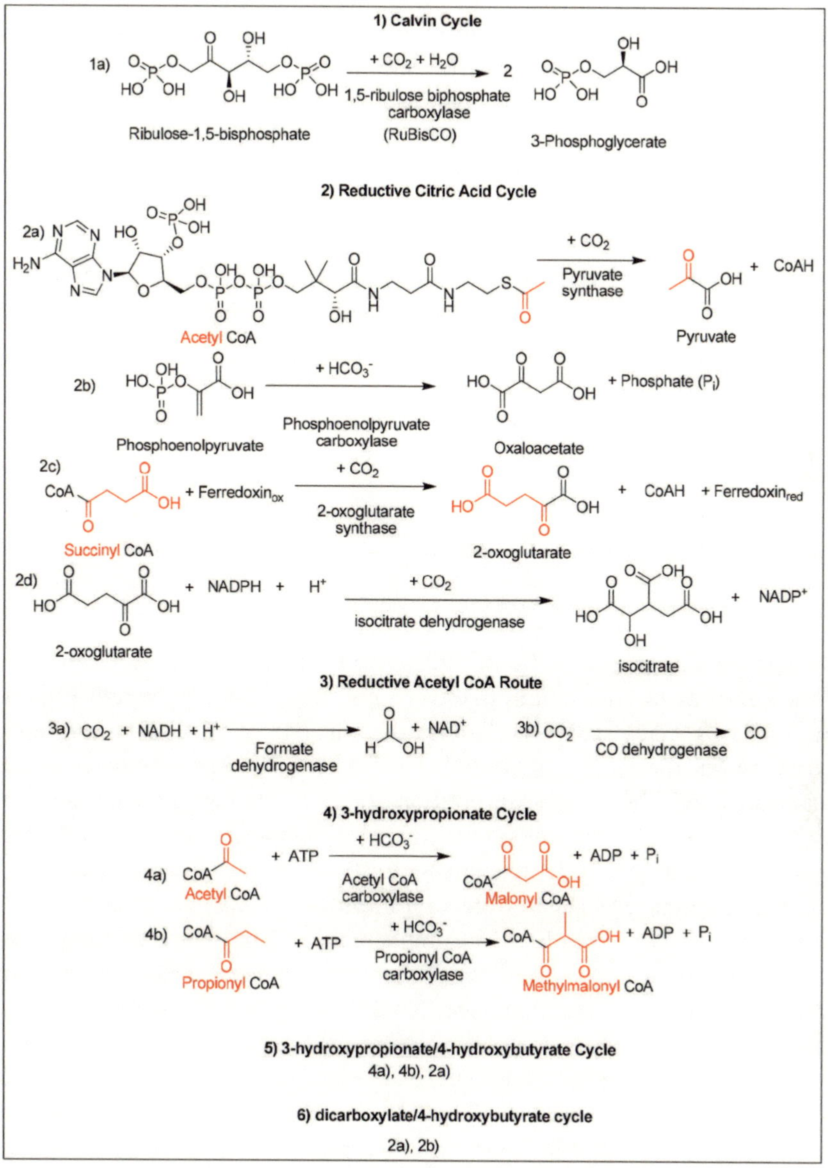

Figure 6. Enzymatic-catalysed reactions of CO_2 incorporation involved in the six main cellular metabolic routes [33].

In all the conversion steps (**Figure 6**) the source of the C_1 synthon is either CO_2 itself or HCO_3 as the intermediate. The preparation of HCO_3 is catalysed by the ubiquitous enzyme carbonic anhydrase (CA):

This enzyme, one of the fastest enzymes known, has a diversity of structures and is catalogued in five different families (α, β, γ, δ and ζ). As an illustrative example, a typical mechanism of a α-CA, which contains in the active site, a Zn(II) centre coordinated with three HIS and one water molecule forming hydrogen bond with the hydroxyl group of THR (activated by GLU) will be described. This concerted interaction, convert water/hydroxide to an enhanced nucleophile towards CO_2, leading to the formation of hydrogencarbonate. Afterwards a new water molecule replaces HCO_3 in the Zn(II) coordination site and a new cycle is initiated [34]. Considering the diverse enzymatic reactions available (**Figure 6**) and the specific instance of CA (**Figure 7**), some concerns should be addressed. From the presented enzymatic-catalysed reactions (**Figure 6**) only a parcel can be used *in vitro* for conversion of CO_2, which is not the case of the reactions involving coenzyme A (CoA), that lead to uncommon/useless products. Other reactions are thermodynamically unfavourable under standard conditions, and pH, ionic strength or temperature tuning should be carried out [35]. It is important to highlight that the possibility of tuning reaction parameters is restrained due to possible inactivation/denaturation of enzyme under specific reaction conditions.

$$H_2O + CO_2 \underset{CA}{\rightleftharpoons} HCO_3^{\ominus} + H^{\oplus}$$

Figure 7. CO_2 conversion to HCO_3 catalysed by carbonic anhydrase.

Other option is the coupling of reactions leading to a net thermodynamic feasible transformation [36]; however, as the reaction system becomes more complex, acceptable operational conditions become usually narrower. Due to the efficiency of CA, the enzymatic reaction associated is, by far, the most represented reaction of CO_2 incorporation in the literature, usually associated with the direct sequestration of CO_2 in the form of solid calcium or magnesium carbonates [34, 37], separation operations [38] and even coupled to other enzymatic CO_2 incorporation reaction in order to increase drastically the concentration of the C_1 synthon in the form of hydrogencarbonate in water and improve the reaction outcome [39]. The high temperatures and inhibitors produced during combustion processes and high commercial value associated with this enzyme, limit the efficiency of CA, nevertheless diverse frameworks were developed in order to overcome operational constraints such as: direct use of carbonic anhydrase mimics [40], support immobilization of the enzyme [41], (to avoid denaturation under harsh conditions), and combination with motion [42] generated by a chemical-engine (to overcome the diffusion constraints associated with immobilization) appear as useful frameworks. Testing carbonic anhydrase from thermophiles [43] and its genetic edition with overexpression in *Escherichia coli* are other alternatives [44].

2.3. Biological

The concluding topic comprises the use of micro-organisms as productive units of added-value products from CO_2 (sequestration and/or utilization). Diverse concepts were explored in the literature such as: (1) microbial electro-synthesis [45, 46], where electrons are supplied from a cathode to micro-organisms, which converts CO_2 into added-value products, usually methane and/or acetate. The electrons are provided by electric current, preferentially from renewable resources. Another framework is (2) microbial carbon capture cell [47], where, different from microbial electro-synthesis, the source of electrons comes from the microbial-assisted degradation of organic compounds from wastewater in the anode with formation of protons, electrons and CO_2. Usually the anode and the cathode are separated by an ion exchange membrane. The cathode receives the generated electrons and, assisted by micro-organisms, converts CO_2 into useful chemicals. Another option available is, (3) microbial electrolytic carbon capture, which similar to microbial electro-synthesis, is used as an external electric power source to increase the potential generated by degradation of organic compounds in the anode assisted by micro-organisms. H_2 and OH^- are generated in the cathode with the anion reacting with CO_2 to obtain hydrogencarbonate [48]. Usually, depending on the microbial cultures on the cathode, it is possible to obtain methane, acetate and other compounds. When acetate is required acetogens are used, nevertheless there is competition associated with the formation of methane, which is inhibited by the addition of compounds, such as 2-bromoethanesulfonate. Other systems available are based solely on the application of light in photo-reactors in order to photosynthetic micro-organisms produce added value compounds. All these options should be optimized in order to be effectively used in real situations.

3. Conclusions and challenges

While renewable sources of energy do not definitely replace conventional fuels, the use of bio-inspired systems for CO_2 capture, sequestration and utilization, constitutes an already open window of opportunity in the context of mitigation of environmental effects associated with excessive anthropogenic GHG emissions. Characteristics such as abundance, low corrosion, biodegradability and possibility to tune interaction with CO_2, constitute clear advantages of the described bio-inspired systems. With bio-molecular frameworks, it is important to overcome the kinetic constraints associated with increase of viscosity when CO_2 is captured. Simultaneously the conception of robust systems requiring low energetics for CO_2 release is important. Additionally, the straightforward use of the specific functionalities of amino acids, saccharides and related bio-polymers on the enhancement of the levels of fixation and utilization of CO_2 is essential. Concerning enzymatic systems is of extreme importance, for stand-alone and especially multi-enzymatic systems, the design/optimization of the system configuration to obtain the target product in high yields or simply sequester CO_2 as a solid. Additional attention should be devoted to the design of robust systems compatible with the real conditions of combustion/processing of gases, addressed to immobilization, design of enzymatic mimics and genetic engineering. Finally, the biological systems available should be improved with application studies in order to overcome robustness and selectivity constraints associated with electro/photochemical systems.

Abbreviations

AABA:	α-Aminobutyric acid
ADP:	Adenosine Diphosphate
AFOLU:	Agriculture, Forestry and Other Land Use
ALA:	Alanine
AMALA:	α-Methyl Alanine
ARG:	Arginine
ASN:	Asparagine
ATP:	Adenosine Triphosphate
BALA:	β-Alanine
t-BuNHGLY:	N-*t*-butylglycine
[C_2OHMIM]:	1-(2-hydroxyethyl)-3-methylimidazolium
CA:	Carbonic Anhydrase
CoA:	Coenzyme A
CYS:	Cysteine
DBU:	1,8-Diazabicyclo[5.4.0]undec-7-ene
DEA:	Diethanolamine
DIGLY:	Diglycine
n-DiPrNHGLY:	N-*n*-dipropylglycine
DMC:	dimethyl carbonate
GABA:	γ-Aminobutyric acid
GHGs:	Green House Gases
GLN:	Glutamine
GLU:	Glutamate
GLY:	Glycine
HIS:	Histidine
HYPRO:	4-hydroxy proline
IPCC:	Intergovernment Panel on Climate Change
LEU:	Leucine
LYS:	Lysine
MDEA:	Methyldiethanolamine
MEA:	Monoetanolamine
[MesylMIM]:	1-mesyl-3-methylimidazolium
MET:	Methionine
[N1111]:	Tetramethylammonium
[N66614]:	trihexyltetradecylammonium
NADP:	Nicotinamide Adenine Dinucleotide Phosphate
[P4444]:	Tetrabutylphosphonium
[P66614]:	trihexyltetradecylphosphonium
PEG:	Polyethyleneglycol
PGA:	Pyroglutamic acid
PHE:	Phenylalanine
i-PrNHAla:	N-*i*-propylalanine
i-PrNHBALA:	N-*i*-propyl-β-Alanine
i-PrNHGLY:	N-*i*-propylglycine

n-PrNHGLY:	N-*n*-propylglycine
PRO:	Proline
RuBisCO:	1,5- Ribulose Bisphosphate Carboxylase
SAR:	Sarcosine
SER:	Serine
TAU:	Taurine
THR:	Threonine
TMG:	1,1,3,3-tetramethylguanidine
TRP:	Tryptophan
VAL:	Valine

Author details

Gonçalo V. S. M. Carrera, Luís C. Branco and Manuel Nunes da Ponte*

*Address all correspondence to: mnponte@fct.unl.pt

LAQV – REQUIMTE – Faculty of Science and Technology – NOVA University of Lisbon, Caparica, Portugal

References

[1] World population clocks – http://www.worldometers.info/world-population/ (accessed on 21/07/2016)

[2] Intergovernmental Panel on Climate Change – Climate Change 2014, Mitigation of Climate Change http://www.ipcc.ch/pdf/assessment-report/ar5/wg3/ipcc_wg3_ar5_full.pdf (accessed on 20/07/2016)

[3] M. K. Mondal, H. K. Balsora, P. Varshney, "Progress and trends in CO_2 capture/separation technologies: A review", *Energy*, 2012, 46, 431–441.

[4] M. S. Shannon, J. E. Bara, "Reactive and reversible ionic liquids for CO_2 capture and acid gas removal", *Separation Sci. Technol.*, 2012, 47, 178–188.

[5] D. J. Heldebrant, C. R. Yonker, P. G. Jessop, L. Phan,"Organic liquid CO_2 capture agents with high gravimetric CO_2 capacity", *Energy Environ. Sci.*, 2008, 1, 487–493.

[6] H. J. Song, S. Park, H. Kim, A. Gaur, J-W. Park, S-J. Lee, "Carbon dioxide absorption characteristics of aqueous amino-acid salt solutions", *Int. J. Greenhouse Gas Control*, 2012, 11, 64–72.

[7] A. H. Liu, R. Ma, C. Song, Z-Z. Yang, A. Yu, Y. Cai, L-N. He, Y-N. Zhao, B. Yu, Q-W. Song, "Equimolar CO_2 capture by N-substituted amino acid salts and subsequent conversion", *Angew. Chem. Int. Ed.*, 2012, 51, 11306–11310.

[8] D. Guo, H. Thee, C. Y. Tan, J. Chen, W. Fei, S. Kentish, G. W. Stevens, G da Silva, "Amino acids as carbon capture solvents: Chemical kinetics and mechanism of the glycine + CO_2 reaction", *Energy Fuels*, 2013, 27, 3898–3904.

[9] S. C. C. Wei, G. Putxy, P. Feron, "Amino acid salts for CO_2 capture at flue gas temperatures", *Energy Procedia*, 2013, 37, 485–493.

[10] S. Yan, Q. He, S. Zhao, H. Zhai, M. Cao, P. Ai, "CO_2 removal from biogas by using green amino acid salts: Performance evaluation", *Fuel Process. Technol.*, 2015, 129, 203–212.

[11] S. Mazinani, R. Ramazani, A. Samsami, A. Jahanmiri, B. Van der Bruggen, S. Darvishmanesh, "Equilibrium solubility, density, viscosity and corrosion rate of carbon dioxide in potassium lysinate solution", *Fluid Phase Equilibria*, 2015, 396, 28–34.

[12] S. Shen, Y. n Yang, Y. Wang, S. Ren, J. Han, A. Chen, "CO_2 absorption into aqueous potassium salts of lysine and proline: Density, viscosity and solubility of CO_2", *Fluid Phase Equilibria*, 2015, 399, 40–49.

[13] N. A. Rahim, N. Ghasem, M. Al-Marzouqi, "Absorption of CO_2 from natural gas using different amino acid salt solutions and regeneration using hollow fiber membrane contactors", *J. Nat. Gas Sci. and Eng.*, 2015, 26, 108–117.

[14] Z. W. Chen, R. B. Leron, M-H. Li, "Equilibrium solubility of carbon dioxide in aqueous potassium L-asparaginate and potassium L-glutaminate solutions", *Fluid Phase Equilibria*, 2015, 400 20–26.

[15] W. Li, X. Zhang, B. Lu, C. Sun, S. Li, S. Zhang, "Performance of a hybrid solvent of amino acid and ionic liquid for CO_2 capture", *Int. J. Greenhouse Gas Control*, 2015, 42 400–404.

[16] S. Shen, Y. n Yang, Y. Bian, and Y. Zhao, "Kinetics of CO_2 absorption into aqueous basic amino acid salt: Potassium salt of lysine solution", *Environ. Sci. Technol.*, 2016, 50, 2054–2063.

[17] X. Wang, N. G. Akhmedov, D. Hopkinson, J. Hoffman, Y. Duan, A. Egbebi, K. Resnik, B. Li, "Phase change amino acid salt separates into CO_2-rich and CO_2-lean phases upon interacting with CO_2", *Applied Energy*, 2016, 161, 41–47.

[18] A. P. Hallenbeck, A. Egbebi, K. P. Resnik, D. Hopkinson, S. L. Anna, J. R. Kitchin, "Comparative microfluidic screening of amino acid salt solutions for post-combustion CO_2 capture", *Int. J. Greenhouse Gas Control*, 2015, 43, 189–197.

[19] S. Saravanamurugan, A. J. Kunov-Kruse, R. Fehrmann, A. Riisanger, "Amine-functionalized amino acid-based ionic liquids as efficient and high-capacity absorbents for CO_2", *ChemSusChem*, 2014, 7, 897–902.

[20] G. V. S. M. Carrera, N. Jordão, M. M. Santos, M. Nunes da Ponte, L. C. Branco, "Reversible systems based on CO_2, amino acids and organic superbases", *RSC Adv.*, 2015, 5, 35564–35571.

[21] B. Lv, G. Jing, Y. Qian, Z. Zhou, "An efficient absorbent of amine-based amino acid-functionalized ionic liquids for CO_2 capture: High capacity and regeneration ability", *Chem. Eng. J.*, 2016, 289, 212–218.

[22] B. Lv, Y. Xia, Y. Shi, N. Liu, W. Li, S. Li, "A novel hydrophilic amino acid ionic liquid [C_2OHMIM][Gly] as aqueous sorbent for CO_2 capture", *Int. J. Greenhouse Gas Control*, 2016, 46, 1–6.

[23] J. Li, Z. Dai, M. Usman, Z. Qi, L. Deng, "CO_2/H_2 separation by amino-acid ionic liquids with polyethyleneglycol as co-solvent", *Int. J. Greenhouse Gas Control*, 2016, 45, 207–215.

[24] H. Thee, N. J. Nicholas, K. H. Smith, G. da Silva, S. E. Kentish, G. W. Stevens, "A kinetic study of CO_2 capture with potassium carbonate solutions promoted with various amino acids: Glycine, sarcosine and proline", *Int. J. Greenhouse Gas Control*, 2014, 20, 212–222.

[25] S. Shen, X. Feng, S. Ren, "Effect of arginine on carbon dioxide capture by potassium carbonate solution", *Energy Fuels*, 2013, 27, 6010–6016.

[26] Y. Gao, F. Zhang, K. Huang, J-W. Ma, Y-T. Wu, Z-B. Zhang, "Absorption of CO_2 in amino acid ionic liquid (AAIL) activated MDEA solutions", *Int. J. Greenhouse Gas Control*, 2013, 19, 379–386.

[27] J. Sun, W.Cheng, Z. Yang, J. Wang, T. Xu, J. Xin, S. Zhang, "Superbase/cellulose: An environmentally benign catalyst for chemical fixation of carbon dioxide into cyclic carbonates", *Green Chem.*, 2014, 16, 3071–3078.

[28] A. H. Tamboli, A. A. Chaugule, H. Kim, "Highly selective and multifunctional chitosan/ionic liquids catalyst for conversion of CO_2 and methanol to dimethyl carbonates at mild reaction conditions", *Fuel*, 2016, 166, 495–501.

[29] G. V. S. M. Carrera, N. Jordão, L. C. Branco, M. Nunes da Ponte, "CO_2 capture systems based on saccharides and organic superbases", *Faraday Discuss.*, 2015, 183, 429–444.

[30] A. F. Eftaiha, F. Alsoubani, K. I. Assaf, W. M. Nau, C. Troll, A. K. Qaroush, "Chitin-acetate/DMSO as a supramolecular green CO_2-phile", *RSC Adv.*, 2016, 6, 22090–22209.

[31] H. Sehaqui, M. E. Gálvez, V. Becatinni, Y. C. Ng, A. Steinfeld, T. Zimmermann, P. Tingaut, "Fast and reversible direct CO_2 capture from air onto all-polymer nanofibrillated cellulose-polyethylenimine foams", *Environ. Sci. Technol.*, 2015, 49, 3167–3174.

[32] H. Xie, S. Zhang, S. Li, "Chitin and chitosan dissolved in ionic liquids as reversible sorbents of CO_2", *Green Chem.*, 2006, 8, 630–633.

[33] J. Shi, Y. Jiang, Z. Jiang, X. Wang, X. Wang, S. Zhang, P. Han, C. Yang, "Enzymatic conversion of carbon sioxide", *Chem. Soc. Rev.*, 2015, 44, 598–6000.

[34] B. K. Kanth, J. Lee, S. P. Pack, "Carbonic anhydrase: Its biocatalytic mechanisms and functional properties for efficient CO_2 capture process development", *Eng. Life Sci.*, 2013, 13, 422–431.

[35] S. Xia, B. Frigo-Vaz, X. Zhao, J. Kim, P. Wang, "Biocatalytic carbon capture via reversible reaction cycle catalyzed by isocitrate dehydrogenase", *Biochem. Biophys. Res. Comm.*, 2014, 452, 147–150.

[36] S. Xia, X. Zhao, B. Frigo-Vaz, W. Zheng, J. Kim, P. Wang, "Cascade enzymatic reactions for efficient carbon sequestration", *Biores. Technol.*, 2015, 182, 368–372.

[37] I. M. Power, A. L. Harrison, G. M. Dipple, "Accelerating mineral carbonation using carbonic anhydrase", *Environ. Sci. Technol.*, 2016, 50, 2610–2618.

[38] L. A. Neves, C. Afonso, I. M. Coelhoso, J. G. Crespo, "Integrated CO_2 Capture and enzymatic bioconversion in supported ionic liquid membranes", *Separat. Purific. Technol.*, 2012, 97, 34–41.

[39] Y. Wang, M. Li, Z. Zhao, W. Liu, "Effect of carbonic anhydrase on enzymatic conversion of CO_2 to formic acid and optimization of reaction conditions", *J. Mol. Catal. B: Enzymatic*, 2015, 116, 89–94.

[40] J. H. Satcher, Jr., S. E. Baker, H. J. Kulik, C. A. Valdez, R. L. Krueger, F. C. Lightstone, R. D. Aines, "Modeling, synthesis and characterization of zinc containing carbonic anhydrase active site mimics", *Energy Procedia*, 2011, 4, 2090–2095.

[41] M. Vinoba, M. Bhagiyalakshimi, S. K. Jeong, S. C. Nam, Y. Yoon, "Carbonic anhydrase immobilized on encapsulated magnetic nanoparticles for CO_2 sequestration", *Chem. Eur. J.*, 2012, 18, 12028–12034.

[42] M. Uygun, V. V. Singh, K. Kaufmann, D. A. Uygun, S. D. S. de Oliveira, J. Wang, "Micromotor-based biomimetic carbon dioxide sequestration: Towards mobile microscrubbers", *Angew. Chem. Int. Ed.*, 2015, 54, 12900–12904.

[43] M. E. Russo, G. Olivieri, C. Capasso, V. De Luca, A. Marzocchella, P. Salatino, M. Rossi, "Kinetic study of a novel thermo-stable α-carbonic anhydrase for biomimetic CO_2 capture", *Enzyme Microb. Technol.*, 2013, 53, 271–277.

[44] B. H. Jo, J. H. Seo, H. J. Cha, "Bacterial extremo-α-carbonic anhydrase from deep-sea hydrothermal vents as potential biocatalysts for CO_2 sequestration.", *J. Mol. Catal. B: Enzymatic*, 2014, 109, 31–39.

[45] S. A. Patil, J. B. A. Arends, I. Vanwonterghem, J. van Meerbergen, K. Guo, G. W. Tyson, K. Rabaey, "Selective enrichment establishes a stable performing community for microbial electrosynthesis of acetate from CO_2", *Environ. Sci. Technol.*, 2015, 49, 8833–8843.

[46] C. W. Marshal, D. E. Ross, E. B. Fichot, R. S. Norman, H. D. May, "Long-term operation of microbial electrosynthesis systems improves acetate production by autotrophic microbiomes", *Environ. Sci. Technol.*, 2013, 47, 6023–6029.

[47] X. Hu, J. Zhou, B. Liu, "Effect of algal species and light intensity on the performance of an air-lift-type microbial carbon capture cell with an algae-assisted cathode", *RSC Adv.*, 2016, 6, 25094–25100.

[48] L. Lu, Z. Huang, G. H. Rau, Z. J. Ren, "Microbial electrolytic carbon capture for carbon negative and energy positive wastewater treatment", *Environ. Sci. Technol.*, 2015, 49, 8193–8201.

Chapter 6

Synergistic Effect on CO_2 Capture by Binary Solvent System

Quan Zhuang and Bruce Clements

Additional information is available at the end of the chapter

http://dx.doi.org/10.5772/65763

Abstract

CO_2 absorption into a binary solvent system was studied using a batch-mode gas/liquid absorption apparatus. The binary system composed of potassium carbonate (K_2CO_3) and piperazine (PZ) showed a strong synergistic effect, whereby the binary solvent performed better than either of the individual solvents for CO_2 absorption. The other pairs of solvents tested (K_2CO_3/monoethanolamine (MEA) and K_2CO_3/NaOH) showed no synergistic effects. The results indicate that this synergistic effect only occurs with specific pairs of solvents. The mechanism for the synergistic effect is postulated that the activated CO_2 on PZ migrates to K_2CO_3, or a more reactive intermediate complex between K_2CO_3 and PZ is formed.

Keywords: post-combustion, carbon capture, binary solvent, synergy effect, piperazine, potassium carbonate, CO_2 absorption

1. Introduction

There has been a growing concern over greenhouse gas emissions as they are considered to be the direct cause of global warming [1, 2]. Postcombustion capture technology is widely being studied for capturing CO_2 produced in power generation plants [3–5]. Compared with other CO_2 capture technologies such as oxy-fuel combustion and integrated gasification combined cycle (IGCC), postcombustion capture is regarded as the most probable technology to be first employed when carbon capture becomes a reality in the near future in terms of technology readiness level, flexibility, and economics [6]. Postcombustion capture technology uses liquid solvents to make efficient contact with CO_2-containing flue gas, during which CO_2 interacts and reacts with the solvent and is removed from the flue gas stream. After absorption, the CO_2-laden solvent undergoes a regeneration operation, releasing pure CO_2 which is then compressed, transported, and sequestered. The regenerated solvent, now at lean state, is returned to start the next cycle of CO_2 capture. The whole operation is a continuous process. The same or similar technologies have been applied for decades for natural gas purification and syngas CO_2 separation [7–9]. For greenhouse gas CO_2 mitigation applications, commercial solvents such as amine, potassium carbonate, and methanol are currently being tested, however, improved solvents are required to reduce the cost and increase the efficiency of postcombustion capture systems. At the moment, solvents that are being developed for CO_2 capture include nonconventional amines, aqueous ammonia, amino acids, ionic liquids, and mixtures of two or more solvents, i.e., hybrid systems [10, 11].

Potassium carbonate is known to be used in industrial CO_2 separation processes, such as Benfield and Catcarb [12], as the main solvent with or without proprietary additives. It has advantages over amines: lower cost, lower heat of absorption, thermal stability, nonvolatile, less corrosiveness, low toxicity, and environmentally friendly. A major downside for using K_2CO_3, however, is its slow absorption rate and low CO_2 absorption capacity, resulting in poor CO_2 mass transfer rate relative to amines. The way to overcome the aforementioned shortcomings of K_2CO_3 is to add promoter, i.e., a hybrid solvent. Hybrid solvent systems have the potential to perform better than the individual components alone. Physicochemical properties of different solvents can supplement each other. Synergistic effect or cooperative effect of hybrid solvents has been found in applications in other areas such as extraction and coal swelling [13, 14]. The mechanisms of the synergistic effects are suggested to be engendered by thermodynamics and hydrogen bonding.

We have been studying CO_2 absorption using an aqueous potassium carbonate solvent solution with the addition of other solvents in an attempt to improve CO_2 absorption performance. In this chapter, we report results of a synergistic effect that became apparent during these studies. When small amount of piperazine (PZ) is added to the potassium carbonate solution, both CO_2 absorption rate and capacity are significantly enhanced, exceeding the mathematical sum of the CO_2 absorption rate and the capacity of the individual solvents.

Piperazine itself is an active absorbent for CO_2 [15]. For some engineering reasons, it has only been used as an additive or a promoter to other common CO_2 capture amines [16]. With amine solvents, piperazine has shown promotional effect. For instance, the CESAR-1 solvent is an aqueous blend of AMP (2-amino-2-methyl-1-propanol) and PZ which showed a reduction of about 20% in the regeneration energy and 45% in the solvent circulation rate compared to those of MEA-based CO_2 capture process under similar process condition [17].

There have been some reports on the promotional/synergistic effect on CO_2 capture by K_2CO_3 and PZ [18]. This study builds upon previous achievements and provides convincing experimental evidence of the synergistic effect.

http://dx.doi.org/10.5772/65763

2. Experimental

A batch-mode liquid-gas absorption apparatus was constructed in CanmetENERGY, Ottawa. A schematic and a photo of the apparatus are shown in **Figure 1**. All of the connections within the system are vacuum-proof. The volume of the four-neck flask is 690 ml. The solute gas used in the experiment is a mixture of CO_2 and air (49 v% of CO_2). CO_2 absorption tests were carried out at 21°C (room temperature). The flask was placed in a water bath to maintain a constant temperature (CO_2 absorption is exothermic). First, the flask was purged by the solute gas for 10 min. Then all of the valves of the flask were closed, leaving the gas in the flask at ambient pressure. After this, 10 ml of solvent was introduced into the flask by opening the two valves of the funnel, and then closing them quickly so that the flask becomes a closed system with gaseous solute in contact with liquid solvent. The liquid was agitated by a magnetic stirrer at 350 rpm (there was no difference on the CO_2 absorption results with rpm in the range of 300–400). When the CO_2 was absorbed, the pressure in the flask decreased. This pressure was monitored with a solid state pressure sensor/transducer (PX209-30V15GI) from Omega. A monotonous pressure declining curve was obtained, revealing the CO_2 absorption kinetics (rate of decline) as well as capacity (the final level-off of the decline).

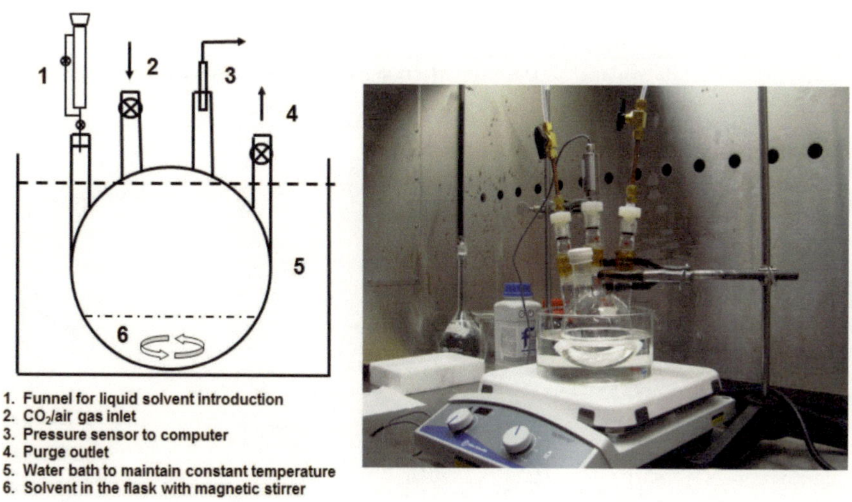

Figure 1. Batch mode gas-liquid absorption apparatus.

The solvents used and their concentrations in aqueous solution are shown in **Table 1**. In the test, the primary solvent was aqueous potassium carbonate, K_2CO_3. Other solvents were used as secondary promoters to see if there was a synergistic effect between the primary and secondary solvents. The hybrid solvents were obtained by mixing the individual solvents (shown in **Table 1**) with certain ratio (quantity in ml). Water was added to adjust the effective concentration and the final volume in a test.

Three test series were completed, one for each of the secondary solvents. These included:

Test Series 1—K_2CO_3 (primary solvent) with PZ (secondary solvent)

- 7 ml K_2CO_3/3 ml H_2O (K_2CO_3 represents its solution in **Table 1**)
- 3 ml PZ/7 ml H_2O (PZ represents its solution in **Table 1**)
- 7 ml K_2CO_3/3 ml PZ

Test Series 2—K_2CO_3 (primary solvent) with MEA (secondary solvent)

- 7 ml K_2CO_3/3ml H_2O
- 3 ml MEA/7 ml H_2O (MEA represents its solution in **Table 1**)
- 7 ml K_2CO_3/3 ml MEA

Test Series 3—K_2CO_3 (primary solvent) with NaOH (secondary solvent)

- 7 ml K_2CO_3/3 ml H_2O
- 3 ml NaOH/7 ml H_2O (NaOH represents its solution in **Table 1**)
- 7 ml K_2CO_3/3 ml NaOH

Solvent	Molecular. formula	Density (g/cm^3)	Molar mass (g/mol)	Concentration used (% wt)	Structure
Potassium carbonate	K_2CO_3	2.43	138.21	33	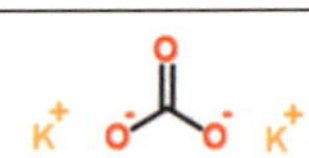
Piperazine (PZ)	$C_4H_{10}N_2$	1.98	86.14	16	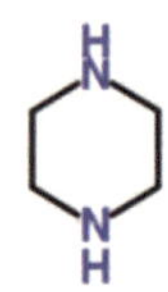
Ethanolamine (MEA)	C_2H_7NO	1.01	61.08	15	
Sodium hydroxide	NaOH	2.13	40.00	15	Na^+ OH^-

Table 1. Properties of chemicals and solvents used in the experiment.

3. Results and discussion

The CO_2 absorption results for test series 1 are shown in **Figure 2**. After the solvent was introduced into the flask filled with CO_2/air, the chemisorption occurred as demonstrated by the pressure decrease. From the results in **Figure 2**, it can be seen that K_2CO_3 showed a slow absorption rate and low absorption capacity. Piperazine's CO_2 absorption rate was faster and had higher capacity. When the two solvents were mixed, the binary solvent system absorbed more CO_2 at an even faster rate. The mathematical sum of the individual CO_2 absorption curves of the K_2CO_3 and piperazine (the sum of the green curve and the light blue curve) is shown in **Figure 2** as well (dark blue line). It is clear that the binary solvent system performed much better for CO_2 absorption than the mathematical sum of the individual solvents. The two curves (orange and purple in **Figure 2**) showing the CO_2 absorption results of the binary solvent system from two different tests under the same conditions indicate that the apparatus worked very well with a high degree of repeatability.

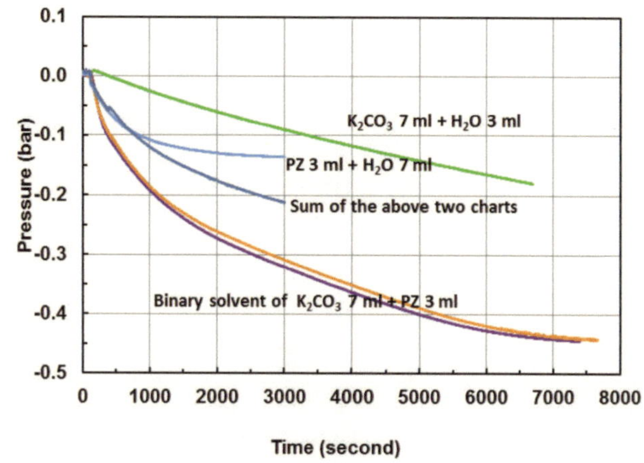

Figure 2. Test series 1—CO_2 absorption with binary solvent system of K_2CO_3 and piperazine.

The test results of the binary solvent system of K_2CO_3 and MEA are shown in **Figure 3**. The component solvents of K_2CO_3 and MEA were of similar effectiveness for CO_2 absorption. The binary solvent system showed only a slight synergistic effect.

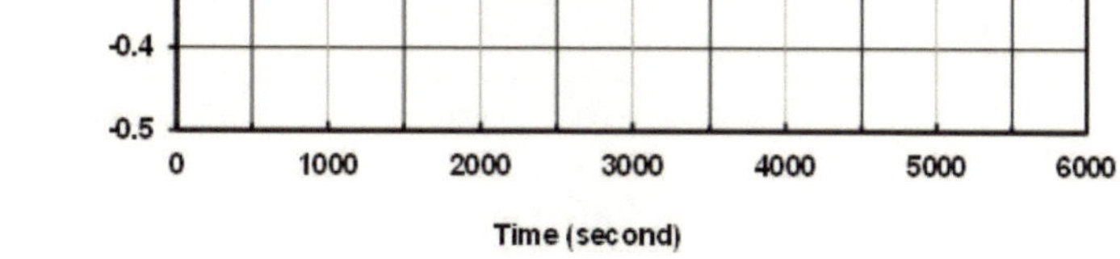

Figure 3. Test series 2—CO_2 absorption with binary solvent system of K_2CO_3 and MEA.

In order to investigate the necessary and/or sufficient conditions for the synergistic effect of a stronger CO_2 solvent with a milder solvent (e.g., PZ with K_2CO_3), the binary solvent system of K_2CO_3 with NaOH was tested (**Figure 4**). It can be seen from **Figure 4** that, although NaOH is a much stronger CO_2 solvent than K_2CO_3, the binary solvent system of K_2CO_3 and NaOH does not show any synergistic effect.

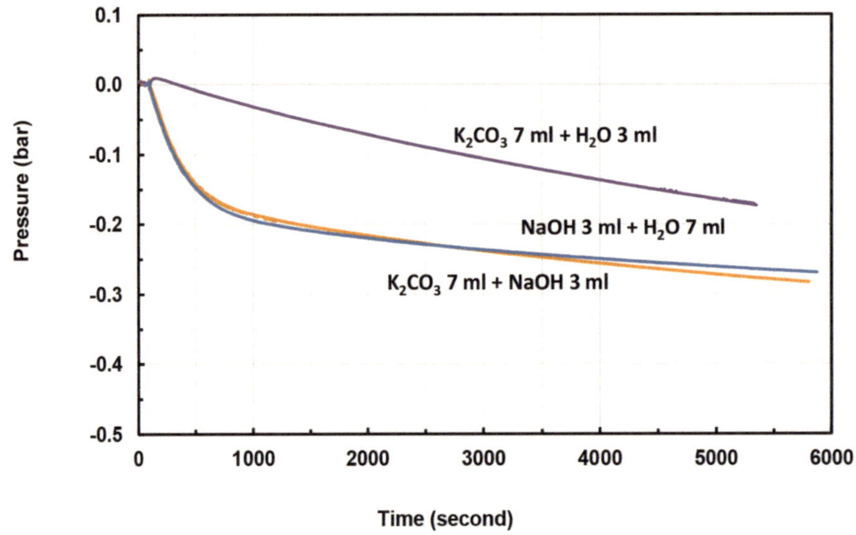

Figure 4. Test series 3—CO_2 absorption with binary solvent system of K_2CO_3 and NaOH.

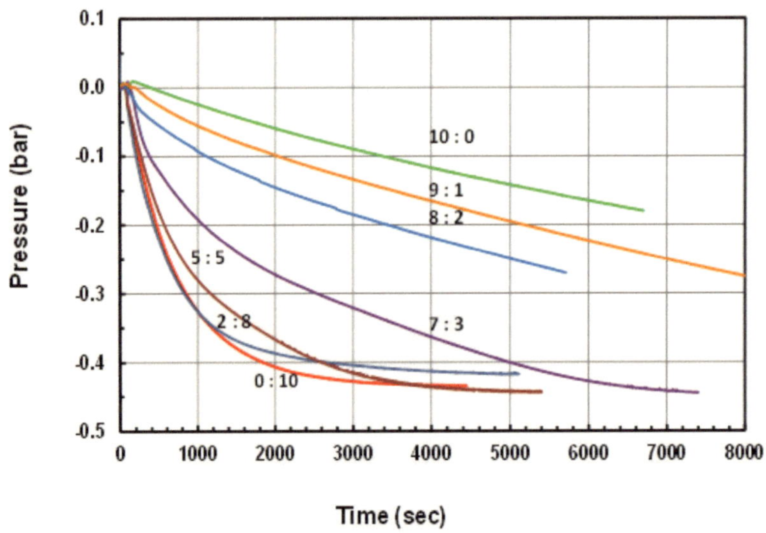

Figure 5. The CO_2 absorption by binary solvent versus the ratio of K_2CO_3:PZ.

http://dx.doi.org/10.5772/65763

Therefore, it is a necessary but not a sufficient condition for a binary solvent system with different CO_2 absorption capacities and kinetics to generate synergistic effect. Among the three pairs, only the binary solvent of K_2CO_3 and PZ showed a positive synergistic effect on CO_2 absorption.

As shown by our experiment (**Figure 2**) and others [19], PZ is a stronger and faster CO_2 solvent than K_2CO_3. When the ratio of K_2CO_3 and PZ was varied, the CO_2 absorption curves shifted from the curve of K_2CO_3 to the curve of PZ, as shown in **Figure 5**. The binary solvent systems between the two pure solvents exhibit synergistic effect. Illustrated in **Figure 6** is the synergistic performance of the binary solvent as well as the relationships with the two pure solvents (this is only a general illustration).

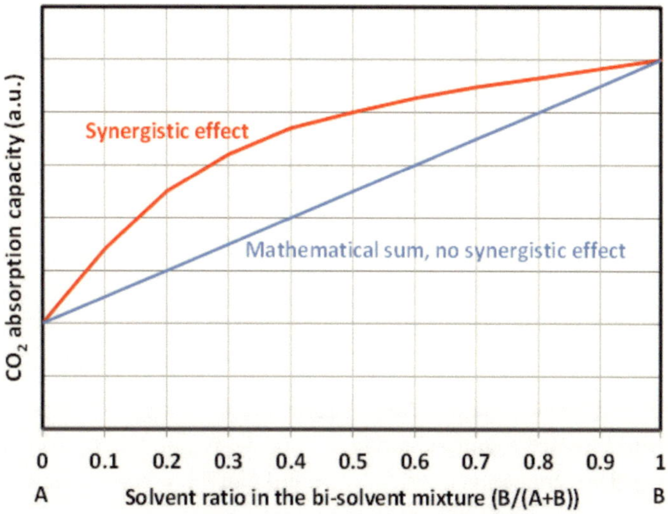

Figure 6. Illustration of synergistic effect by a binary solvent system, e.g., K_2CO_3/PZ.

PZ is an expensive solvent. Whether or not it is suitable, alone, as a CO_2 capture solvent is still being explored in terms of thermal stability, corrosiveness and cost, etc. [19]. As shown by this study, it is promising to apply a binary solvent of K_2CO_3 and PZ at a ratio that maximizes the synergistic effect on CO_2 capture. Savings from operating at this condition could be realized in terms of solvent cost, reduction of the absorber and regenerator sizes due to the improved CO_2 absorption rate and capacity. More effective solvents would require smaller absorbers and regenerators, leading to lower capital costs.

J. Tim Cullinane and Gary T. Rochelle have reported the promotional effect of K_2CO_3 and PZ by kinetics [18]. They concluded that the promotional effect comes from the kinetics of the two individual solvents and that the two solvents absorb CO_2 independently. These cannot explain the observations of this study. The promotional or synergistic effect of PZ to K_2CO_3 has been suggested to occur through an intermediate formed between CO_2 (aq) and PZ [20–22]. This hypothesis, however, still needs to be verified experimentally. Our results indicate that there may be a more interactive mechanism affecting the hybrid solvent performance. Having a binary solvent system with one solvent more effective than the other is a necessary condition for the synergistic effect (the pairs of K_2CO_3 and PZ, K_2CO_3, and NaOH), but not a sufficient condition (K_2CO_3 and NaOH). There must be other reasons behind the synergistic effect. Here we postulate two mechanisms:

- CO_2 transition (or spill over or migration): CO_2 is reactivated by solvent B forming a labile state $[[B] \cdot [CO_2]_{(aq)}]$, then transfers or migrates to solvent A to finish CO_2 absorption (**Figure 6**). Likely hydrogen bonding is involved.

- Reactive complex intermediate structure between the two solvents: in the CO_2 absorption system, there occur some kind of interactions between the two solvents by hydrogen bonding or local ionic attraction, forming a more reactive intermediate complex [A·B] with improved CO_2 absorption ability.

The factors of electron donner strength, dielectric constants, solubility parameters of the individual absorbent, and hydrogen-bonding/nonhydrogen-bonding may influence the degree of synergistic effects. There needs more research work to capture and characterize the reactive intermediate complex or transition state, to prove or disprove these postulated mechanisms.

4. Conclusion

The idea of combining solvents to improve absorption is effective for piperazine and K_2CO_3. These two solvents interact together and generate a greater absorption than each of the individual solvents. The other solvents, i.e., MEA and NaOH, when mixed with K_2CO_3 did not improve CO_2 absorption, implying that the synergistic effect only occurs selectively between specific pairs of solvents. The solution of 3 ml piperazine with 7 ml potassium carbonate is the optimal ratio that increases CO_2 absorption using the least amount of piperazine. The results of these tests show the possibility of using piperazine and K_2CO_3 solution at an industrial scale. If correctly implemented, it would result in savings in capital by reducing the absorber size compared to use K_2CO_3 alone. The next step for this project is to apply these results within a larger system. The major conclusions from the tests conducted are summarized below:

- A synergistic effect between K_2CO_3 and piperazine was observed.

- This synergistic effect only happens between this specific pair of solvents and is not universal. Other than the thermodynamic reasons behind the effect, there seems to be some additional mechanism that enhances the reaction (potentially a labile $[CO_2]$ formation followed by migration or some more reactive intermediate complex structure formed between the two solvent molecules).

- 3 ml piperazine/7 ml K_2CO_3 ratio is the most effective (faster absorption rate and higher absorption capacity).

http://dx.doi.org/10.5772/65763

Acknowledgements

The project was financially supported by the Canadian Federal Government ecoEII Program. Thanks goes to Mr. Zlatko Lovrenovic, coop student from University of Ottawa, for his contribution to the project. An extra financial support from AirScience Technologies Inc., Montreal, Canada, is gratefully acknowledged.

Author details

Quan Zhuang* and Bruce Clements

*Address all correspondence to: quan.zhuang@canada.ca

Natural Resources Canada, CanmetENERGY, Ottawa, Ontario, Canada

References

[1] Kathryn A. Mumford, Yue Wu, Kathryn H. Smith, Geoffrey W. Stevens, "Review of solvent based carbon-dioxide capture technologies", Front. Chem. Sci. Eng. 2015, 9(2): 125–141.

[2] IPCC Report, 2014 and the previous annual editions.

[3] Yasaman Mirfendereski, "Techno-Economic Assessment of Carbon Capture and Sequestration Technologies in the Fossil Fuel-based Power Sector of the Global Energy-Economy System", Master Thesis, Technische Universität Berlin, May 2008.

[4] Quan Zhuan, Bruce Clements, Ying Li, "From ammonium bicarbonate fertilizer production process to power plant CO_2 capture", Int. J. Greenhouse Gas Control, 2012, 10: 56–63.

[5] Quan Zhuang, Richard Pomalis, Ligang Zheng, Bruce Clements, "Ammonia- based carbon dioxide capture technology: issues and solutions", Energy Procedia, 2010, 4: 1459–1470.

[6] Bruce G. Miller, Clean Coal Engineering Technology, Elsevier, Burlington, USA, 2011.

[7] Salako Abiodun Ebenezer, "Removal of carbon dioxide from natural gas for LNG production", Institute of Petroleum Technology Norwegian University of Science and Technology, 2005.

[8] Chao Chen, "A technical and economic assessment of CO_2 capture technology for IGCC power plants", Ph.D. Thesis, Carnegie Mellon University, 2005.

[9] Barry Burr and Lili Lyddon, "A comparison of physical solvents for acid gas removal", GPA 2008.

[10] Young Eun Kim, Jeong Ho Choi, Sung Chan Nam, Yeo Il Yoon, "CO_2 absorption capacity using aqueous potassium carbonate with 2-methylpiperazine and piperazine", J. Ind. Eng. Chem., 2012, 18: 105–110.

[11] Gary T. Rochelle, Frank Seibert, "CO_2 capture by absorption with potassium carbonate", DOE Final Report, December 2007.

[12] B. Amerijiafari, F. Grange, H. Ford, "Alternative concepts for supplying carbon dioxide for enhanced oil recovery projects", DOE Final Report, 1980.

[13] Yongseung Yun, Eric M. Suuberg, "Cooperative effects in solvent swelling of a bituminous coal", Energy Fuels, 1998, 12: 798–800.

[14] Izuru Matsubayashi, Yuko Hasegawa, "Thermodynamic consideration for solvent effects in the synergistic extraction of europium (III) with pivaloyltrifluoroacetone and 1,10-Phenonthroline", Anal. Sci., 2001, 17: 221–223.

[15] Francis Bougie, Maria C. Iliuta,"CO_2 absorption in aqueous piperazine solutions: experimental study and modeling", J. Chem. Eng. Data, 2011, 56: 1547–1554.

[16] Derks, P.W.J., "Carbon Dioxide Absorption in Piperazine Activated N-Methyldiethanolamine", PhD thesis, University of Twente, The Netherlands, 2006.

[17] Eva Sanchez Fernandez, "Novel process designs to improve the efficiency of postcombustion carbon dioxide capture", Ph.D. Thesis, Universidad Complutense de Madrid, 2013.

[18] J. Tim Cullinane, Gary T. Rochelle, "Carbon dioxide absorption with aqueous potassium carbonate promoted by piperazine", Chem. Eng. Sci. 2004, 59: 3619–3630.

[19] David H. Van Wagener, Gary T. Rochelle, Eric Chen, "Modeling of pilot stripper results for CO_2 capture by aqueous piperazine", Intl. J. Greenhouse Gas Control, 2013,12: 280–287.

[20] R. Ramazania, S. Mazinanib, A. Jahanmiria, B. Van der Bruggen, "Experimental investigation of the effect of addition of differentactivators to aqueous solution of potassium carbonate: absorption rate and solubility", Intl. J. Greenhouse Gas Control, 2016, 45: 27–33.

[21] A.L. Shrier, P.V. Danckwerts, "Carbon dioxide absorption into amine-promoted potash solutions", Ind. Eng. Chem. Fundam. 1969, 8: 415–423.

[22] G. Astarita, D.W. Savage, J. Longo, "Promotion of CO_2 mass transfer in carbonate solutions", Chem. Eng. Sci. 1981, 36: 581–588.

Section 3

Soil Carbon Sequestration

Chapter 7

Maximizing Soil Carbon Sequestration: Assessing Procedural Barriers to Carbon Management in Cultivated Tropical Perennial Grass Systems

Jon M. Wells, Susan E. Crow, Manyowa N. Meki,
Carlos A. Sierra, Kimberly M. Carlson,
Adel Youkhana, Daniel Richardson and
Lauren Deem

Additional information is available at the end of the chapter

http://dx.doi.org/10.5772/66741

Abstract

The natural capacity of the terrestrial landscape to capture and store carbon from the atmosphere can be used in cultivated systems to maximize the climate change mitigation potential of agricultural regions. A combination of inherent soil carbon storage potential, conservation management, and rhizosphere inputs should be considered when making landscape-level decisions about agriculture if climate change mitigation is an important goal. However, the ability to accurately predict soil organic carbon accumulation following management change in the tropics is currently limited by the commonly available tools developed in more temperate systems, a gap that must be addressed locally in order to facilitate these types of landscape-level decisions. Here, we use a case study in Hawaii to demonstrate multiple approaches to measuring and simulating soil carbon changes after the implementation of zero-tillage cultivation of perennial grasses following more than a century of intensive sugarcane cultivation. We identify advancements needed to overcome the barriers to potential monitoring and projection protocols for soil carbon storage at our site and other similar sites.

Keywords: carbon sequestration, soil carbon, climate change mitigation, perennial grasses, zero-tillage

1. Introduction

As global demand for agriculture and bioenergy increases, so does the need to understand and predict not only the amount of food or energy that can be produced in large-scale agricultural systems, but also the environmental impacts associated with changes in agricultural land use and management. The effects of land-use change on soil carbon sequestration are poorly understood, particularly for novel bioenergy feedstocks. Maximization of carbon capture in agricultural systems through successive sequestration of photosynthetically fixed biomass carbon into soils for long-term storage has great potential to offset greenhouse gases in the atmosphere and mitigate climate change. However, the potential to sequester carbon in soil across cultivated landscapes for the purpose of climate change mitigation remains largely untapped, in part due to the complexity of soil carbon stabilization processes. In this chapter, we focus on a heterogeneous landscape in central Maui, a Hawaiian island, where sugarcane was intensively cultivated for over a century using preharvest burns and deep tillage. In 2011, a ratoon harvest system with zero-tillage management replaced sugarcane cultivation at select sites. Our objective was to identify current gaps in knowledge within the Maui system that diminish efforts toward accurate prediction of carbon capture and storage across this, and other similar landscapes in transition.

1.1. Factors controlling soil carbon stocks

1.1.1. Soil texture and mineralogy

Soil texture, particularly clay concentration, is commonly thought to predominantly influence soil organic carbon storage and therefore percent clay is commonly used as a modulator in simulation models like CENTURY [1] and Roth C [2] to help project carbon sequestration. However, other researchers investigating soil texture and soil carbon [3] found improved water holding capacity in silt-dominated soils and subsequently improved plant productivity, and thus suggest that silt may have greater effects on soil carbon sequestration than clays. Water holding capacity regulates oxygen supply and thus affects microbial decomposition [4]. In some tropical and subtropical soils, Fe-oxide cementations defy standard protocols for dispersion during texture determination and require specialized methodology to attain accurate clay concentrations that are not yet widely recognized in the literature [5]. Torn et al. [6] concluded that geological timescales were the strongest controlling factor of soil carbon change, but that was based on the stages of mineral weathering and the direct organomineral interactions that result in carbon stabilization. Specifically, the concentration of poorly or noncrystalline clay minerals can be a stronger factor controlling soil organic carbon storage than net primary productivity on millennial [6] and decadal [7] time scales. Although percent clay can be an adequate modulator for many systems, greater detail of information on soil texture and mineralogy often is needed in others such as systems of volcanic origin, arid regions, and subtropic/tropical ecosystems to improve model simulations of soil carbon accumulation.

1.1.2. Soil carbon stabilization, equilibrium and saturation

In reality, soil texture, mineralogy, climate, gross productivity and carbon allocation, land management, soil biota and their carbon use efficiency, and stabilization mechanisms such as

http://dx.doi.org/10.5772/66741

physical protection by aggregation and organomineral interactions act together to control soil carbon stocks. In the context of managing land to maximize soil carbon capture, the concepts of soil carbon stabilization, saturation, and equilibrium are critical because these processes dictate how quickly soil carbon will increase, the level of soil carbon reached, and whether the accumulated soil carbon stock will be resilient to future disturbance. In 2002 and 2004 reviews, Six et al. [8, 9] found (1) physiochemical soil characteristics control the maximum soil carbon stabilization capacity of soils; (2) microaggregates (<250 μm) are better at long-term soil carbon stabilization compared to macroaggregates (>250 μm); and (3) macroaggregate turnover is a strong driver of soil organic carbon stabilization across soil types and disturbance, with decreased macroaggregate turnover promoting increased long-term microaggregate stabilization. Soil aggregates, therefore, play multiple roles in soil carbon accumulation and should be protected and promoted with management decisions such as shifts to zero-tillage and minimal disturbance regimes.

Plant inputs also are an important factor in carbon stabilization, saturation, and equilibrium. In natural grassland systems of North America, litter mass losses contributed to soil carbon quickly at first through microbial decomposition, as well as more slowly through litter fragments moving into the mineral soil profile [10]. In tropical perennial grass systems, deep root inputs may have a greater influence on soil carbon accumulation than surface processes [11] because the aboveground biomass is removed. In cultivated landscapes, increases in organomineral complexes in the deeper soil profile under no tillage compared to conventional tillage led to a 16% increase of organic carbon [12]. A recent study of conservation agriculture in grasslands also found exchangeable calcium as the strongest single predictor of soil carbon in the top 10 cm [13], likely due to the positive effect of Ca^{2+} on soil aggregation in arid systems. Improving our assessment and understanding of soil carbon storage requires increasing our understanding of carbon stabilization while continuing to test and update conceptual models, especially those that span disciplines [14].

Advancements in technology as well as recent research findings support the move to mechanistic models of soil carbon processes. For example, Schmidt et al. [15] provide a succinct but wide-ranging source of reasoning behind the need for better observation-based and mechanistically driven conceptual frameworks. Lehmann and Kleber [16] argue for the need of soil science and interrelated disciplines to progress to a new model of soil organic matter and its interactions in the soil ecosystem. Their soil continuum model (SCM) is an attempt to reconcile three current conceptual models of the fate of organic debris in soils: (1) humification, or the classic belief in the synthesis of large recalcitrant molecules from decomposition products; (2) selective preservation, the assumption that preferential mineralization leaves intrinsically stable compounds; and (3) progressive decomposition, the concept of faunal and microbial size processing of plant inputs into smaller molecules. They argue that humic terminology should be relinquished; instead of suggesting that humic substances are a distinct category of organic matter in soils, they should be considered an analytical process of alkaline extracts. Further, recent advances in nuclear magnetic resonance (NMR) imaging, termed comprehensive multiphase NMR, has given one of the first analytical looks into the *in situ* soil-water interface and shows a complex mix of microbial and plant biopolymers with no evidence for cross-linked humic material [17]. The study also describes notable findings in relation to the soil-water interface that suggests carbon storage locations depend on the form of soil organic carbon. These varied mechanisms of carbon stabilization could contribute to the disconnect between measured and modeled soil carbon as different carbon inputs may be mechanistically stored in different ways, thus creating nonlinearity between plant inputs and soil carbon sequestration.

1.1.3. *Soil and crop management*

Landscapes with soil properties that favorably control carbon stabilization could be preferentially transitioned into conservation management to improve carbon storage in certain cultivated agricultural systems. Conservation agriculture practices—including minimal tillage, residue management, and plant cover—affect the carbon cycle in agricultural systems, thereby altering ecosystem service provision [18]. In bioenergy production, where high primary productivity and maximum carbon capture in plant biomass are the primary goal, soil carbon sequestration can be a desirable secondary outcome with potent climate mitigation potential. Specifically, fast growing deep-rooted perennial grasses have the potential to input large amounts of carbon deep into the soil profile that can be protected by aggregate formation or organomineral interactions. A review of bioenergy crop "management swing potential" illustrates how management changes can swing the greenhouse gas emissions balance of agricultural production systems in positive or negative directions [19], which could offset negative carbon emissions from harvest and planting. Mutuo et al. [20] also discuss tropical agroforestry as another potential means to sequester carbon, finding large aboveground potential (60 Mg C/ha) but low belowground storage (25 Mg C/ha). However, they only investigated the top 20 cm of soil; investigation of the full soil profile may have revealed agroforestry increasing deeper, and potentially longer-term, soil carbon stocks. Anderson-Teixeira et al. [21] also found significant increases in belowground biomass of fast growing perennial crops compared to corn. Though soil carbon was not directly investigated in their study, it is expected that higher biomass in deeper depths likely increased deep soil carbon stocks compared to the typically shallow rooting of row crops like corn and soy. Getting carbon to deeper depths, minimizing its disturbance, and allowing physical and chemical protection mechanisms to remain intact are goals that conservation agriculture can help to achieve.

1.1.4. *Soil bulk density and profile depth*

When measuring soil carbon stocks, soil bulk density (i.e., mass per volume) and the whole soil profile (i.e., depth and development features) are critical to determining accurate total carbon stocks. Soil bulk density and profile depth have the most direct and simple effect on measured soil carbon stock, and more importantly accurately assessing the change in soil carbon stock postland use or management shift. In cultivated systems, determination of cumulative soil carbon using equivalent soil mass (ESM) methods accounts for changes in soil bulk density caused by management changes in compaction or tillage [22–24]. A recent study of a tropical forested system by Crow et al. [25] illustrates how different soil carbon measurement techniques (bulk density vs. ESM) can lead to conflicting carbon stock interpretations, especially in transitions between land management and crop type. By not accounting for compaction during land use change, average soil C change was overestimated by 14.9%, a difference that could have led to vastly different management decisions [25]. Though mineral soil profiles can be several meters deep, many studies have only investigated surface soils even though change can be profound at depth [26]. More soil measured, even with lower carbon concentration at deeper depths, results in larger carbon stocks and, increasingly, deep soil sampling (to at least 1 m, if not more) are commonplace.

1.2. Simulation modeling and monitoring soil carbon change

In bioenergy feedstock production, where the environmental goal is to displace fossil fuels and promote carbon negative or neutral activity, maximizing carbon capture and sequestration in soils is important. As the societal costs of climate change increase, the economic value of carbon will also increase on a global scale [27, 28]. If carbon sequestration in a bioenergy feedstock production system can be fairly monetized, it could offset costs of establishing bioenergy production sites and help reduce uncertainty in an industry currently closely tied to fluctuating oil prices. Future management plans for soil carbon capture should include both environmental and economic sustainability factors.

The challenge of accurately measuring and projecting of carbon storage in agricultural systems increases as novel crops, large spatial scales, and heterogeneous soils and landscapes are utilized. Simulation modeling allows, with minimal on-site data collection, prediction of the potential of an area for climate mitigation or carbon monetization. Such models, calibrated to the specific processes controlling soil carbon accumulation at each site, can provide insight to land managers making landscape-level carbon decisions. With the precise tools, adaptive management plans may be made if monitored soil carbon stocks meet or miss simulated carbon potentials. Using simulation and projection modeling as a tool to investigate the effectiveness of carbon stock assessment methods (e.g., understanding the necessary sample quantities, spatial arrangements across a landscape, etc.) also helps determine the number and spatial distribution of samples needed to accurately quantify soil carbon stock change over time.

With subsequent advances in understanding of soil processes and representation in ecosystem and Earth system models, the potential exists to improve estimates of soil sequestration and projection models. Although many site-specific studies of soil organic carbon stocks have been completed, there remains uncertainty in the predominant soil processes that influence soil carbon storage and how these processes apply across heterogeneous landscapes of varying soil, temperature, rainfall, management, and other conditions to achieve carbon sequestration. Overly simplified models that consider only plant inputs as the driver of soil carbon are not accurate. Simple clay modifiers also are not effective in accurately modeling carbon storage for many soils, with noncrystalline mineral modifiers showing increased accuracy for some Andisol systems [29].

Several soil carbon models were developed for a specific cropping system, which makes it difficult to compare their performance against other models. Moreover, most models have not been fully parameterized and effectively tested for lack of adequate field measured SOC data, which is also crucial for the verification of model outputs. Soil subcomponent models that consider soil carbon dynamics and multiple pools, such as CENTURY, have shown to be reasonably good in simulating changes in SOC stocks, it is, however, important to note that the C pool compartments are only conceptual, and have not been verified experimentally. Fundamentally, these models are based on outdated and oversimplified concepts of soil carbon formation. For example, static transfer rates among pools in the CENTURY model breaks down conceptually through time, and as demonstrated in this case study, have been shown empirically to be dynamic in certain systems. Moving toward more empirical models, resembling the SCM [16], that employ mechanisms like aggregate formation, organomineral interactions, and soil microbial biomass, among others, could help to better describe and model soil carbon cycles in the soil microbiome.

2. Ratoon harvest and zero-tillage management: a bioenergy case study in central Maui

Located in the central valley of Maui, Hawaii, between Haleakala and West Maui mountains, the Hawaiian Commercial & Sugar Company (HC&S) produced sugar from irrigated cane beginning in 1870 (**Figure 1**). The 36,000 acres of the HC&S plantation span large gradients of elevation, temperature, wind, and rain, which in turn generated high soil heterogeneity. From the start, HC&S preburned their sugarcane fields to reduce extraneous foliage and increase the percent sugar of collected material, which in turn improved the efficiency of their sugar extraction process. After the burn, cane stalks were mechanically ripped from the soil with their associated root bulb for processing; the fields underwent deep soil ripping (40 cm), and then were left barren until being replanted several weeks later. HC&S, as the last remaining large-scale agriculture company in the Hawaiian Islands, began transitioning in 2016 from sugarcane to diversified agriculture that will include large areas of perennial grasses for bioenergy feedstock and/or cattle forage. In this case study, the focus was on the heterogeneous landscape and identifying factors with predominant control on soil carbon stocks and accumulation following an experimental change from past intensive cultivation to ratoon harvest, zero-tillage management of tropical perennial grasses.

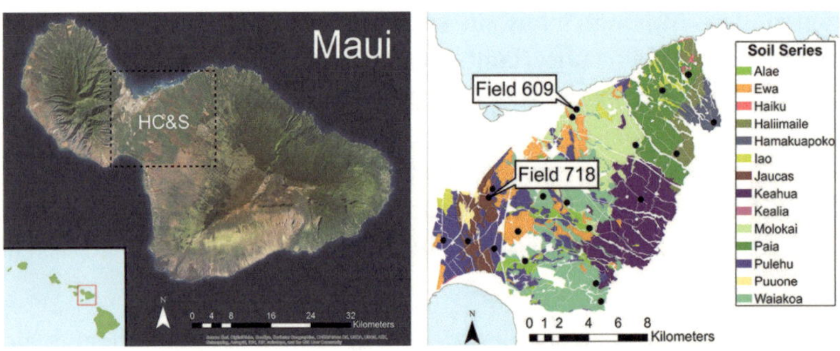

Figure 1. Maui with the main Hawaiian Islands inset (left) and Hawaiian Commercial & Sugar Company fields and associated soil series (right). Soil series data from: Soil Survey Staff, Natural Resources Conservation Service, United States Department of Agriculture. Web Soil Survey. Available online at http://websoilsurvey.nrcs.usda.gov/. Accessed [07/30/2016].

http://dx.doi.org/10.5772/66741

2.1. Geospatial representation of soil parameters and baseline soil stocks

HC&S contains 14 soil series identified by the U.S. Soil Taxonomic system (**Figure 1**) that reflect heterogeneous soil properties across large areas of the plantation. As a first step to improve model simulations of soil carbon accumulation at the plantation scale, primary soils data (including GPS locations, horizon depths, bulk density, soil texture, pH, total carbon and nitrogen concentration, and organic carbon concentration) were collected for 20 map units across HC&S. These 20 map units represent 7 soil orders, 10 soil series, and ~77% of total plantation area. Data from these 20 fields also provide a baseline of soil carbon stocks under more than a century of intensive cultivation of sugarcane.

Raster interpolation was used to investigate geospatial relationships between soil organic carbon and potential factors that affect soil carbon sequestration. The field data were analyzed in ESRI ArcGIS using ordinary spherical kriging from the 3D analyst toolbox. Geospatial patterns emerged in soil texture data in both the percent sand and clay (**Figure 2**) but none were apparent for percent silt, which was approximately 50% for most of the soils tested (data not shown). The wetter, higher elevations going up Haleakala Mountain showed high levels of clay, while the west side of the plantation had more sand dominant soils. Standard protocols were used for the textural classification and therefore the percent clay may be underestimated and percent sand may be overestimated for some of the soils; however, these observation are consistent with greater clay development in wetter, upland soils and sandier clays in low lying areas that were subject to sea-level rise. pH did not show clear trends in space (results not shown), although we did find that a majority of the soils across the plantation were very basic (pH 7–8), with the most acidic areas slightly under pH 6. This is likely due to soil parent material (e.g., basic igneous rock in the Keahua series and calcareous sand deposits in the Jaucas series) in parts of the plantation and high application rates of lime as needed for productivity throughout.

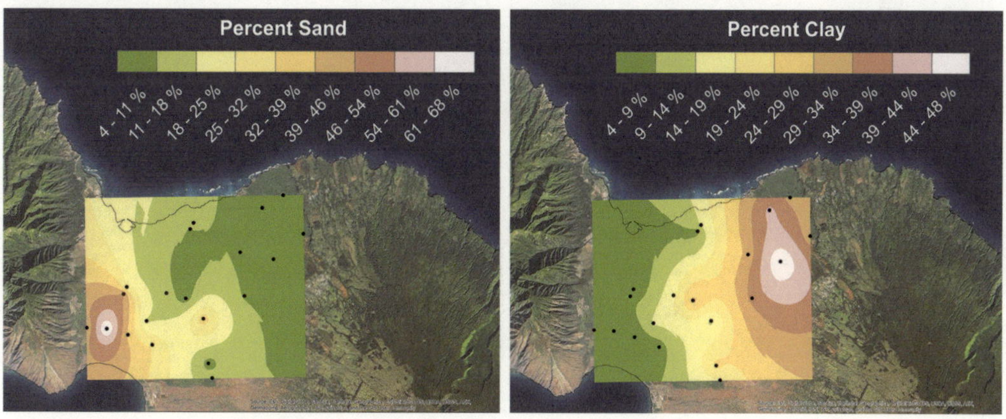

Figure 2. Simple spherical kriging of texture data gathered from 20 fields across the HC&S plantation: percent sand (left) and percent clay (right). Interesting geospatial patterns of sand and clay appear but do not well explain the patterns found in organic carbon stocks across HC&S.

The equivalent soil mass (ESM) method as described in the introduction (Section 1.1.4) was used to calculate a carbon stock for each of the 20 fields and the same methods described above were used to generate a geospatial representation of baseline soil carbon stock for the plantation (**Figure 3**). Four arbitrary cumulative soil reference masses were chosen based on the sampling scheme and typical cumulative soil masses found by the ESM method in the 20 fields data. The ESM reference masses chosen were 2500, 5000, 7500, and 10,000 Mg/ha, with each ESM reference mass interpolated similarly to the texture data using ordinary spherical kriging in ArcGIS. These masses roughly equate to a 0.6 m depth, although the exact depth is different for every soil. There were no strong geospatial patterns for soil carbon stock in the shallowest soils (ESM 2500), with spatial patterns emerging only after the inclusion of the deeper soil profile. Long periods of deep tillage and a monoculture cropping system with similar amounts of litter and shallow root inputs are possible causes for the lack of geospatial difference in the surface soil carbon stocks. However, the final pattern that emerged in ESM 10,000, where higher carbon stocks appear to be in wetter northeastern fields along the windward side of Haleakala and western fields toward the West Maui Mountains, does not have a simple explanation.

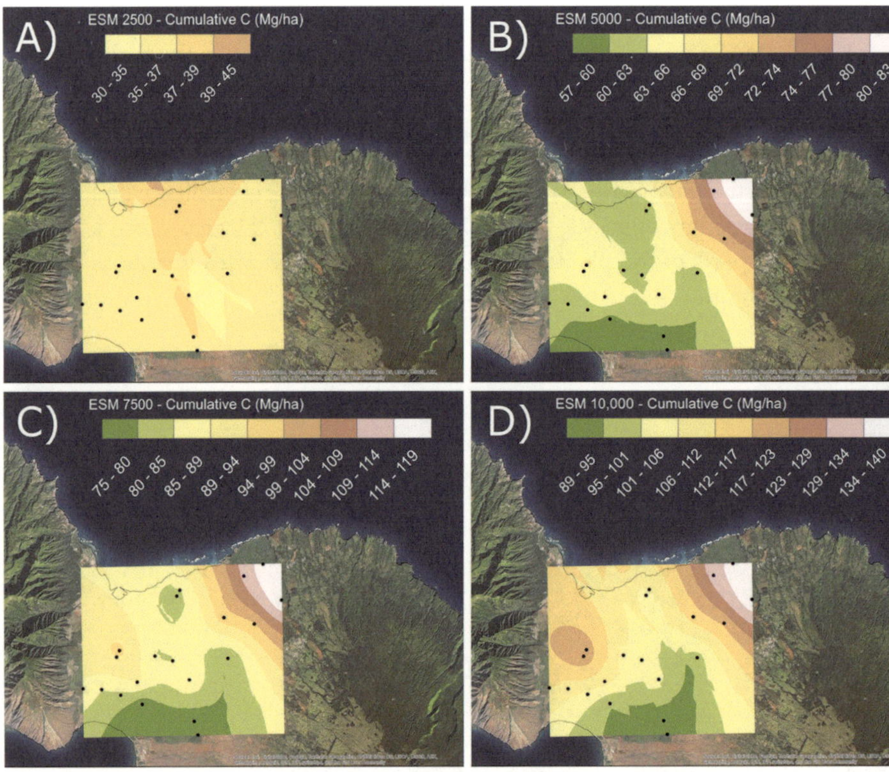

Figure 3. Simple spherical kriging done in ArcGIS estimates the distribution of baseline carbon across the HC&S plantation at four ESM reference masses: (A) 2500 Mg/ha; (B) 5000 Mg/ha; (C) 7500 Mg/ha; (D) 10,000 Mg/ha. Carbon stocks are represented on differing scale bars to illustrate changes in geospatial patterns as more of the depth profile is considered. ESM 10,000 (full profile) shows highest carbon accumulation toward the West Maui Mountains and northeast toward Haleakala Mountain, possibly following a climatic gradient of rainfall and soil weathering.

Soil texture and pH did not align overall with the patterns in soil organic carbon (**Figure 3**), although higher clay in the northeast areas of HC&S does generally align with higher carbon storage. Some of the higher clay areas are also known to have volcanic ash deposits and the soils have andic properties (e.g., Haliimaile and Hamakuapoko series in the northeast) or are classified as Andisols (e.g., Alae series spread throughout in small areas). Poorly or noncrystalline minerals derive from volcanic ash deposits, thereby confounding the influences of clay concentration and mineralogy. Future work will include exploring modified methods to quantify various forms of iron and aluminum oxides as possible drivers of organomineral sorption, water stable aggregates as a representation of physical protection, and comparison of climate data like temperature and rainfall as factors controlling clay weathering. Nonetheless, this is the first geospatial look at soil carbon stocks at this location and represents an initial attempt at soil carbon stock measurement and monitoring in this highly cultivated tropical perennial grass system. Higher sampling density of the plantation will likely be needed to corroborate these potential geospatial patterns.

2.2. Measured soil carbon stocks

2.2.1. Validation of geospatial interpolation of baseline values

Experimental plots were established at multiple locations across the plantation to investigate aspects of potential bioenergy production including: (a) growth characteristics of multiple novel feedstock crops, (b) water use efficiency and stress management, (c) the effects of elevation/wind/rain gradients, (d) emissions of greenhouse gases from soils, and (e) soil carbon sequestration based on management and soil properties. Soil carbon stock changes over time were measured at each site from baseline to year 3 postmanagement change. This case study focuses on soil carbon stocks at two-field plots established in the HC&S commercial fields 718 (Pulehu series; fine-loamy, mixed, semiactive, isohyperthermic Cumulic Haplustoll) and 609 (Molokai series; very-fine, kaolinitic, isohyperthermic, Typic Eutrotorrox) (**Figure 1**) to validate the geospatial interpolation method for determining baseline carbon stocks. Comparison of the baseline cumulative soil carbon data collected at field 718 (~126 Mg/ha at ESM 10,000) and 609 (~111 Mg/ha at ESM 10,000) (**Figure 4**) shows fairly close agreement with the plantation-practice baseline interpolation of the 20 field's dataset that indicate 117–123 Mg/ha and 101–106 Mg/ha of cumulative carbon at fields 718 and 609, respectively (**Figure 3d**). However, further comparison of both pit sampling and core sampling at identical locations will be needed to confirm agreement between the two ESM calculations.

2.2.2. Change over time

Focusing specifically on field 718 and a high performing hybrid energy cane (*Saccharum Officinarum* × *Saccharum Robustum*) feedstock, soil carbon stocks after conversion from intensive sugarcane cultivation to annual ratoon harvested energy cane exhibited substantial sequestration during the first 3 years (**Figure 5**). For this comparison, ESM data was calculated from soil cores dug using hand augers, with samples divided into 20 cm soil depth increments to 1.2 m. Within the collected soil profile, as represented in the ESM of 18,000 Mg/ha (roughly equivalent to 1 m depth, but all cores were different), high levels of cumulative

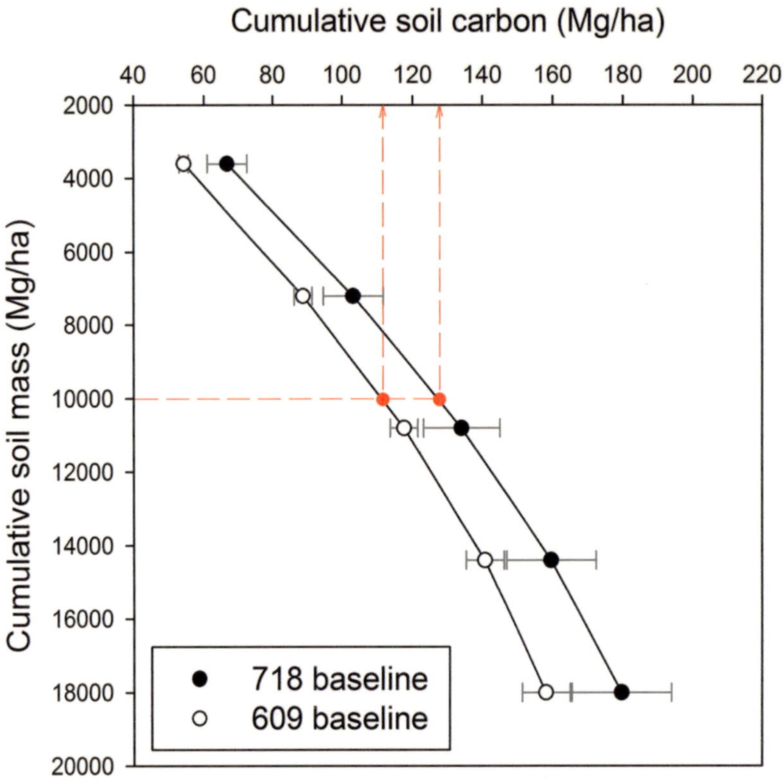

Figure 4. ESM data from baseline measurements of fields 718 and 609 are depicted with red line representing the 10,000 Mg/ha soil reference mass. When cumulative carbon from fields 718 and 609 are compared to the 20 field's interpolation, decent agreement between spatial model and measured ESM values are found.

carbon sequestration are apparent under conservation agriculture. It is known that following the shift from intensive cultivation to conservation practice, initial soil carbon accumulation rates will be high and decrease over time as a system saturates and reaches a new equilibrium. In this case study, the average annual cumulative soil carbon over the first 3 years was calculated as an estimate of soil carbon sequestration potential after the implementation of conservation agriculture in the surface soils and the deepest soil mass depth (ESM 3600 and 18,000; **Figure 5**). Net carbon sequestration is expected to continue at decreasing rates if conservation agriculture is maintained, but, in the first 3 years, the mean soil carbon sequestration at field 718 was 2.34 ± 1.03 Mg C/ha/yr in the surface soils and 12.75 ± 2.76 Mg C/ha/yr in the deeper soil profile over the 3-year experiment. These gains are in the range of recently reported rates of 3.9 Mg C/ha/yr in the surface soils of a tropical Napier grass system [30] and 5.0 Mg C/ha/yr in the top 1 m of a subtropical sorghum study [31]. Fluctuations in soil carbon stock occurred during the transition to conservation agriculture in this tropical perennial system. Rapid carbon sequestration in the first 2 years from baseline while the crops were establishing the below ground system and rhizopshere, with a slight reduction in carbon storage in year 3 (**Figure 5**). These data may indicate increasing but also naturally oscillating soil carbon stocks under improved soil management.

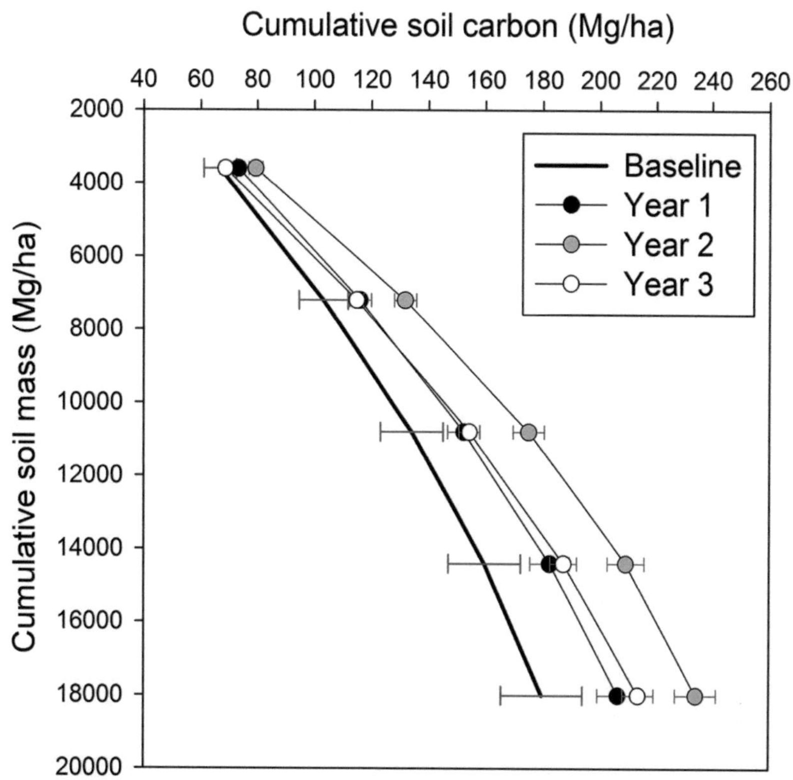

Figure 5. Cumulative soil carbon as measured by equivalent soil mass (ESM) methods at field 718 under ratoon harvest energy cane and zero-tillage. Increases in cumulative carbon at 5 ESM reference masses are shown by a shifting to the right from baseline. An unexplained drop in year 3, especially in the lowest reference mass, may show large natural fluctuation in carbon stocks.

To scale-up carbon storage to other areas of the plantation, we compared soil properties of the Pulehu series found at field 718 with nearby soils (**Table 1**). Specifically, the Paia, Waiakoa, and Ewa soil series (Mollisols) were chosen as adjacent areas with similar soil properties to field 718. However, important differences in detailed soil properties (e.g., percent clay) among the four soil series, suggest that only the plantation area under the original Pulehu soil should be used for scaling up (**Table 1**). Simple extrapolation of surface soil and deep soil carbon sequestration potentials found at field 718 (2.34 ± 1.03 and 12.75 ± 2.76 Mg C/ha/yr, respectively) across areas of the Pulehu series (1763 ha, representing ~11% of the HC&S plantation) equates to a prospective soil carbon sequestration potential of 4.1 ± 1.8 Gg C/yr that could be taken from the atmosphere and stored in similar surface soils compared to 22.5 ± 4.9 Gg C/yr that could be stored throughout the deeper soil profile in the first 3 years of transition to conservation agriculture. However, these initial carbon sequestration rates are expected to decrease with time as the soil's potential for carbon storage is saturated. Geospatial differences in carbon sequestration due to soil and environmental heterogeneity across the plantation also make these estimates rather uncertain, but these findings clearly indicate that deep (≥1 m) carbon sampling is important when considering landscape level soil carbon stocks as inclusion of the deep soil profile increased carbon stocks several fold.

Properties: Sites sampled	Pulehu $n = 4$	Ewa $n = 2$	Paia $n = 2$	Waiakoa $n = 4$
taxonomy (NRCS)[1]	Cumulic haplustolls	Aridic haplustolls	Torroxic haplustollus	Torroxic haplustolls
Parent material[1]	Igneous alluvium	Basaltic alluvium	Igneous residuum	Igneous residuum
Clay mineral type[1]	Mixed	Kaolinite	Iron oxide	Kaolinite
Bulk density (g/cm^3)	1.27 ± 0.06	1.19 ± 0.15	1.15 ± 0.03	1.32 ± 0.07
Soil porosity (%)	51.99	55.09	56.60	50.2
Texture	Clay loam	Silty clay loam	Silty clay	Silty clay loam
Clay (%)	5.85 ± 1.16	11.22 ± 3.35	35.90 ± 3.78	22.64 ± 3.10
Silt (%)	51.91 ± 9.96	75.15 ± 6.5	57.53 ± 4.97	66.61 ± 1.07
Sand (%)	42.24 ± 9.84	13.66 ± 3.14	6.58 ± 1.20	10.75 ± 2.33
Soil pH	7.45 ± 0.28	7.20 ± 0.2	7.60 ± 0.13	6.87 ± 0.53
SOC (%)	1.44 ± 0.16	1.99 ± 0.78	1.72 ± 0.20	1.28 ± 0.12
Total nitrogen (%)	0.12 ± 0.01	0.13 ± 0.03	0.17 ± 0.03	0.1 ± 0.01
C/N (using organic C)	12.43	15.27	10.12	12.30

[1]Data taken from NRCS Soilweb database, all other data collected during 20 field sampling using NRCS soil sampling protocols.

Table 1. Descriptive data of similar Mollisol soils series in Hawaii at HC&S (sampled during 20 fields experiment).

2.3. Model comparison: ALMANAC versus three pool transfer model

Finally, in an effort to project past the 3 years of data and to better understand the mechanisms that control how carbon is entering and moving through the soil system, we performed a physical separation of soil pools (i.e., density fractionation with sonication to disrupt an aggregated fraction) and subsequent simulation and projection model in SoilR [32] using the surface soils of field 718 and the ESM 3600 carbon accumulation values. The fractionation method used was based on Golchin et al. [33], in which sodium polytungstate (SPT) was used to increase the extraction density with free light, occluded light, and dense fractions sequentially separated by 1.6 g/L SPT solution. The free light fraction, which represents fresh plant inputs like roots and litter, was separated from the soil through light agitation by hand followed by centrifugation and aspiration. To obtain the occluded light fraction, which represents carbon that has been physically protected by soil aggregation, the soil slurry was sonicated with 400 kJ/mL to disrupt aggregates, and the released occluded carbon was captured by centrifugation and aspiration. Finally, the dense fraction was quantified as the soil that remained. The weights of recovered fractions and the percent carbon of each fraction were measured and then used to calculate the distribution of carbon between the pools. Strong decreases in litter/root inputs and aggregate protected carbon, as represented by the free light and occluded light fractions, respectively, were found. However, large increases of carbon in the mineral-rich dense fraction drove increases in carbon stocks in the surface soils (**Figure 6**). As this data-driven model is only a representation of surface soils, it will be important to

http://dx.doi.org/10.5772/66741

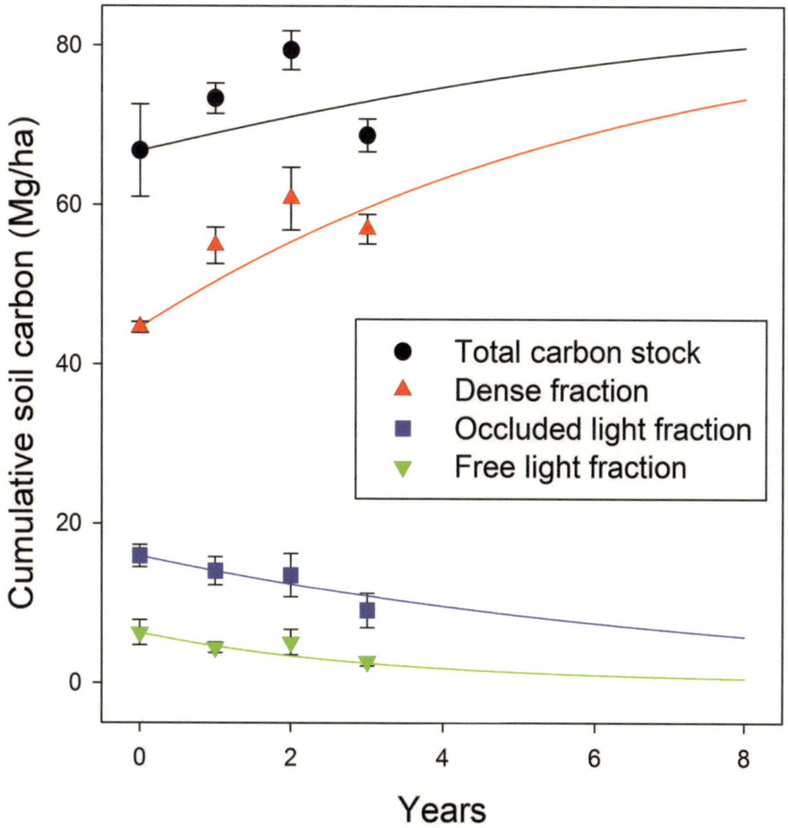

Figure 6. Three-pool model completed in SoilR (R project package, [32]) using density fractionation and measured ESM soil stock data from field 718 under ratoon harvest energy cane and zero-tillage. An ESM reference mass of 3600, which represents the shallowest mass soil depth was used for C stocks (**Figure 5**). Pool fraction data was determined from 0 to 20 cm depth samples (same cores ESM carbon stocks were calculated from).

further investigate the lower soil mass depths to see if mineral-driven carbon sorption is prevalent throughout the depth profile. A model of the entire soil profile (ESM 18,000) will be completed to project forward changes in total carbon stock as more of the depth profile is density fractionated.

A second model simulation of carbon storage was completed using the ALMANAC software [34]. As a crop model, ALMANAC uses a broad range of inputs to model soil carbon compared to SoilR (**Figure 7**). In the ALAMANC model, based on the field experiment and expected future management, the plough layer was set to 20 cm, energy cane was ratooned for 4 years and was then killed, ripped, harrowed, and replanted. These operations were based on farmer practice of periodically ploughing-back their conservation-tilled lands to alleviate problems of drainage, pests, and soil compaction [35]. The model was then run for a total of 25 years, with 10 years of preruns to stabilize model input variables prior to 2011. From 2011, the cumulative soil carbon of the surface soil (ESM 3600) and full soil profile (ESM 18,000) at field 718 were projected out to 2025 (**Figure 8**). Projected soil carbon stocks from the surface

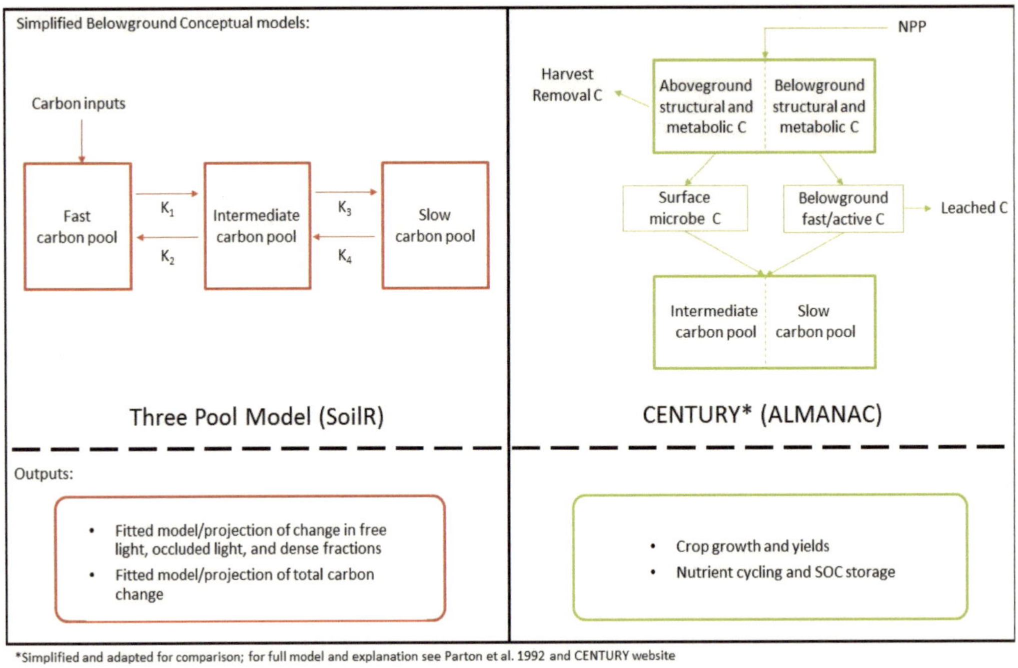

Figure 7. Conceptual representations of the processes involved in both the three pool SoilR model and the ALMANAC crop model. The empirical SoilR model requires soil carbon and density fractionation data to estimate fluxes between soil carbon pools, while the ALMANAC process model focuses on soil, crop, management, and weather data to estimate soil carbon stocks using the CENTURY soil carbon submodel for belowground carbon estimates. Comparing these two vastly different approaches helps to identify areas of improvement for carbon stock modeling in the future.

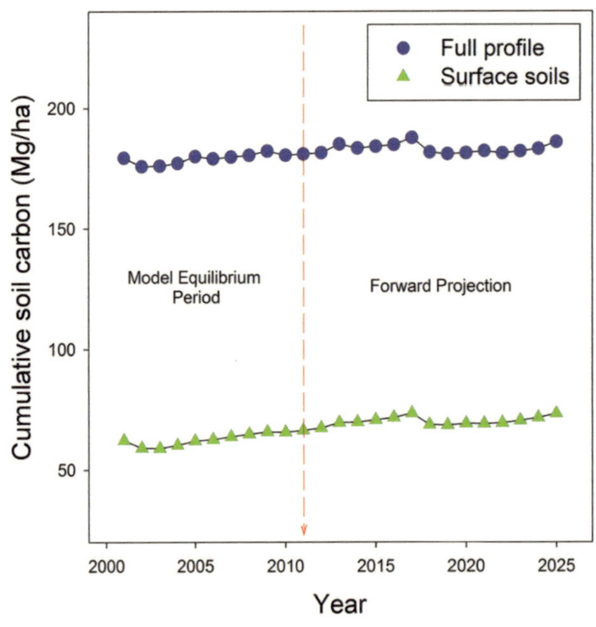

Figure 8. ALMANAC projection of surface soils (ESM 3600) and full soil profile (ESM 18,000). Ten years (2001–2010) were completed as a model equilibrium period before forward projection from 2011 for 15 further years.

soils for both the three pool SoilR model and the ALMANAC model were then compared to the measured data (**Figure 9**).

To compare modeled outputs with averaged yearly soil carbon increase data, the year 5 modeled outputs from SoilR and ALMANAC were compared and scaled across the Pulehu soil series in the same manner as the measured data. Increases of 1.9 and 1.1 Mg C/ha/yr were found in the surface soils by the SoilR and ALAMANAC models, respectively. Scaling these values with the same area of similar Pulehu soils (1763 ha) shows a soil carbon sequestration estimate of 3.3 and 1.9 Gg C/yr for SoilR and ALAMANC models, respectively. This estimate covers approximately 11% of the HC&S plantation area. As the SoilR model was of the surface soil only, it will be important to repeat this exercise with the deeper soil profile, especially considering that surface soils have shown the least differences under conservation agriculture. In contrast, the ALMANAC model was projected forward from a carbon baseline derived from the full depth profile, with year 5 showing an increase of only 0.8 Mg C/ha/yr when the deep soil profile is considered. Scaling up the ALMANAC output at year 5 gave a comparatively low value of 1.3 Gg C/yr sequestration potential across Mollisol soils at HC&S. Importantly, comparison of measured data and multiple models allows us to test different assumptions used, with further data collection and analysis expected to help improve and

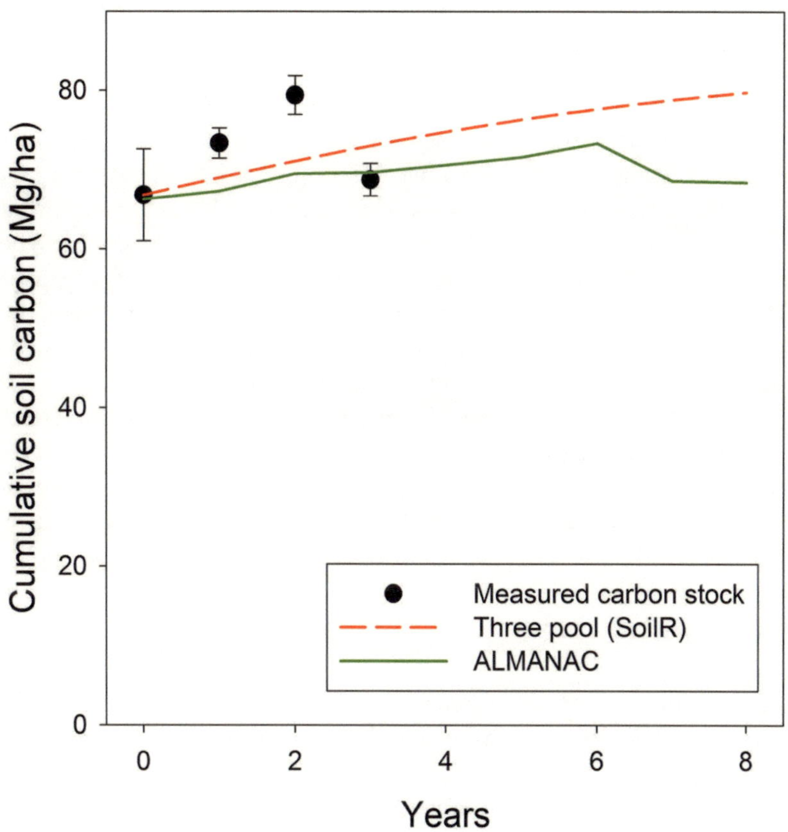

Figure 9. Two models (ALMANAC and SoilR) projected forward from surface soil (ESM 3600) for 8 years compared to measured data.

refine not only the model end products, but also the underlying assumptions that are relied on to understand soil carbon sequestration and storage in the tropics.

3. Conclusion

Carbon storage potential on the plantation scale in productive tropical cultivated systems is high when transitioning into conservation agriculture, especially when the deep soil profile is considered. Deep soil carbon increases, attributable to fast growing deep-rooted perennial grasses, led to a high belowground carbon sequestration potential (4.1 ± 1.8 and 22.5 ± 4.9 Gg C/yr in the surface and deep soils, respectively) if the average yearly increase of soil carbon at field 718 is generalized across similar Pulehu soils at the HC&S plantation. Taking modeled soil carbon stocks in the surface soils and scaling similarly gave 3.3 Gg C/yr from the SoilR model and 1.9 Gg C/yr from the ALMANAC model. However, much lower estimates were found when the deep soils were projected forward (only 1.3 Gg C/yr estimated by the ALMANAC model). As SoilR requires more fraction data from the deeper profile, it will be interesting to see if SoilR consistently has a greater estimate and ALMANAC a lower estimate of soil carbon sequestration. Nonetheless, these results point to a large potential for carbon storage through conversion to conservation agriculture; if monetized, these carbon storage potentials could be a huge prospective boon to the value of HC&S as a bioenergy site with proper soil and harvest management.

As indicated by the density fractionation results from commercial field 718, soil carbon storage is likely driven by mineral sorption in our system. However, there is need to refine our understanding of mineral changes across such a heterogeneous landscape, with continued density fractionation and iron/aluminum oxide measurements expected to better explain variations in soil carbon storage across HC&S. The estimated sequestration potentials will also need further improvement through comparison of our other field trials to determine if geospatial variations in soil texture and mineralogy will affect total carbon sequestration potentials across this heterogeneous landscape. Through continued and more detailed mineralogy and the addition of detailed climate, net primary productivity, and belowground biomass data, we expect to uncover relationships between the many factors controlling soil carbon sequestration in this system. Determination of better metrics and relationships of soil properties to carbon sequestration across the heterogeneous landscape of HC&S will enable more accurate projection of carbon sequestration potentials in Hawaii and other similar tropical perennial systems.

Acknowledgements

We thank Mae Nakahata and Hawaii Commercial & Sugar workers for their help with access, research priorities, and field work, and Richard Ogoshi and Andrew Hashimoto for field trial design and implementation and funding support. Preparation of this chapter was supported by the USDA-ARS, Grassland, Soil and Water Research Laboratory, and Texas A&M AgriLife Research, Temple, TX, through Specific Cooperative Agreement: 58-6206-1-053, and is funded by the U.S. Navy, Office of Naval Research (ONR). This work was also partly supported by ONR Grant N00014-12-1-0496 and USDE Award DE-FG36-08GO88037 to the University of Hawaii.

Author details

Jon M. Wells[1], Susan E. Crow[1*], Manyowa N. Meki[2], Carlos A. Sierra[3], Kimberly M. Carlson[1], Adel Youkhana[1], Daniel Richardson[1] and Lauren Deem[1]

*Address all correspondence to: crows@hawaii.edu

1 Natural Resources and Environmental Management Department, University of Hawaii at Manoa, Honolulu, HI, USA

2 Texas A&M AgriLife Research, Blackland Research and Extension Center, Temple, TX, USA

3 Max Plank Institute for Biogeochemistry, Jena, Germany

References

[1] Parton WJ, Hartman M, Ojima D, Schimel D. Daycent and its land surface submodel: description and testing. Global and Planetary Change. 1998;19(1–4):35–48.

[2] Jenkinson DS, Adams DE, Wild A. Model estimates of CO_2 emissions from soil in response to global warming. Nature. 1991;351(6324):304–6.

[3] Augustin C, Cihacek L. Relationships between soil carbon and soil texture in the Northern great plains. Soil science. 2016;181(8):386–92.

[4] Xu X, Shi Z, Li D, Rey A, Ruan H, Craine JM, et al. Soil properties control decomposition of soil organic carbon: Results from data-assimilation analysis. Geoderma. 2016;262:235–42.

[5] Silva JHS, Deenik JL, Yost RS, Bruland GL, Crow SE. Improving clay content measurement in oxidic and volcanic ash soils of Hawaii by increasing dispersant concentration and ultrasonic energy levels. Geoderma. 2015;237–238:211–23.

[6] Torn MS, Trumbore SE, Chadwick OA, Vitousek PM, Hendricks DM. Mineral control of soil organic carbon storage and turnover. Nature. 1997;389(6647):170–3.

[7] Crow SE, Reeves M, Schubert OS, Sierra CA. Optimization of method to quantify soil organic matter dynamics and carbon sequestration potential in volcanic ash soils. Biogeochemistry. 2015;123(1):27–47.

[8] Six J, Conant RT, Paul EA, Paustian K. Stabilization mechanisms of soil organic matter: Implications for C-saturation of soils. Plant and Soil. 2002;241(2):155–76.

[9] Six J, Bossuyt H, Degryze S, Denef K. A history of research on the link between (micro) aggregates, soil biota, and soil organic matter dynamics. Soil and Tillage Research. 2004;79(1):7–31.

[10] Cotrufo MF, Soong JL, Horton AJ, Campbell EE, Haddix ML, Wall DH, et al. Formation of soil organic matter via biochemical and physical pathways of litter mass loss. Nature Geosci. 2015;8(10):776–9.

[11] Rasse DP, Rumpel C, Dignac MF. Is soil carbon mostly root carbon? Mechanisms for a specific stabilisation. Plant and Soil. 2005;269(1):341–56.

[12] Plaza C, Courtier-Murias D, Fernández JM, Polo A, Simpson AJ. Physical, chemical, and biochemical mechanisms of soil organic matter stabilization under conservation tillage systems: A central role for microbes and microbial by-products in C sequestration. Soil Biology and Biochemistry. 2013;57:124–34.

[13] O'Brien SL, Jastrow JD, Grimley DA, Gonzalez-Meler MA. Edaphic controls on soil organic carbon stocks in restored grasslands. Geoderma. 2015;251–252:117–23.

[14] Marín-Spiotta E, Gruley KE, Crawford J, Atkinson EE, Miesel JR, Greene S, et al. Paradigm shifts in soil organic matter research affect interpretations of aquatic carbon cycling: Transcending disciplinary and ecosystem boundaries. Biogeochemistry. 2014;117(2):279–97.

[15] Schmidt MWI, Torn MS, Abiven S, Dittmar T, Guggenberger G, Janssens IA, et al. Persistence of soil organic matter as an ecosystem property. Nature. 2011;478(7367):49–56.

[16] Lehmann J, Kleber M. The contentious nature of soil organic matter. Nature. 2015; 528(7580):60–8.

[17] Masoom H, Courtier-Murias D, Farooq H, Soong R, Kelleher BP, Zhang C, et al. Soil organic matter in its Native State: Unravelling the most complex biomaterial on earth. Environmental Science & Technology. 2016;50(4):1670–80.

[18] Palm C, Blanco-Canqui H, DeClerck F, Gatere L, Grace P. Conservation agriculture and ecosystem services: An overview. Agriculture, Ecosystems & Environment. 2014;187:87–105.

[19] Davis SC, Boddey RM, Alves BJR, Cowie AL, George BH, Ogle SM, et al. Management swing potential for bioenergy crops. GCB Bioenergy. 2013;5(6):623–38.

[20] Mutuo PK, Cadisch G, Albrecht A, Palm CA, Verchot L. Potential of agroforestry for carbon sequestration and mitigation of greenhouse gas emissions from soils in the tropics. Nutrient Cycling in Agroecosystems. 2005;71(1):43–54.

[21] Anderson-Teixeira KJ, Masters MD, Black CK, Zeri M, Hussain MZ, Bernacchi CJ, et al. Altered belowground carbon cycling following land-use change to perennial bioenergy crops. Ecosystems. 2013;16(3):508–20.

[22] Davidson EA, Ackerman IL. Changes in soil carbon inventories following cultivation of previously untilled soils. Biogeochemistry. 1993;20(3):161–93.

[23] Gifford RM, Roderick ML. Soil carbon stocks and bulk density: Spatial or cumulative mass coordinates as a basis of expression? Global Change Biology. 2003;9(11):1507–14.

[24] Wendt JW, Hauser S. An equivalent soil mass procedure for monitoring soil organic carbon in multiple soil layers. European Journal of Soil Science. 2013;64(1):58–65.

[25] Crow SE, Reeves M, Turn S, Taniguchi S, Schubert OS, Koch N. Carbon balance implications of land use change from pasture to managed eucalyptus forest in Hawaii. Carbon Management. 2016:1–11.

[26] Richter Dd, Billings SA. 'One physical system': Tansley's ecosystem as Earth's critical zone. New Phytologist. 2015;206(3):900–12.

[27] Dietz S, Stern N. Endogenous growth, convexity of damage and climate risk: How nordhaus' framework supports deep cuts in carbon emissions. The Economic Journal. 2015;125(583):574–620.

[28] Burke M, Craxton M, Kolstad CD, Onda C, Allcott H, Baker E, et al. Opportunities for advances in climate change economics. Science. 2016;352(6283):292–3.

[29] Shirato Y, Hakamata T, Taniyama I. Modified rothamsted carbon model for andosols and its validation: changing humus decomposition rate constant with pyrophosphate-extractable Al. Soil Science and Plant Nutrition. 2004;50(1):149–58.

[30] Sumiyoshi Y, Crow SE, Litton CM, Deenik JL, Taylor AD, Turano B, et al. Belowground impacts of perennial grass cultivation for sustainable biofuel feedstock production in the tropics. GCB Bioenergy. 2016: Advance online publication.

[31] Dou F, Wight JP, Wilson LT, Storlien JO, Hons FM. Simulation of Biomass Yield and Soil Organic Carbon under Bioenergy Sorghum Production. Plos One. 2014;9(12):e115598.

[32] Sierra CA, Müller M, Trumbore SE. Models of soil organic matter decomposition: The Soilr package, version 1.0. Geoscientific Model Development. 2012;5(4):1045–60.

[33] Golchin A, Oades J, Skjemstad J, Clarke P. Study of free and occluded particulate organic matter in soils by solid state ^{13}C Cp/MAS NMR spectroscopy and scanning electron microscopy. Soil Research. 1994;32(2):285–309.

[34] Meki MN, Kiniry JR, Youkhana AH, Crow SE, Ogoshi RM, Nakahata MH, et al. Two-year growth cycle sugarcane crop parameter attributes and their application in modeling. Agronomy Journal. 2015;107(4):1310–20.

[35] Causarano HJ, Doraiswamy PC, McCarty GW, Hatfield JL, Milak S, Stern AJ. EPIC modeling of soil organic carbon sequestration in croplands of iowa All rights reserved. Journal of Environmental Quality. 2008;37(4):1345–53.

Chapter 8

Relationship Between Mineral Soil Surface Area and Carbon Sequestration Rate for Biosolids Added to Soil

Dongqi Wen, Wenjuan Zhai and Kenneth E. Noll

Additional information is available at the end of the chapter

http://dx.doi.org/10.5772/65862

Abstract

Biosolid degradation in soil comprises important biological and geochemical processes that operate in the soil matrix and on the soil surface. The microbial ecology is assumed to be associated with mineral soil surface area because of the large surface area of soil. Biological degradation rates for 27 fields (10°C and 10% moisture) ranged from 0.01 to 0.30 yr^{-1} and were determined by applying a degradation rate model (DRM). A 1-year-long laboratory study was also conducted to determine biosolid microbial degradation rates (21°C and 20% moisture) for soils from eight of the fields. Changes in degradation rates were correlated with changes in mineral soil surface area (1–10 m^2/g) with larger degradation rates associated with soils with larger surface areas. The annual soil sequestration rate was calculated to increase from 1 to 6% for field conditions and from 4 to 14% for laboratory conditions when the soil total surface area increased from 1 to 10 m^2/g. Therefore, land application of biosolids is an effective way to enhance carbon sequestration in soils and reduce greenhouse gas (GHG) emissions.

Keywords: carbon sequestration, biosolids, biological degradation, mineral soil surface area

1. Introduction

This chapter evaluates the relationship between carbon sequestration rates for biosolids added to soils and soil surface area to provide a better understanding of the variables that control sequestration. Biosolids are nutrient-rich organic materials formed as a result of anaerobic digestion of primary and secondary sludge from wastewater treatment plants. Each year 7.1 Mt of biosolids are generated (dry tons) in the United States [1]. Previous studies have accounted for effects of temperature and moisture on carbon sequestration rate but have not included adjustments for soil surface area. We hypothesize that due to the large surface area of soils, biological processes that operate on the soil surface are potentially important to the sequestration rate of biosolids in soil. A degradation rate model (DRM) is used to predict the portion of biosolids added to soil that is sequestered (residual microbial biomass). The quantification of the biomass yield is especially important and is a unique feature of the DRM because it provides the ability to separate the soil organic carbon (SOC) into two components: (i) carbon (C) in biosolids that has not been degraded and (ii) C in residual microbial biomass produced during the microbial degradation process (sequestered carbon). Biomass is being developed as biosolids are consumed. One of the most important DRM characteristics is that it can be used to represent the pattern of biomass accumulation from multiple applications of biosolids to soil.

A basic relationship is developed that uniquely predicts changes in the sequestration of organic matter added to soils as a function of mineral surface area that provides a better understanding of the important variables that control sequestration and allows the application of technologies based on their ability to increase the rate of C sequestration. Results demonstrate that soil surface area is much more profound as an indicated of carbon sequestration in soils than previously indicated, and therefore soil surface area is an essential parameter in assessing sequestration rates of organic material added to soils.

The United Nations Framework Convention on Climate Change (UNFCCC) declares that greenhouse gas (GHG) emissions result from anthropogenic activities and recommends C counting as a necessary step toward reduction of such emissions [2]. The land application of biosolids is an effective way to increase SOC. The ability for soils to biologically degrade biosolids and sequester carbon (C) is recognized as one method to mitigate greenhouse gas emissions [2, 3]. Jarecki and Lal [4] also suggested that application of biosolids is an important management practice to increase soil C sequestration in agricultural soils. Net C sequestration rates from biosolids applied to soil have been reported to be between 1 and 3 Mg ha^{-1} yr^{-1} with biosolid application rate between 56 and 71 Mg ha^{-1} yr^{-1} [5].

The world's degraded soils (1216 Mha) and agricultural soils (4961 Mha) both have high potential for C sequestration. Historical data show that 40 Pg of SOC have been lost in these soils. Considering these soils have capacity to sequester C, it is important to realize that there is a way to reverse the SOC depletion process. The total potential of soil C sequestration is around 0.6–1.2 Pg C yr^{-1}, in which the world cropland could sequester C at the rate of 0.4–0.6 Pg C yr^{-1} [6] and the desertification control has the C sequestration potential of around 0.2–0.6 Pg C yr^{-1}. Conant et al. [7] pointed out that the grassland also has relative high potential of C sequestration, which can be included in desertification control. These data imply that about 0.9 ± 0.3 Pg C yr^{-1} additional C could be sequestered in soils [8].

Efforts to improve C sequestration in agricultural soils focus on changes of management practices such as tillage/no tillage, irrigation, farm machinery, and other similar strategies [9]. Yet, C sequestration associated with improved management practices has not been investigated comprehensively because terrestrial C sequestration is a complex function of plant species, type of soil, regional climatic conditions, and topography in addition to management practices [10, 11]. These intricate details are further amplified by the need for long-term studies required to evaluate dynamic processes involved in the better understanding of a quantitative dynamic process for C sequestration in agricultural soils. The present study overcomes this difficulty by focusing on a specific and clearly defined system: repeated soil application of biosolids under conditions of variable application rates.

Biosolids applied to soil are aerobically transformed into inorganic C, which is released to the atmosphere and into humid substances (biomass) that are available for sequestration. Microbial decomposition of biosolids occurs over a period of years not decades and therefore is susceptible to depletion. Decomposition of accumulated biomass is much slower, and consequently this material has a significant potential as a repository for excess atmospheric C. Through the process described above, biosolid degradation rate plays a key role in governing the SOC dynamics. The rate of biosolid decomposition is one of the key processes governing the dynamic of C sequestration. The DRM is used in this chapter to quantify the degradation rate for biosolids and yield for residual microbial biomass from repeated application of biosolids to soil and provides an easy quantitative method for evaluating C sequestration.

2. Method and materials

2.1. Degradation rate model (DRM)

To assess the dynamic of SOC sequestered process and better understand C sequestration in agriculture soil, Zhai et al. [12] developed a degradation rate model (DRM)l to describe the degradation rate process and determine the yield for residual microbial biomass (sequestered carbon) from repeated application of biosolids to soil. The DRM [12] was used in this study to provide the field biological degradation rate for biosolids applied to selected fields from 1972 to 1985 in Fulton County, Illinois [5] (see **Table 1** and **Figure 1**).

This site is located approximately 300 km southwest of Chicago. The climate of the site is continental with an annual average air temperature of 10°C and annual precipitation of 1013 mm. The monthly mean moisture content of soil is near 10%. The pH of surface spoils was neutral to alkaline with variable soil texture [5]. The biosolids applied to 41 fields were in liquid phase with average organic carbon 23.2% [5]. The advantage of using the Tian et al. [5] database is its long duration and the inclusion of repeated measurements of biosolids applied (including the organic constituent) and resulting soil organic carbon (SOC) gain. Forty-one fields were

Type of biosolids	Total solids¥	Volatile solids¥	Organic carbon¥	Organic N*	NH_4-N*	C/N*	Total Fe*	Total Al*
			(%)				(%)	
Liquid	2.9–73.0	14.6–47.6	8.5–27.6	2.38	1.54	6.21	3.98	1.13

¥ Values vary based on biosolids applied from the year 1972 to 1985 in these properties, for detailed information, see Tian et al. database [5].

* Values vary based on biosolids applied from the year 1972 to 1985 in these properties, an average was shown here; for detailed information, see Tian et al. database [5].

Table 1. Properties of biosolids applied in this study (values are from Tian et al. database) [5].

Figure 1. Biosolids applied on selected field in Fulton County, IL [13].

divided into three groups based on soil type. Group I consisted of 20 fields of "coarse" mine spoil soils, primarily Lenzburg (fine-loamy, mixed, active, calcareous, mesic Haplic Udarents) and Lenzwheel soil series [14]. Group II is primarily Rapatee (fine-silty, mixed, superactive, nonacid, mesic Mollic Udarents) soil series and contains nine fields of "fine" mine spoil soils, [14, 15]. Group III contains 12 fields of various non-mined soils that were degraded by intensive cultivation or overgrazing [14, 15] (**Figure 2**).

The DRM was developed by employing pertinent information from Tian et al. [5] on the long-term application of biosolids to soil in 41 fields with variable application periods ranging from 8 to 34 years (1972 to 2006). The model is based on a mass balance between the amount

Figure 2. Typical soil samples from group I, II, and III applied in this study.

of biosolid carbon applied to soil and the amount of SOC present in the soil plus the carbon evolved as CO_2. The mass balance could be written as follows:

$$\text{Applied biosolids carbon} \xrightarrow{\text{first-order kinetic}} \text{biosolids remaining} + \text{Carbon sequestered} + CO_2\,\text{emissions} \tag{1}$$

The biosolid decomposition rate in the DRM is described by first-order kinetics:

$$\frac{dC}{dt} = -kt \tag{2}$$

where:

C = the carbon concentration present in the biosolids (Mg/Ha);

k = the first-order degradation rate (yr^{-1});

t = biosolid decomposition time (yr).

Based on the first-order kinetics, the accumulated residue could be calculated by the following equation:

$$\text{residue} = \left(1 + f^1 + f^2 + f^3 + \cdots + f^n\right)\left(\text{SEQequation\_num} * MERGEFORMAT3\right) \tag{3}$$

where

f is the fraction left after 1-year decay or $f = \frac{C_t}{C_0} = e^{-kt}$, with $t = 1$.

To develop the DRM model, a curve fitting approach was applied that compared field measurements of SOC to calculations of SOC. Curve fitting, the measured values of SOC for each year with model-generated values of SOC using trial and error, produced a best-fit average degradation rate for biosolid degradation and biomass yield. The DRM is based on quantification of both the degradation rate for the biosolids and the yield for residual microbial biomass and provides an easy quantitative method for evaluating residual microbial biomass. One of the most important DRM characteristics is that it uses one degradation rate constant to adequately represent the pattern of accumulation from multiple applications of biosolids to soil. The DRM can be applied to estimate (1) the biosolid degradation as a function of time, (2) the SOC portion due to biosolids remaining, and (3) the residual microbial biomass (C sequestered). To apply the DRM, the appropriate biomass yields and degradation rate that are estimated from curve fitting are needed [15]. It should be noticed that the microbial biomass yield is considered as constant (35–40%) determined by curve fitting results. It is an important assumption when computing biomass (C sequestered) during the biosolid degradation process in each field.

Figure 3 shows the C flow simulated in the DRM model. A wide range of factors control the rate of sequestration of carbon from biosolids and their residence time in soil.

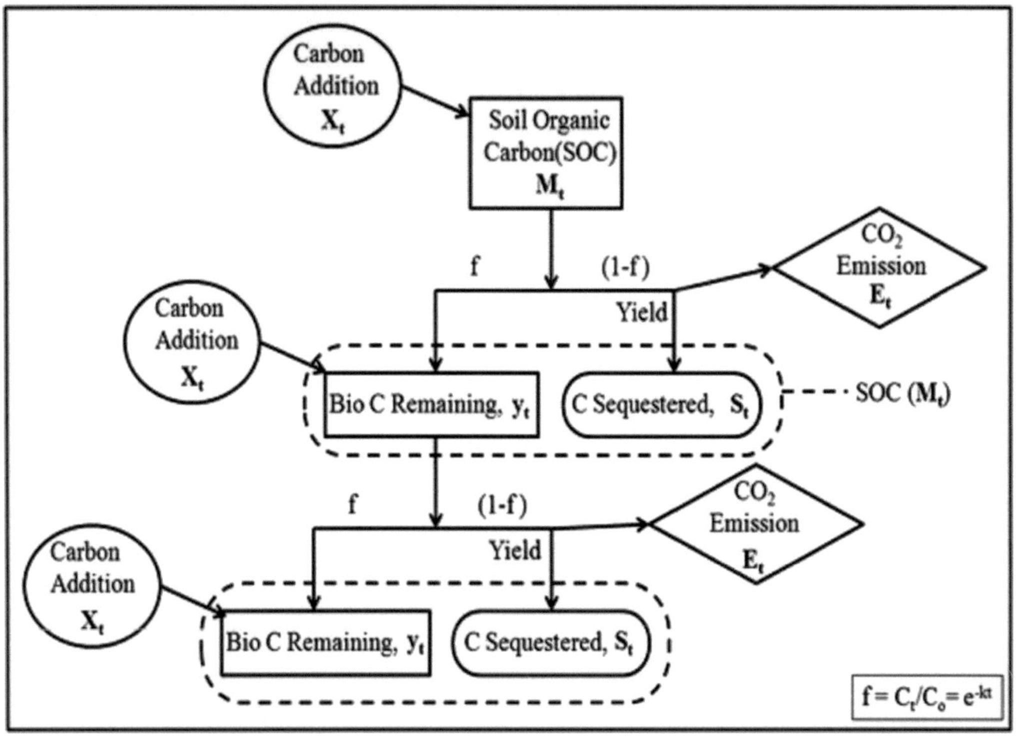

Figure 3. Flow diagram of the DRM as a function of C addition [12].

The DRM is presented as follows:

$$y_t = \left(y_{t-1} + X_{t-1}\right) \times f \tag{4}$$

$$S_t = \left(\frac{(1-f)}{f}\right) \times \text{Yield} \times \sum_1^t y_t \tag{5}$$

$$M_t = y_t + S_t = y_t + \left(\frac{(1-f)}{f}\right) \times \text{Yield} \times \sum_1^t y_t \tag{6}$$

$$E_t = \frac{(1-f)}{(f)} \times (1 - \text{Yield}) \times y_t \tag{7}$$

where $f = e^{-k}$;

X_t is the biosolid carbon application amount (mg) at time t (day);

y_t is the biosolid carbon remaining at time t (mg);

S_t is the biomass carbon mass or sequestered carbon accumulation at time t (mg);

M_t is mass of SOC at time t (mg);

E_t is the annual C-CO_2 emission at time t (mg).

2.2. Soil particulate surface area

Tian et al. [5] classified the 41 fields evaluated in this study as coarse (group I), fine (group II), and mixed (group III) but provided no additional information concerning physical differences in the type of soils in each group. Therefore, a sieve analysis based on the mass fraction of soil that passes through a specified screen size was used to determine the mass distribution of the coarser, larger-sized particles, and a hydrometer was used to determine the size distribution of the finer particles [16]. Twenty-seven fields were strategically selected from the 41 fields to have a wide range of mineral surface areas (see **Table 2**). Of the 27 fields selected, nine were from group I, seven from group II, and 11 from group III.

A soil texture analysis was used to determine the physical characteristics of the soils. [16]. Quantitatively, soil texture denotes the proportion of sand (0.05–2 mm diameter), silt (0.002–0.05 mm diameter), and clay (less than 0.002 mm diameter) that occur in a given soil.

Particulate surface area size distributions were estimated from the mass size distributions to determine the effect of surface area on biosolid degradation. It was assumed that the soil particles are spherical with a smooth surface and the number of soil particles was estimated by Eq. (8) [15]:

$$N = M / \left(\rho \cdot 4/3 \cdot \pi \cdot r^3\right) \tag{8}$$

where:

N = the number of soil particles;

r = soil particle radius (cm);

M = the mass of the soil particle (g);

ρ = soil density [17] (g/cm^3).

The soil particulate surface area (S) was estimated by Eq. (9):

$$S = 4\pi r^2 \cdot N \tag{9}$$

where:

N = the number of soil particles;

r = soil particle radius (m);

S = particle surface area (m^2).

Eqs. (8) and (9) were applied to calculate mineral soil surface based on known mass distribution and average particle size for sand, silt, and clay [18].

2.3. Laboratory experiment using soil respirator

A yearlong laboratory study using a soil respirator was conducted to determine the degradation rate constants for eight of the 27 fields under laboratory conditions (21°C and 20% moisture). The fields were selected strategically from the 41 fields to represent different mineral

Soils	Organic materials	G	Field no.	Environment conditions		C^{*}	K1 $	K2¥	D	SFC	Soil texture
				Fields	Lab	Mg ha^{-1}	Yr^{-1}	Yr^{-1}	g cm^{-3}	m^{2} g^{-1}	
The soil samples from this study (eight soil samples)	Biosolids	I	F32	10°C and 10%	21°C and 20%	8.93	0.02	0.11	2.6	1.29	Sandy loam
	Biosolids	I	F39	10°C and 10%	21°C and 20%	11.95	0.05	0.12	2.6	1.73	Sandy loam
	Biosolids	III	F10	10°C and 10%	21°C and 20%	12.85	0.06	0.14	2.6	2.29	Sandy loam
	Biosolids	I	F15	10°C and 10%	21°C and 20%	15.44	0.11	0.24	2.6	2.42	Sandy loam
	Biosolids	III	F37	10°C and 10%	21°C and 20%	11.75	0.14	0.35	2.6	6.55	Clay loam
	Biosolids	II	F47	10°C and 10%	21°C and 20%	18.02	0.15	0.42	2.6	8.54	Clay loam
	Biosolids	II	F45	10°C and 10%	21°C and 20%	12.95	0.15	0.38	2.6	7.63	Clay loam
	Biosolids	II	F43	10°C and 10%	21°C and 20%	15.63	0.20	0.47	2.6	7.65	Clay loam
The soil samples from this study (19 soil samples)	Biosolids	I	F8	10°C and 10%	N/A	11.59	0.01	N/A	2.6	1.12	Loamy sand
	Biosolids	III	F16	10°C and 10%	N/A	12.30	0.01	N/A	2.6	1.27	Loamy sand
	Biosolids	III	F22	10°C and 10%	N/A	14.74	0.02	N/A	2.6	1.14	Sandy loam
	Biosolids	III	F23	10°C and 10%	N/A	8.64	0.03	N/A	2.6	1.27	Loamy sand
	Biosolids	I	F4	10°C and 10%	N/A	9.01	0.03	N/A	2.6	2.29	Loamy sand
	Biosolids	III	F35	10°C and 10%	N/A	15.39	0.06	N/A	2.6	1.66	Sandy loam
	Biosolids	I	F2	10°C and 10%	N/A	10.46	0.06	N/A	2.6	2.34	Sandy loam
	Biosolids	III	F34	10°C and 10%	N/A	10.80	0.07	N/A	2.6	2.41	Sandy loam
	Biosolids	I	F30	10°C and 10%	N/A	18.00	0.10	N/A	2.6	2.66	Sandy loam
	Biosolids	I	F28	10°C and 10%	N/A	17.97	0.12	N/A	2.6	4.04	Loam
	Biosolids	I	F7	10°C and 10%	N/A	11.64	0.13	N/A	2.6	4.80	Loam
	Biosolids	II	F9	10°C and 10%	N/A	9.71	0.01	N/A	2.6	4.58	Silt loam
	Biosolids	III	F36	10°C and 10%	N/A	11.98	0.10	N/A	2.6	5.15	Loam

Soils	Organic materials	G	Field no.	Environment conditions		C*	K1 $	K2¥	D	SFC	Soil texture
				Fields	Lab	Mg ha^{-1}	Yr^{-1}	Yr^{-1}	g cm^{-3}	m^2 g^{-1}	
	Biosolids	III	F21	10°C and 10%	N/A	11.32	0.13	N/A	2.6	7.83	Clay loam
	Biosolids	III	F31	10°C and 10%	N/A	16.55	0.17	N/A	2.6	7.64	Clay loam
	Biosolids	II	F41	10°C and 10%	N/A	11.50	0.19	N/A	2.6	7.89	Clay loam
	Biosolids	II	F42	10°C and 10%	N/A	13.90	0.20	N/A	2.6	8.45	Clay loam
	Biosolids	II	F40	10°C and 10%	N/A	15.48	0.25	N/A	2.6	10.09	Silt clay loam
	Biosolids	III	F20	10°C and 10%	N/A	10.93	0.30	N/A	2.6	10.02	Silt clay loam
Terry et al. [21] soils	Sludge	N/A	Fincastle	N/A	21°C and 20%	5.15	N/A	0.35	2.6	4.81	Silt loam
	Sludge	N/A	Chalmers	N/A	21°C and 20%	5.15	N/A	0.29	2.6	4.66	Silt loam
	Sludge	N/A	Tracy	N/A	21°C and 20%	5.15	N/A	0.12	2.6	2.13	Sandy loam

G, group no.;
C*, annual carbon application rate;
K1$, degradation rate under field conditions;
K2¥, degradation rate under lab conditions;
D, soil density; SFC, mineral soil surface area.

Table 2. Variation in the environmental conditions (temperature and moisture content), annual application rate (C), degradation rate under both field (K1) and laboratory (K2) conditions, soil density(D), mineral soil surface area (SFC), and soil texture for soils evaluated.

surface areas. This approach facilitates the objective of establishing an association between total mineral surface area and degradation rate.

For the laboratory experiment, 10 g of air-dried soil from each field was added to 500 ml Erlenmeyer flasks plus additional water to provide the desired moisture content (0.2 g of water per gram of soil). The flasks were connected to a scrubber system consisting of a series of flasks containing concentrations of sulfuric acid, sodium hydroxide, and water to provide humidified, CO_2-free air to pass over the surface of each soil sample. The flow rate of humidified, CO_{CO}-free air was controlled at 9 ml/min for each soil sample, which was incubated in the dark at 21°C for 360 days. The evolved CO_2 was absorbed in 200 ml of a 0.5 NaOH solution. Periodic replacement of NaOH solutions guaranteed the accuracy of CO_2 production rate measurements for each sample. Duplication experiments were conducted. The CO_2 amount was measured by back-titration with 1.0M HCl after the CO_2 was stabilized by precipitation with 1.5M $BaCl_2$ solutions [19]. At the beginning and the end of the 1-year incubation period, samples of soil taken from the flasks were analyzed to determine SOC concentrations using the Walkley-Black method [20]. This method used excess dichromate ion to oxidize the SOC and titrate the dichromate residual after oxidization with ferrous ion.

Terry et al. [21] also conducted a yearlong laboratory experiment to evaluate the biological degradation of synthetic biosolids (with the decomposition and degradation rate similar to real biosolids) using three different soil types under controlled conditions (**Table 2**). Terry's paper reports on the emissions from biosolids added to soil with little analysis of the results. Terry's CO_2 emission database was reanalyzed in this study to provide the laboratory degradation rate of biosolids, based on first-order kinetics. Results from Terry's experiment were compared with results from this study.

3. Results

3.1. Sequestrate rates determined from DRM application

DRM provides C sequestration rates based on first-order kinetics and estimates the SOC concentration for each field in Tian et al. [5] database. After annual biosolid application for 15–22 years to each field, the modeled SOC concentrations and amount of C sequestered were estimated by applying DRM. **Figure 4** provides example plots for two of the selected fields [15]. The SOC measured and estimated with the DRM matched very good base on the coefficient of determinations between them (average coefficient of determination is 0.94), indicating that the curve fitting technique was able to provide biomass yield and degradation rate. Therefore, DRM model is allowed to provide acceptable estimates of the measured SOC values and sequestration amounts [15].

The slow rates of biosolid degradation resulted in large increases in SOC; this is because of the presence of biosolids that have not reacted [15]. The peak of the SOC concentration occurred in the mid-1980s after 12 years of annual biosolid application. After the year 1985, the accumulation of biosolids stopped in increasing and the biosolid application declined. An increased SOC was caused by the accumulation of stored biosolids when the biosolid

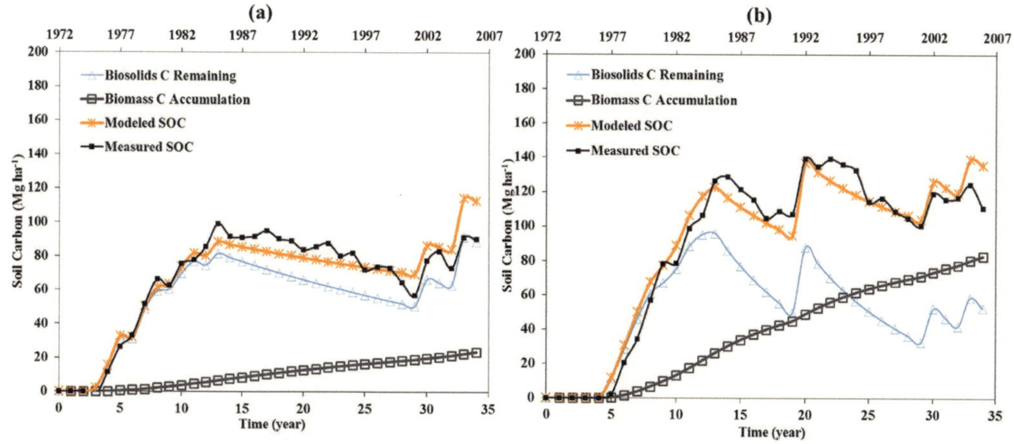

Figure 4. Example of the variation in biosolids remaining, measured and modeled soil organic carbon (SOC), and accumulated biomass. (a) Field 32 and (b) field 47.

accumulation exceeded degradation during the early stage of biosolid application. A steady state was approached when of decomposition converged on the amount of biosolid application [15]. This result is supported by Hamaker's study [22], which developed a mathematical model to predict the cumulative levels of pesticides in soil. The study indicated that when pesticides application rate equals to its decomposition rate, a steady state was approached. Also, Jastrow et al. [23] suggested carbon sequestration occurs when a positive disequilibrium sustained between C input and C degraded over some period of time. A new steady-state system would eventually be achieved when the amount of degradation converged on the amount of application. Jastrow's [23] result explains the increasing of SOC during the early stages of biosolid addition in this study, and Hamaker's [22] finding corresponds to achieving a steady state between biosolid degradation and biosolid application.

Analyses of DRM simulation results (**Figure 5**) for the group I and II in the Tian et al. [5] database indicate that higher biosolid degradation rates occur with finer soils. Several long-term agroecosystem studies also indicate that SOC accumulation increases with increase in C input [23, 24]. To assess the relationship between soil type, biosolid degradation rate, and biosolid C application rate, the fields were divided into coarse and fine soils [5]. It can be observed that the microbial degradation rate was larger for fine than for coarse soil type based on the separation of the regression lines. The error bars in **Figure 5** represent one standard deviation for each average biosolid application rate for related fields in Tian's database [15]. The average difference in the degradation rate between the linear regression lines was near 0.10 yr^{-1} (**Figure 5**). This represents the difference in the average degradation rate of biosolids when applied on coarse and fine soils. It can be observed in **Figure 5** that there is a linear relationship between biosolid degradation rate and biosolid carbon application rate for both coarse and fine soils.

Figure 5 and **Table 2** also identify eight of the 41 soils that were selected for a laboratory experiment using a soil respirator (21°C and 20% moisture content). The fields were selected strategically from the 41 fields to represent different mineral surface areas.

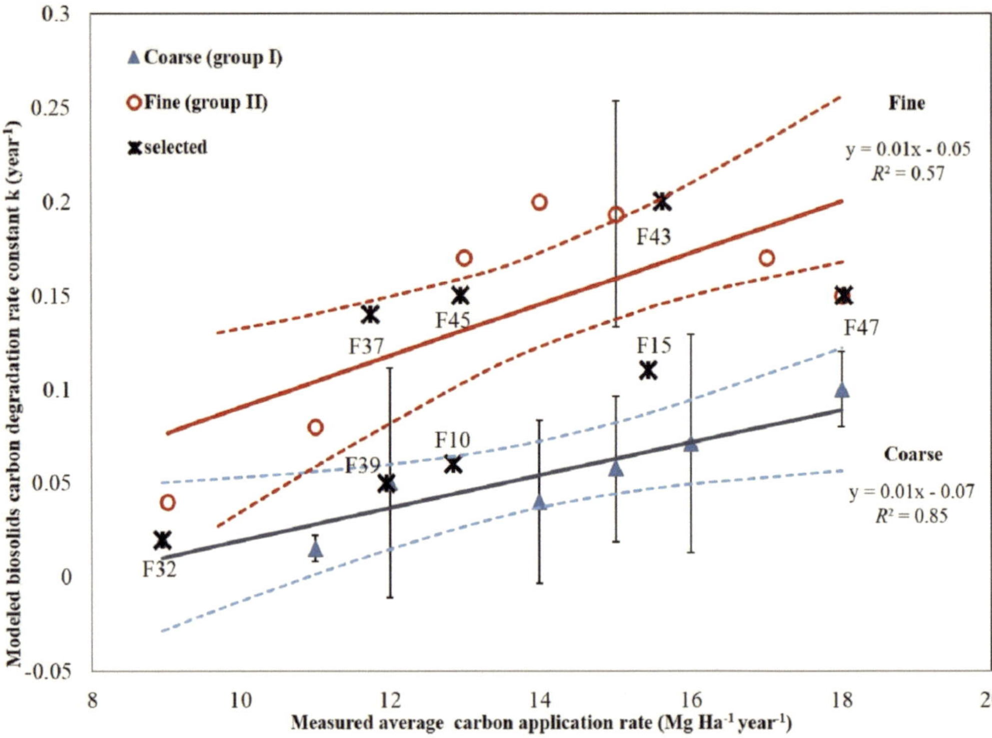

Figure 5. Biosolids degradation rate as a function of average annual carbon application rate separated into coarse and fine soil types (group I and II). Also, identified are the eight soil samples used in this study with 95% confidence interval shown for each soil group.

The DRM was used to determine rates of C sequestration for the eight soils based on Eqs. (2)–(7) (see **Figure 6**). There was a marked increase in the ratio of C sequestered to C application rate up to the year 1985 due to accumulation of the C from the conversion of biosolids to new biomass [15]. Beyond 2025, the ratio of C sequestration to C application rate shows the almost same trend as indicated in **Figure 6**. In the short term, lower biosolid degradation rates result in less microbial production and produce a smaller increase in C sequestration [15], i.e., F32. In the long term, the total amount of biosolid application determined the amount of sequestered C since all of the applied biosolids may undergo degradation. Under aerobic conditions, it may take a long time, i.e., 20–100 years, to sequester 35–40% biosolid C based on known degradation rate k (0.20 and 0.02 yr^{-1}) with 95% biosolid conversion (see **Table 2** and **Figure 6**) [15].

3.2. Relationship between measured soil surface area and DRM-simulated degradation rate

Table 2 provides a summary of the information used in this study. The table identified the 27 soils evaluated in this study (field numbers) that were selected from the 41 field samples

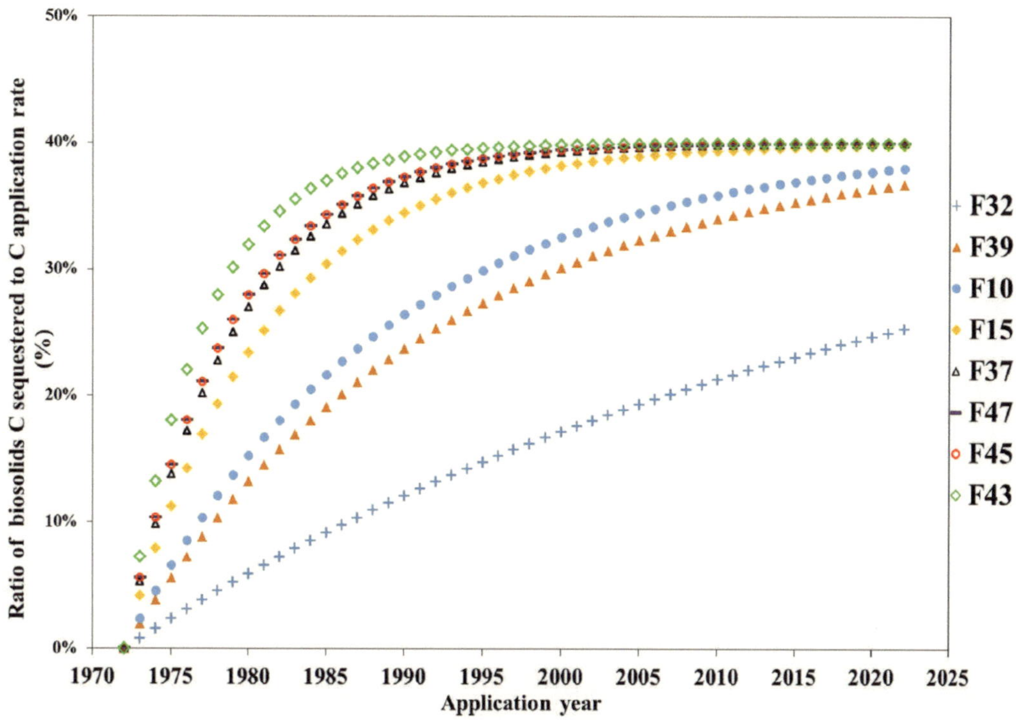

Figure 6. Percent of biosolids converted to biomass (carbon sequestration) between 1972 and 2025 for the eight fields. The variation between fields is due to difference in the degradation rate (see **Table 1**).

in the [5] database. The table also identifies the eight soils evaluated in the laboratory soil respirator experiment and three soils from Terry's laboratory study [21]. The table provides (1) the experiment conditions (temperature, moisture content, annual carbon application rate), (2) the degradation rate calculated with the DRM model for field samples and measured for the laboratory studies, and (3) the surface area and soil texture for each soil.

Figure 7 provides the relationship between the total mineral soil surface area and the DRM-simulated degradation rate for the 27 fields in **Table 2**. The figure indicates that the degradation rate increased when there was an increase in the total surface area. The degradation rate varied from 0.01 to 0.30 yr^{-1} when the soil total surface area varied from 1.1 to 10.1 m^2/g of soil. A multiple linear regression analysis was conducted between the degradation rate for the 27 soils and the variables of total mineral soil surface area and the annual biosolid application rate. The analysis performed by SPSS (version 22.0, 2015) indicates that the total mineral soil surface area is a significant indicator of degradation rate with a high coefficient of determination ($R^2 = 0.87$), but annual application rate is not statistically significant ($p = 0.34 > 0.05$). The equation for the relationship between soil surface area and degradation rate (average-field environmental conditions of 10°C and 10% moisture) is:

$$y = 0.02x + 0.01 \tag{10}$$

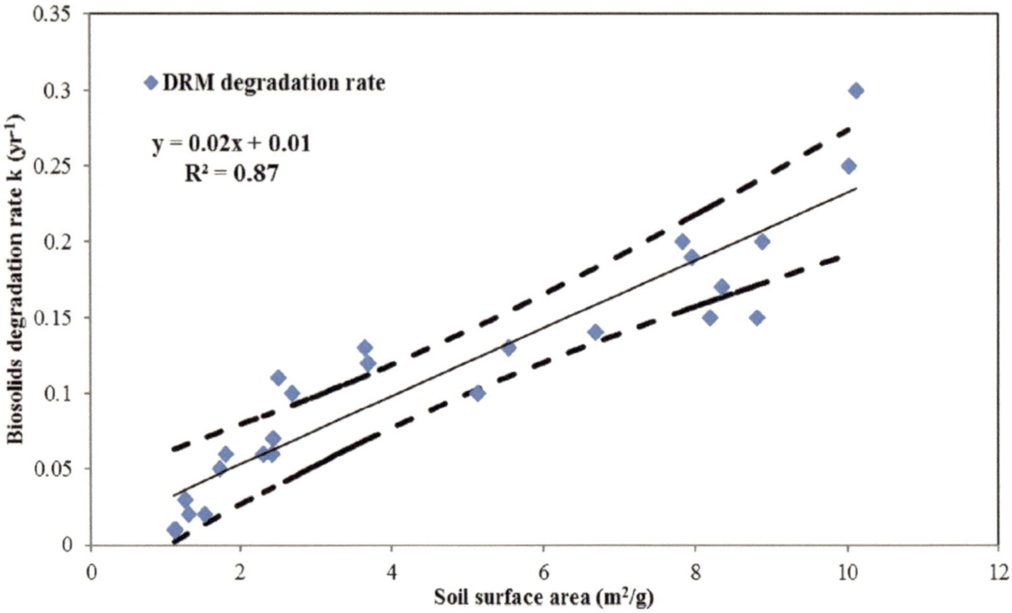

Figure 7. Biosolids degradation rate as a function of surface area for total of 27 soils from different fields with 95% confidence interval.

where:

y = the degradation rate (yr^{-1});

x = the total mineral soil surface area (m^2/g).

Soil texture and mineral surface area (**Table 2**) are closely related with higher mineral soil surface area associated with finer soil texture. Additionally, many studies have demonstrated that decomposition rates are related to soil texture [25, 26]. However, soil texture represents a range of soil surface areas, and therefore surface area provides a more definitive parameter to relate to organic degradation rates in soil.

Historically soils with a finer texture have been associated with a higher retention of applied biomass (carbon sequestration) [27]. Application of the DRM model demonstrates that higher degradation rates are associated with higher long-term sequestration rates and also with larger soil surface areas. [15]. And that soil surface area represents a more definitive parameter to relate to organic degradation rates in soil than soil texture.

3.3. Laboratory biosolid degradation rates

Figure 8 and **Table 2** provide the laboratory biosolid C degradation rates (incubated at 21°C and 20% moisture) determined from the yearlong soil respirator experiment for the eight soil samples based on first-order kinetics. The slopes of the regression lines in **Figure 8** represent the average first-order degradation rate for the biosolids in the soil (from Eq. (1) where

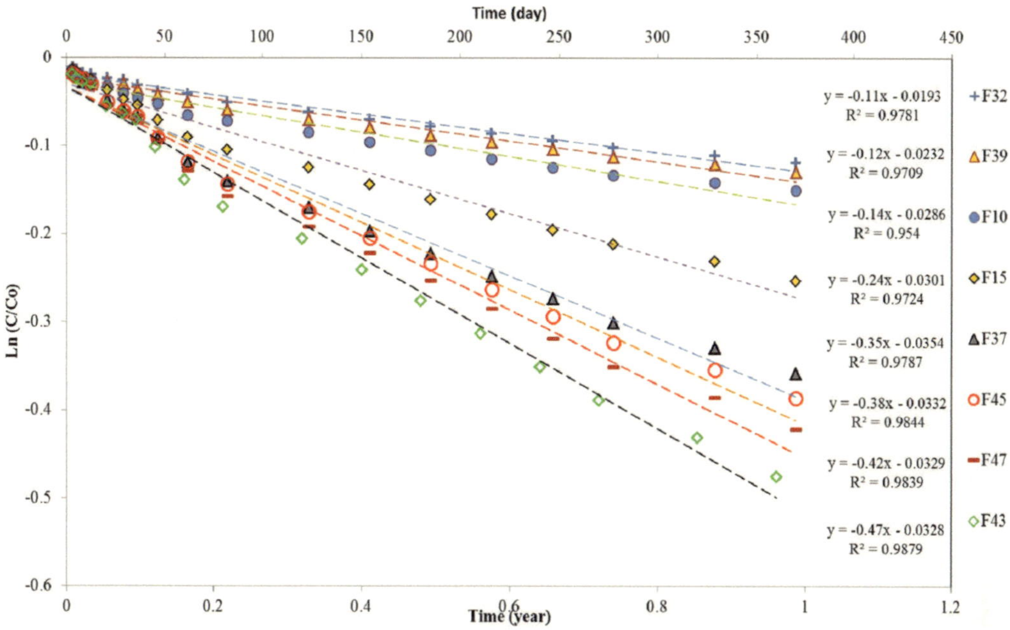

Figure 8. Variation in laboratory degradation rate (yr^{-1}, slope of trend line) for eight soil samples based on first-order kinetics ($ln(C/C_0 = -kt)$).

ln(C/Co) = −*Kt*). The difference between biosolid carbon remaining and the original biosolid organic carbon (C/Co) was determined by measuring evolved CO_2 concentration once per week during the incubation period. CO_2 evolution is an indication of biological decomposition and is used as an index of biosolid C degradation [28, 29]. At the end of 360 days of incubation, between 11 and 40% of the original biosolid organic carbon was evolved as CO_2 from the eight soil samples. The variation in the slopes for the eight soils is due to differences in the soil surface area as shown in **Table 2**. Soils with more surface area had higher degradation rates and therefore larger slopes.

Analyses of DRM simulation results for the eight fields indicate that the field degradation rates for the eight fields varied between 0.02 and 0.20 yr^{-1} and were much lower than the laboratory degradation rates that varied between 0.11 and 0.47 yr^{-1} (**Table 2**).

Figure 9 and **Table 2** provide the laboratory degradation rates determined for the biosolids added to the three soils from Terry et al. [21]. The biosolid degradation rate varied from 0.19 to 0.35 yr^{-1}. The data are for synthetic biosolids incubated from 28 to 336 days at 21°C and 20% moisture. The synthetic sludge applied to Terry's experiment was in the liquid phase with volatile solids similar to biosolids applied to 41 fields in Illinois. The synthetic biosolids had an organic carbon percent (22.3%) similar to biosolids used in this study (23.2%) [5]. Decomposition of the biosolids was initially very rapid with 54–63% of the total C in the biosolids removed during the first 28 days followed by a slow decomposition for the period from 28 to 336 days. This is because fresh biosolids were applied that had

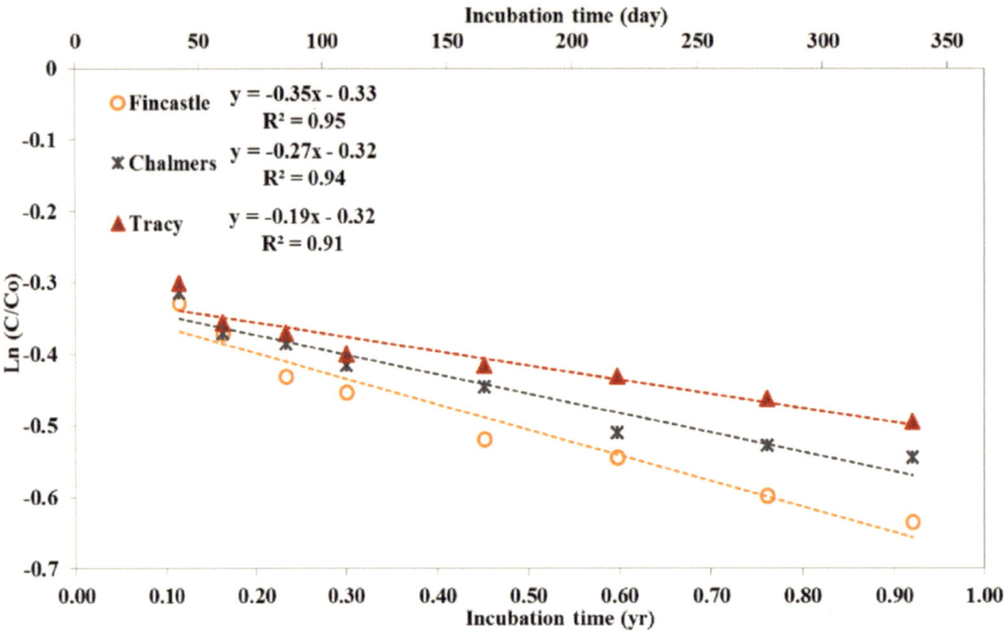

Figure 9. Variation in degradation rate (yr^{-1}, slope of trend line) for three soils based on first-order kinetics ($ln(C/C_0 = -kt)$) and Terry et.al [21] laboratory decomposition data (incubated for 28–336 days at a temperature of 21°C and moisture content of 20%).

not undergone short-term volatile losses. For the eight soil samples used in our laboratory study, it is assumed that the rapid fraction was consumed before 1985 and there was no rapid degradation phase [30].

Figure 10 compares the soil surface area and degradation rates for the field and laboratory studies. The annual average-field environmental conditions were estimated to be 10°C and 10% moisture content [5], and the laboratory conditions were 21°C and 20% moisture content. **Figure 10** indicates that degradation rates for the 27 fields varied between 0.02 and 0.30 yr^{-1} and were much lower than the laboratory degradation rates that varied between 0.11 and 0.47 yr^{-1}. The decomposition of biosolids at the field site with environmental conditions of 10°C and 10% moisture content is calculated to be only 37% of that under laboratory conditions of 20°C and 20% moisture content.

3.4. C sequestration

Figure 11 shows the relationship between increased annual percentage of applied biosolids that can be sequestered and total mineral soil surface area. The DRM was applied to estimate biomass C sequestered in the soil and gases C emitted to the atmosphere based on mass balance described in Eq. (1). The annual percentage of biosolids converted to biomass is determined then. Eqs. (4)–(6) were used to determine the C sequestration values with assumed 40% biomass yield. Eq. (9) was applied to estimate the total mineral soil surface area for different

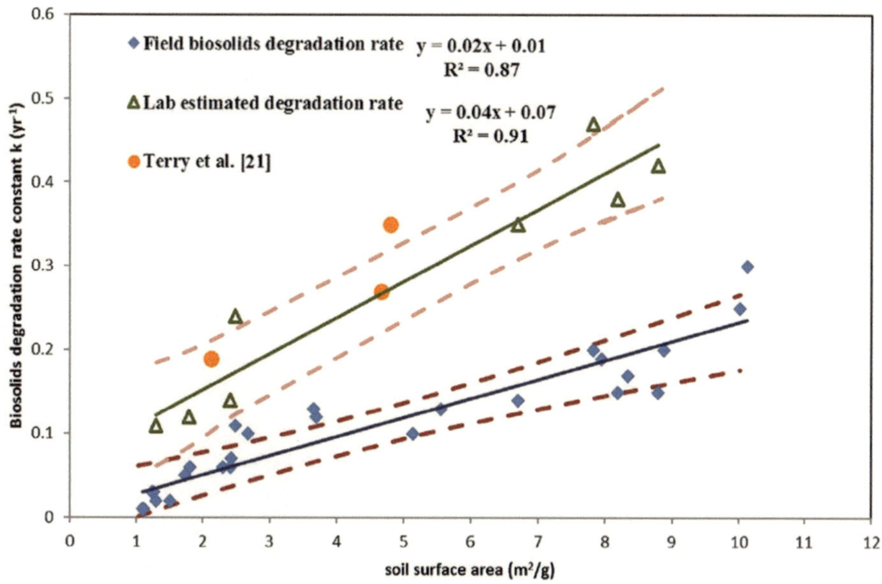

Figure 10. A comparison of field modeled (10°C and 10% moisture content) and laboratory measured biosolids (21°C and 20% moisture content) degradation rates for different fields with 95% confidence interval as a function of soil surface area.

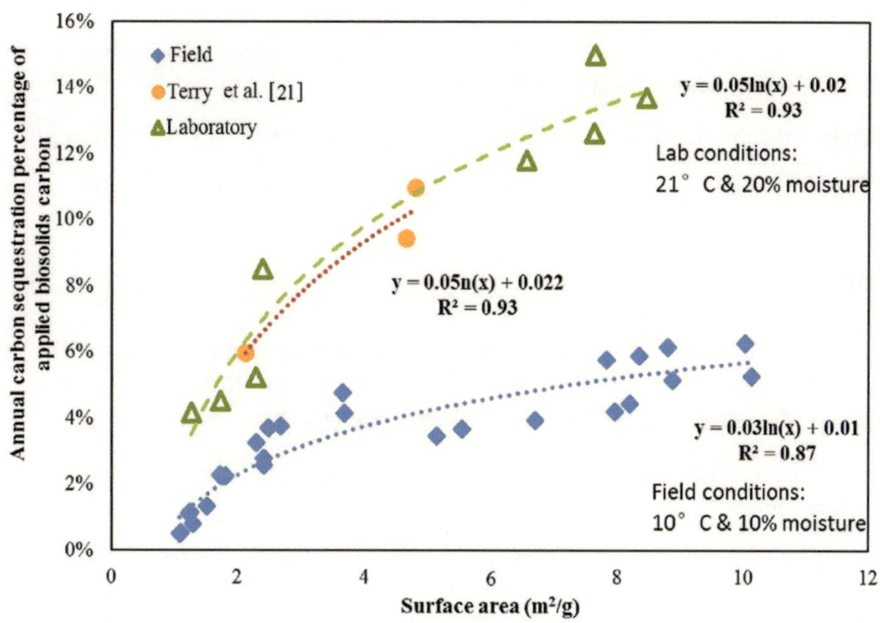

Figure 11. Annual percentages of applied biosolids converted to sequestered carbon as a function of mineral soil surface area of 27 soil samples for field (10°C and 10% moisture) and 11 laboratory soil sample (21°C and 20% moisture) conditions. Sequestered % is based on first-order kinetics and a biomass yield of 40%.

degradation rates [15]. The sequestration rates for Terry et al. [21] soils were computed by using degradation rates determined from **Figure 8** and for the eight fields from **Figure 9**. For Terry's [21] soils the increase in total surface area was 2.7 m^2/g and produced an increase in the annual sequestration rate from 6 to 11%. For the eight soil samples, the increase in total surface area was 7.2 m^2/g and produced an annual increase sequestration rate between 4 and 14%. For the soils from the 27 fields, the increase in total surface area was 9.9 m^2/g and this produced an increase in the annual sequestration rate from 1 to 6%. Applying biosolids is much more effective in enhancing C sequestration than other agriculture methods such as applying animal manure or plant materials [31, 32].

4. Conclusion

Applying biosolids to soils with fine texture contributes to the reduction of GHG more effectively than applying to coarser soils. Importantly, the present study indicated that land application of biosolids is an appropriate way to enhance C sequestration in soils and contribute to the reduction of greenhouse gas emissions and that soil total surface area as well as temperature and moisture affect the rate of biosolid degradation and the rate of C sequestration.

Author details

Dongqi Wen*, Wenjuan Zhai and Kenneth E. Noll

* Address all correspondence to: dwen1@hawk.iit.edu

Civil, Architecture and Environmental Engineering, Illinois Institute of Technology, Chicago, IL, USA

References

[1] Lal R. Potential of desertification control to sequester carbon and mitigate the greenhouse effect. Storing Carbon in Agricultural Soils: A Multi-Purpose Environmental Strategy. 2001; **51**: 35–72. DOI:10.1007/978-94-0172089-1_3

[2] Brown S, Leonard P. Building carbon credits with biosolids recycling, part II. BioCycle. 2004; September: 25–29.

[3] Post WM, Peng TH, Emanuel WR, King AW, Dale VH, DeAngelis DL. The global carbon cycle. American Scientist. 1990; **78(4)**: 310226. Retrieved March 27, 2016, from, http://www.jstor.org/stable/10.2307/29774118?ref=search-gateway:904fadd8268013f480e76dc0aa2560de

[4] Jarecki MK, Lal R. Crop management for soil carbon sequestration. Critical Reviews in Plant Sciences. 2003; **22(6)**: 471–502. DOI:10.1080/713608318

[5] Tian G, Granato TC, Cox AE, Pietz RI, Carlson CR, Abedin Z. Soil carbon sequestration resulting from long-term application of biosolids for land reclamation. Journal of Environment Quality. 2009; **38(1)**: 61–74. DOI:10.2134/jeq2007.0471

[6] Lal R. World cropland soils as a source or sink for atmospheric carbon. Advances in Agronomy 2000; **71**: 145–191. DOI:10.1016/s0065–2113(01)71014-0

[7] Conant R, Puastian K, Elliot E. Grassland management and conversion into grassland: effects on soil carbon. Applied Ecology.2001; **11**: 343255. DOI:10.3334/cdiac/tcm.005

[8] Lal R. Soil carbon sequestration to mitigate climate change. Geoderma. 2004; **123(1–2)**: 1–22. DOI:10.1016/j.geoderma.2004.01.032

[9] Paustian K, Brenner J, Killian K, Cipra J. State level analyses of C sequestration in agricultural soils. In: Kimble JM, Lal R, Follett RF, editors. Agricultural practices and policies for carbon sequestration in soil. Boca Raton, FL: CRC Press; 2001. p. 193–204.

[10] Parton WJ. Abiotic section of ELM. In: Grassland simulation model. New York: Springer-Verlag; 1978. p. 31–53.

[11] Parton WJ, Schimel DS, Cole CV, Ojima DS. Analysis of factors controlling soil organic matter levels in great plains grasslands1. Soil Science Society of America Journal. 1987; **51(5)**: 1173. DOI:10.2136/sssaj1987.03615995005100050015x

[12] Zhai W, Moschandreas DJ, Tian G, Venkatesan D, Noll KE. Degradation rate model formulation to estimate soil carbon sequestration from repeated biosolids application. Soil Science Society of America Journal. 2014; **78(1)**: 238. DOI:10.2136/sssaj2013.05.0180

[13] Metropolitan Water Reclamation District of Greater Chicago [/internet]. 2016. Available from: http://www.mwrd.org/irj/portal/anonymous/biosolids [Accessed: 2016-09-15]

[14] Natural Resources Conservation Service, Shul SE. Soil survey of Fulton County, Illinois, part 1. [Internet]. Soil survey of Fulton County, Illinois, part 1. USDA; 1997. Available from: http://www.nrcs.usda.gov/internet/fse_manuscripts/illinois/il057/0/fulton_il.pdf

[15] Wen D, Zhai W, Moschandreas D, Tian G, Noll K. Relationship between mineral soil surface area and the biological degradation of biosolids added to soil. Agriculture. 2015; **6(1)**: 1–11. DOI:10.3390/agriculture6010001

[16] Liu C, Evett JB. Soil properties: testing, measurement, and evaluation 6th ed. Upper Saddle River, NJ: Pearson/Prentice Hall; 2009.

[17] Hillel D. Fundamentals of soil physics. New York: Academic Press;1980.

[18] Soil Conservation Service. Textural soil classification [Internet]. Textural soil classification USDA; 1987. Available from: http://www.wcc.nrcs.usda.gov/ftpref/wntsc/h&h/training/soilsother/soil-usda-textural-class.pdf

[19] Stotzky G. Microbial respiration. In: Chemical and microbiological properties. Madison: American Society of Agronomy; 1965. p. 1550–1572.

[20] Walkley A, Black IA. An examination of the Degtjareff method for determining soil organic matter, and a proposed modification of the chromic acid titration method. Soil Science. 1934; **37(1)**: 2928. DOI:10.1097/00010694-193401000-00003

[21] Terry RE, Nelson DW, Sommers LE. Carbon cycling during sewage sludge decomposition in soils. Soil Science Society of America Journal. 1979; **43(3)**: 494. DOI:10.2136/sssaj1979.03615995004300030013x

[22] Hamaker JW. Mathematical prediction of cumulative levels of pesticides in soil. In: Gould R.F, editor. Advances in Chemistry Organic Pesticides in the Environment. Washington D.C.: ACS; 1967. p. 122–131. DOI:10.1021/ ba-1966-0060.ch010

[23] Jastrow JD, Amonette JE, Bailey VL. Mechanisms controlling soil carbon turnover and their potential application for enhancing carbon sequestration. Climatic Change. 2007; **80(1–2)**: 5–23. DOI:10.1007/s10584-006-91782

[24] Paustian K. Soil organic matter in temperate agroecosystems: long-term experiments in North America. In: Collins HP, Paul EA, editors. Management controls on soil carbon Boca Raton: CRC Press; 1997. p. 15–49.

[25] Ladd J, Oades J, Amato M. Microbial biomass formed from 14C, 15N-labelled plant material decomposing in soils in the field. Soil Biology and Biochemistry. 1981; **13(2)**: 119–126. DOI:10.1016/0038-0717(81)90007-9

[26] Schimel D, Coleman D, Horton K. Soil organic matter dynamics in paired rangeland and cropland toposequences in North Dakota. Geoderma. 1985; **36(3–4)**: 201–214. DOI:10.1016/0016-7061(85)900022

[27] Sorensen L. Carbon-nitrogen relationships during the humification of cellulose in soils containing different amounts of clay. Soil Biology and Biochemistry. 1981; **13(4)**: 313221. DOI:10.1016/0038-0717(81)90068-7

[28] Paul EA, Van Veen JA. The use of tracers to determine the dynamic nature of organic matter. 11th ed. Edmonton, Alberta, Canada: International Science Congress; 1978. p. 61–102.

[29] Gilmour JT, Gilmour CM. A Simulation Model for Sludge Decomposition in Soil. Journal of Environment Quality, **9(2)**: 194–201.DOI:10.2134/jeq1980.00472425000900020006x

[30] Gilmour JT, Clark MD, Daniel SM. Predicting long-term decomposition of biosolids with a seven-day test. Journal of Environment Quality. 1996; **25(4)**: 766. DOI:10.2134/jeq1996.00472425002500040016x

[31] Pinck LA, Allison FE, Sherman MS. Maintenance of soil organic matter: I. Inorganic soil colloid as a factor in retention of carbon during formation of humus. Soil Science. 1949; **68**: 463–478.

[32] Gerzabek MH, Haberhauer G, Kirchmann, H. Soil organic matter pools and carbon-13 natural abundance in particle size fractions of a long-term agricultural field experiment receiving organic amendments. Soil Science Society of America Journal 2001; **65**: 352–358.

Section 4

Carbon Storage and Utilization

Chapter 9

CO_2 Conversion to Chemicals and Fuel for Carbon Utilization

Wonjun Cho, Hyejin Yu and Yonggi Mo

Additional information is available at the end of the chapter

http://dx.doi.org/10.5772/67316

Abstract

Recent direction dealing with climate change has changed more to focus on carbon utilization rather than the direct carbon capture and storage. Conceptually converting CO_2 to sellable chemicals or fuels should be more benign to environment by substituting the fossil raw materials like oil, natural gas, or coal. Instead of converting CO_2 fully to valuable chemicals or fuels, it is much easier to employ a portion of CO_2 with existing raw materials in many natural gas conversion processes. Dimethyl ether (DME) and gas-to-liquids (GTL) are most prominent processes that can be modified to accommodate CO_2 as a reacting raw material. There are already several successful technology developments in using CO_2-rich natural gas for DME and liquid fuels, although they are not yet fully reached the commercialized level. This chapter highlights recent developments in utilizing CO_2-containing natural gas and landfill gas to yield valuable chemicals and fuels like diesel or DME.

Keywords: CO_2, chemicals, fuel, DME, GTL

1. Introduction

Carbon dioxide (CO_2) is widely used in process industries as its own form or feedstock for the production of bulk chemicals including methanol, dimethyl ether (DME), synthetic fuel, etc. Although chemical activity of CO_2 is low, CO_2 can be activated enough for chemical reactions with the presence of suitable catalysts and the appropriate operating conditions. The most of reactions involving CO_2 as reactants, except CO_2 insertion reactions, are based on reductions in which hydrogen or electrons are added. Cost-effectiveness of industrial manufacturing using CO_2 is heavily dependent on energy consumption required for the reaction, and hence it is very important to provide energy required for CO_2 conversion in a most economic and effective manner. Various options for improving economics of reactions using CO_2 are possible, for example, reducing inherent energy requirement for the reaction from the development of new catalysts, saving energy cost through strategic utilization of renewable energy sources, etc. Another important aspect in the utilization of CO_2 from the viewpoint of carbon capture sequestration and utilization (CCSU) is provision of co-reactants for CO_2 utilization because of huge quantity of CO_2 available.

Methanol is the key feedstock for C1 chemistry as it is used for producing formaldehyde, acetic acid, chloromethane, and other chemicals for chemical industries. Attention has been paid recently to methanol as a clean synthetic fuel because it can be converted to hydrogen-rich gas via the steam reforming which can be utilized in the fuel cell systems for generating electricity.

A breakthrough in the synthesis of methanol was made from the observation as methanol can be favorably synthesized at the presence of CO_2. The main reactions during hydrogenation of CO_2 are as follows:

$$CO_2 + 3\,H_2 \rightarrow CH_3OH + H_2O \tag{1}$$

$$CO_2 + H_2 \rightarrow CO + H_2O \tag{2}$$

The first reaction is a direct synthesis of methanol from CO_2, while the second reaction is reverse water gas shift one. These two reversible reactions are both exothermic. It should be noted that the cost of such CO_2 hydrogenation reaction is higher than that of reaction based on CO and CO_2 mixture. However, the CO_2 hydrogenation for methanol synthesis provides opportunity for utilizing CO_2 gases, which are often wasted in process industries.

Industrial catalysts for methanol synthesis are available for gases containing H_2 and CO, which normally come with small quantity of CO_2 presented. Therefore, it is required to develop a new catalyst for methanol synthesis that can deal with CO_2-rich feed.

A colorless, nontoxic, and environmentally benign DME is widely used as a solvent and propellant in aerosol products, of which physical properties are close to those of liquefied petroleum gas (LPG). As considerably fewer pollutants are generated from the combustion of DME compared to that of conventional diesel fuel, DME is regarded as a sustainable substitute for diesel fuel. DME has a structure of CH_3-O-CH_3 including oxygen between two CH_3s and thus can work as a clean fuel. DME can be produced from a wide range of feedstocks, including natural gas, coal, biomass, and waste plastics. Another benefit for DME is that DME can be the alternative to conventional diesel fuels and LPG because high cetane number can be achieved from DME and physical properties of DME is close to that of LPG. Furthermore, existing infrastructure for transportation and storage of conventional fuels can be readily adopted for DME once it is introduced to the market.

Ammonia (NH_3) and carbon dioxide (CO_2) produced from ammonia plant react as below and the urea and water are produced.

$$2\,NH_3 + CO_2 \rightarrow H_2NCONH_2 + H_2O \tag{3}$$

The intermediate products such as carbamate, non-reacted ammonia and CO_2 are separated and recycled as raw materials. The water generated as byproduct is removed during the process of condensing urea.

http://dx.doi.org/10.5772/67316

2. DME manufacturing using CO_2

DME is an ultra-clean burning alternative to LPG and diesel. It is easy to liquefy, making it a convenient fuel to transport and store. These properties make DME a versatile and promising solution in the worldwide consideration of clean and low-carbon fuels. And DME can be produced from various feedstocks like natural gas, Coal Bed Methane (CBM), shale gas, biomass, coal, and CO_2.

In 2000, Korea Gas Co. (KOGAS) embarked the development of proprietary catalyst and process development with the ultimate goal of producing DME on a commercial scale. Central to the KOGAS DME technology is the one-step DME synthesis from synthesis gas compared to the conventional two-step process including methanol synthesis. Conceptually, the one-step technology offers a possibility to produce DME with a lower capital and production cost.

The KOGAS technology has progressed considerably toward commercialization. The technology has been undergoing extensive testing since 2008 in the 3000 metric tons/year demonstration plant at the KOGAS R&D Center in Incheon, Korea [1]. In 2011, KOGAS completed the Basic Engineering Package (BEP) of 300,000 metric tons/year. In parallel, the Korean Government had initiated market test studies in distributing DME to local end-users.

2.1. DME market (case study in Korea)

The demand and supply dynamics for DME in Korea are analyzed based on potential usage of DME in Korea. There are three market sectors in Korea where DME can be potentially used [2]:

- LPG replacement for domestic and transportation usage by diluting LPG with DME.
- Bunker-C oil replacement for district heating with industrial boilers.
- Diesel fuel replacement for diesel-fueled vehicles.

DME can be blended up to 20–30% with LPG [3]. The LPG replacement for domestic usage is estimated with the assumption that DME will replace 20% of LPG usage. Based on this analysis, it is expected that DME demand will decrease from 254 kilotons in 2013 to 163 kilotons in 2021. This is because LPG demand for domestic usage has been decreasing due to the introduction of more economical LNG for domestic use. It is projected that this trend will continue.

On the other hand, the demand of DME for transportation for LPG fueled vehicles is expected to increase from 265 kilotons in 2013 to 360 kilotons in 2021, when LPG is replaced with a 5-mol% DME/LPG mixture. When LPG-DME blended fuel is used, additional distribution and end-use infrastructures are not required. Thus, it is possible to establish a market for DME in a very short time. KOGAS also successfully carried out field tests for DME-LPG blends.

The potential demand for DME in replacing diesel for diesel vehicles is expected to increase from 105 to 1046 kilotons in 2021. This is because the diesel engine–based vehicles have been steadily increasing, and DME can provide advantages over diesel in terms of air pollution. Particulate matter and NO*x* have been two of the major problems of diesel engines despite their higher energy efficiency.

DME may also command a price premium with respect to diesel due to its cleaner burning properties. The emission requirements for diesel vehicles are getting tighter in many countries, and the use of DME will help vehicle manufactures and end-users comply with the tighter regulations.

Diesel substitute and fuel for power generation could be a big market for DME in the future. Natural gas is a good fuel for power generation, but DME is comparable to natural gas in performance as a fuel for power generation. This has been approved by gas turbine manufacturers, and DME can be an efficient alternative fuel for medium-sized power plants, especially for remote or isolated locations where it is difficult to transport natural gas.

2.2. DME production technology

The KOGAS process represents the newest generation of DME production technology. At the most conceptual level, its distinguishing feature is that DME is synthesized directly from synthesis gas and hence called a "direct" or "one-step" process. By contrast, the conventional process is called the "indirect" or "two-step" process because DME is produced from an intermediate product, methanol. The Toyo process was used as a representative indirect process to compare to KOGAS DME, as Toyo's process is the most established conventional two-step DME technology.

Key technical comparisons between the KOGAS DME and Toyo processes are summarized in **Table 1**. KOGAS DME represents the first commercial-scale (demonstration) plant for KOGAS' process, whereas Toyo's process has several commercial-scale plants in operation. Based on demonstration plant data, KOGAS DME process was shown to be competitive to Toyo in terms of catalyst longevity, operations reliability, and similar number of equipment, which is an indicator of fixed capital costs. A fundamental advantage of the KOGAS DME process is that it needs just one reactor section to convert syngas to DME, whereas Toyo requires two sections. Another advantage is KOGAS' proprietary catalyst, which can utilize high CO_2 content in the reformer feed, allowing it to handle a more diverse and economic source of feed gas. The estimation of energy efficiency for KOGAS process is comparable to that of the Toyo.

	KOGAS DME process	Toyo process
Process development stage	10 metric tons/day DME production plant	30 metric tons/day DME production plant
	Demonstration plant	340 metric tons/day commercial scale
Catalyst life expectancy	Tri-reformer: 1 year	ISOP reforming: 3 years
	DME reactor: 1 year	Methanol conversion: N/A
		Methanol dehydration: N/A
Number of reaction steps	2	3
Tolerance for high CO_2 in NG feed	Can use natural gas with as much as 30-mol% CO_2	Not known
Total number of major equipment	80 plus ASU	90 plus ASU
Process energy efficiency	60%	55%

Table 1. Comparison of KOGAS DME process and Toyo process.

http://dx.doi.org/10.5772/67316

2.3. Description of KOGAS DME process

A schematic process flow diagram (PFD) of KOGAS' commercial scale DME plant is shown in **Figure 1**.

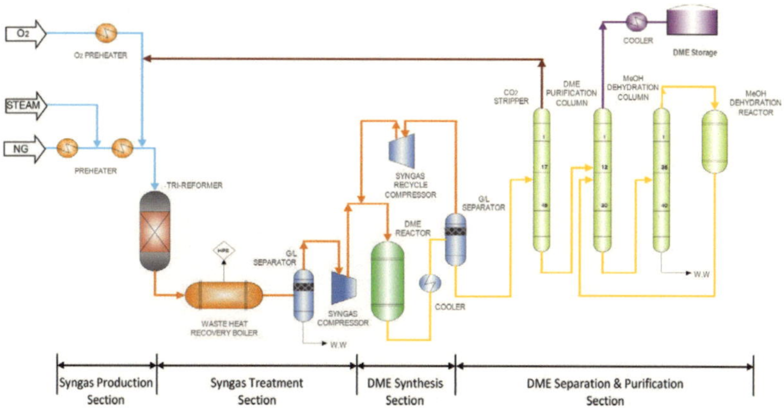

Figure 1. PFD of KOGAS DME plant.

The four major sections of the KOGAS DME process and their functions are as follows:

(1) Reforming section

Synthesis gas, a mixture of H_2 and CO, is produced from natural gas, steam, O_2, and CO_2 using tri-reformer, an adiabatic auto-thermal reformer based on KOGAS' proprietary catalyst, KDN-1. This KOGAS proprietary catalyst involves pre-coating Ce-ZrO_2 onto a commercially available Al_2O_3 substrate before impregnating with Ni.

The tri-reformer consists of a homogeneous section and a fixed-bed catalyst section, in which the pre-reformed natural gas (mainly methane) is reacted with steam, oxygen and carbon dioxide for producing synthesis gas. Maintaining right amounts and ratio of carbon monoxide and hydrogen for the reaction is important. Auto-thermal nature in the reaction compensates heating requirement for reforming reactions with exothermic combustion reactions. The resulting temperature for the exit stream from the tri-reformer is around 1080°C, and the pressure is 3.1 MPa.

The global reactions taking place in the tri-reformer can be summarized as

$$CH_4 + O_2 + CO_2 \rightarrow 3\,H_2 + 3CO + H_2O + \text{Heat} \tag{4}$$

$$2\,CH_4 + \tfrac{1}{2}O_2 + H_2O \rightarrow 5\,H_2 + 2CO \tag{5}$$

The composition of the product syngas (in particular the H_2:CO ratio) is a function of the three key molar feed ratios: steam, oxygen, and CO_2.

Figure 2 shows the tri-reformer reforming reactor and KDN-1 catalyst for reforming process. Compared to other traditional reforming catalysts, the KDN-1 catalyst provides better conversion of CH_4 and CO_2 both initially and over time. The KDN-1 enables the production of syngas with the desired H_2 to CO ratio for optimum performance in the DME synthesis for a wide range of compositions in the natural gas feedstock, including high CO_2 content.

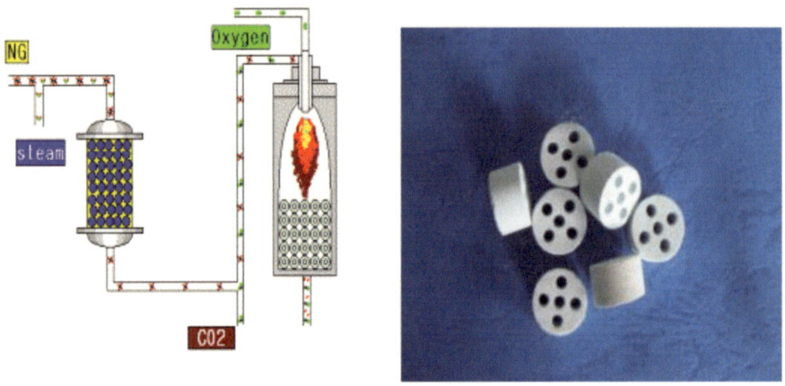

Figure 2. Tri-reformer reforming reactor and KDN-1 catalyst for tri-reformer reactor.

(2) Syngas treatment section

The raw syngas produced by KOGAS' tri-reformer has a carbon dioxide content of around 15 mol%. This CO_2 content must be reduced to around 1.3 mol% to meet the 4 mol% requirements of KOGAS' DME reactor feed. KOGAS BEP design uses a UOP SELEXOL absorption column to remove CO_2 from the raw syngas down to the desired level.

(3) DME synthesis section

The H_2 and CO in the syngas are catalytically reacted to produce DME, and a small amount of methanol and water in a single-step DME reactor goes through the synthesis according to the following set of global reactions:

$$CO_2 + 3\,H_2 \rightarrow CH_3OH + H_2O \tag{6}$$

$$CO + H_2O \rightarrow CO_2 + H_2 \tag{7}$$

$$2\,CH_3OH \rightarrow CH_3OCH_3 + H_2O \tag{8}$$

Figure 3 shows the DME synthesis reactor and KD-540-27 catalyst for DME synthesis. This proprietary reactor consists of a multiple tubular reactor configuration filled with KD-540-27, a hybrid bifunctional catalyst consisting of Cu/ZnO, which is catalyzing methanol synthesis and γ-Al_2O_3 catalyzing methanol dehydrogenation to DME.

In overall, heat removal is made via a cooling jacket for the vertical tubes containing the catalyst with which the temperature of reacting gas is maintained at 260°C.

(4) DME separation & purification section

The stream leaving the DME reactors will contain unreacted syngas, which needs to be separated from the condensable DME, methanol, and water. There is also a significant amount of

http://dx.doi.org/10.5772/67316

CO_2 present. The unreacted syngas needs to be recompressed and recycled to the DME reactor feed, and most of the CO_2 needs to be removed.

Figure 3. DME synthesis reactor and KD-540-27 catalyst for DME synthesis.

Figure 4 shows the KOGAS's 3000-metric tons/year demonstration plant. The KOGAS DME commercial design calls for the exit stream from the DME reactor to be cooled in stages: first with heat recovery against boiler feed water, process condensate, cooling water, chilled condensate, and finally refrigeration to −68°C. The condensing methanol is expected to be sufficient to remove most of the CO_2 in a single step. Therefore, no CO_2 absorption column is required. The DME/methanol/water/CO_2 condensate is sent to a CO_2 stripping column with three sections of packing equivalent to 30 theoretical stages that operates at 4.68 MPa. The stripped CO_2 from the DME exit stream is combined with the CO_2 removed from the raw syngas and sent back to the tri-reformer inlet.

Figure 4. KOGAS' 3000 metric tons/year demonstration plant.

3. Fertilizer (ammonia/urea) manufacturing using CO_2

Manufacturing fertilizers (ammonia/urea) is one of the ways of increasing added value of the natural gas and CO_2. The fertilizers in modern society are inevitable for enhancing agricultural productivity. The main elements used for plant growth are nitrogen, phosphorus, potash, and sulfur. Among them, the nitrogen is the most consumed fertilizer and is easily produced by using nitrogenous in air and natural gas. Phosphorus and potash are provided from minerals, and sulfur is produced as byproducts in refineries and gas plant. Ammonia containing nitrogen is a toxic gas at room temperature and cannot be used directly as fertilizer. Therefore, ammonia should be converted into other chemicals such as urea, diammonium phosphate (DAP) and ammonium nitrate, which are suitable for plant applications. Urea is produced by combining ammonia and carbon dioxide generated from production of ammonia. It is a solid, and it can be granule or prill form. The 80% of ammonia used worldwide as fertilizer is converted into urea. The present chapter deals with manufacturing and selling the urea in granule form by producing ammonia with natural gas.

3.1. Production process and characteristics of facilities

The production process consists of following facilities, and its overall block diagram is represented in **Figure 5**.

- Ammonia plant
- Urea plant
- Granulation unit
- Warehouse
- Utilities & offsite

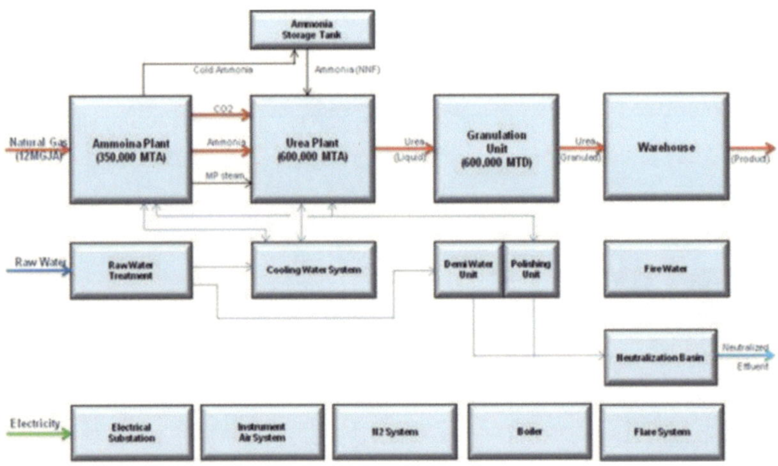

Figure 5. Block diagram of ammonia/urea plant.

http://dx.doi.org/10.5772/67316

3.2. Ammonia plant

Figure 5 shows the block diagram of ammonia/urea plant. This plant manufactures ammonia by use of natural gas and air. The ammonia and carbon dioxide (CO_2) generated as byproduct are used as raw materials for manufacturing urea. For enhancing operational flexibility and the possibility of direct sale of ammonia to customers, the ammonia storage tank is required. Block diagram of ammonia plant is shown in **Figure 6**.

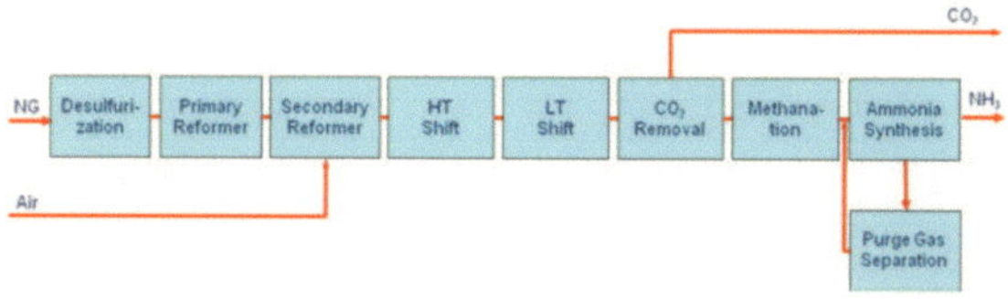

Figure 6. Block diagram of ammonia plant.

The existence of sulfur substances in gas reduces the activity of catalyst. For this prevention, the sulfur substances shall be removed from the gas. And then the produced gas is supplied to reformer. The synthetic gas is generated in passing through various reaction steps of the reformer. In reaction steps, the air is supplied for providing nitrogen needed in producing ammonia. After these reactions, the main substances of hydrogen (H_2), carbon dioxide (CO_2), and nitrogen (N_2) are produced.

The produced CO_2 is separated by using amine in CO_2 removal unit and then sent to urea plant. The CO_2 and CO that are not eliminated in the CO_2 removal unit are removed in methanation unit. N_2 and H_2 are reacted in ammonia reactor, and then the ammonia (NH_3) is finally produced.

3.3. Urea plant

Ammonia (NH_3) and carbon dioxide (CO_2) produced from ammonia plant reacts as below, and the urea and water are produced.

$$2\,NH_3 + CO_2 = H_2NCONH_2 + H_2O \tag{9}$$

The intermediate products such as carbamate, nonreacted ammonia, and CO_2 are separated and recycled as raw materials. The water generated as byproduct is removed during the step of condensing urea. **Figure 7** shows the block diagram of urea plant.

The produced urea liquid in high temperature is solidified by cooling. The urea is solid state in atmospheric conditions, and it can be a form of prill or granule. The size of prill is smaller than granule. The investment cost for granulation processing is higher than that of prilling, but the granulated urea is more suitable as fertilizer. The granulated urea is packed in various sizes of bags for storage in warehouse.

For the operation of above plants, the following utility & offsite facilities are required:

- Water treatment including demineralization unit

- Cooling water system
- Boiler system
- Wastewater treatment system
- IA/PA
- Nitrogen
- Flare system
- Warehouse for Urea
- Ammonia storage tank

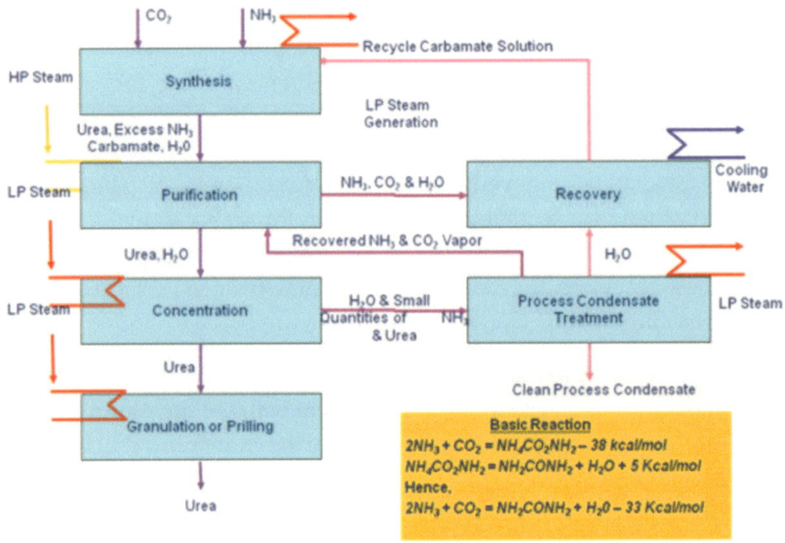

Figure 7. Block diagram of urea plant.

4. Gas-to-liquid using CO_2

4.1. What is GTL?

Gas-to-liquid (GTL) is the processing of converting natural gas to synthetic oil. This synthetic oil will be the fuel or the product based on hydrocarbon. Liquefied Natural Gas (LNG: for city gas and bunkering fuel), Pipeline Natural Gas (PNG: for city gas), and Compressed Natural Gas (CNG: for vehicles) are classified by their respective transportation technology, but GTL in liquid state at room temperature is the long-chain hydrocarbon products identified by transformation technology of chemical conversion. **Figure 8** shows the transformation technology of GTL production.

Coal, natural gas, and biomass are used as raw materials in Fischer-Tropsch (FT) process while the meaning of GTL is based on conversion of natural gas to pure synthetic oil in removing impurities such as sulfur, aromatic compounds, and metal substances. By refining this synthetic oil, it can produce diesel, naphtha, wax, and other liquid compounds based oil or other special products. This transformation technology is based on FT process, which was developed *ca.* 100 years ago. Technology of pre-treatment of gas, reforming, and upgrading process is in the mature stage, but FT process has been in the stage of commercialization. New technology is continuously developed, and the already developed technology is applied to conversion process for enhancing efficiency. Also mini GTL technology to be applied to small-scaled gas fields is being developed. The factors influencing on competitiveness are enumerated as investment cost, operation cost, materials price, plant dimension, and technology enhancing usability of products.

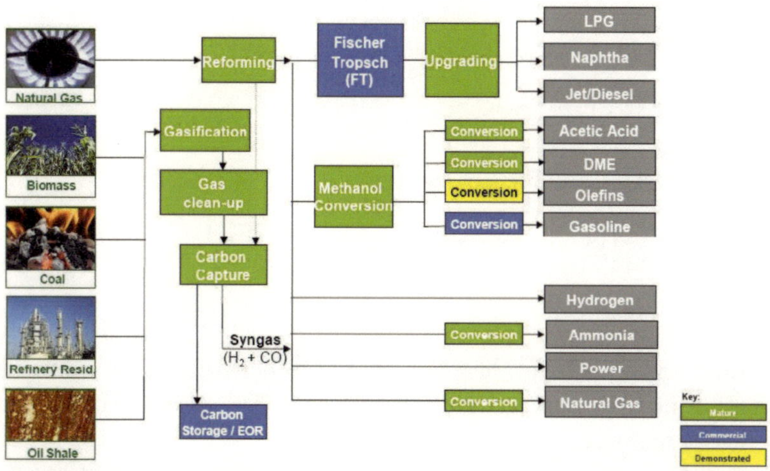

Figure 8. Transformation technology of GTL.

In comparing with history of coal-to-liquid (CTL) process, GTL is relatively a new technology and globally commercialized facilities are actually very rare. **Table 2** shows the global GTL project status (see **Figure 9**).

Even when a plant in Nigeria is completed, total production capacity of GTL is only 260,000 barrels per day. On the contrary, daily consumed oil in the world is 87 million barrels. Thus, GTL production is not subject to restriction of consumption

Project	Country	Scale (bbl/day)	Start-up
Shell Bintulu GTL	Malaysia	14,700	1993
PetroSA Mossgas GTL	South Africa	36,000	1993
Sasol/QP ORYX GTL	Qatar	34,000	2007
Shell Pearl GTL	Qatar	140,000	2011
Chevron Escravos GTL	Nigeria	34,000	2013
Total		258,700	

Table 2. Worldwide GTL project status.

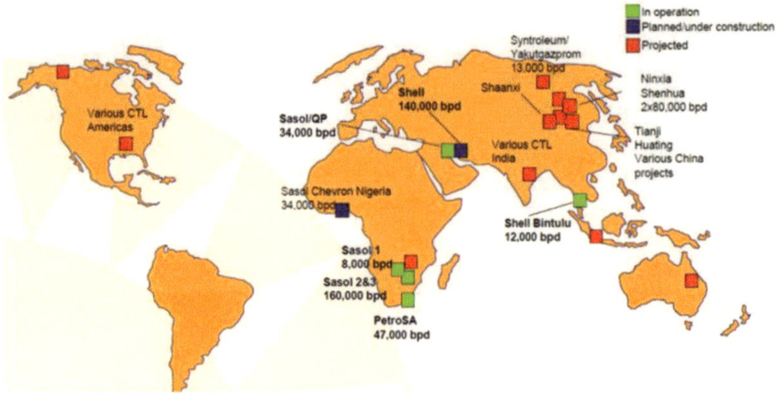

Figure 9. GTL and CTL status and project plans.

4.2. Production processing and facility features

GTL Process is composed of next four main steps.

4.2.1. Step 1: gas pre-treatment (gas clean-up)

The gas pre-treatment in GTL process is generally to dehydrate and remove sulfur compounds, mercury, and hydrocarbon C3+. It is similar to the requirements of LNG process, but it needs no stage of removal of CO_2.

4.2.2. Step 2: synthesis gas generation

The production of synthetic gas covers the conversion into CO and H_2 mixture by autothermal reforming (ATR) and steam carbon dioxide reforming (SCR). The oxygen needed in ATR is supplied by air or pure oxygen. When using the air, there are some advantages by eliminating the cost of air separation unit (ASU) and electricity cost but finally become obstacles to process due to large volume of nitrogen. By using reformed steam methane, we supply necessary oxygen. However, this method has disadvantage of producing synthetic gas of H_2 to CO ratio by 4:1 against the required H_2 to CO ratio by 2:1.

SCR reformer has a homogeneous section and a fixed-bed catalyst section and reacts the pre-reformed natural gas (primarily methane), steam, and carbon dioxide to produce synthesis gas containing the correct amount and ratio of carbon monoxide and hydrogen. The SCR reformer uses exothermic combustion reactions to off-set the endothermic reforming reactions. The resulting exit temperature is around 870°C, and the pressure is 2.5 MPa. The composition of the product syngas (in particular the H_2:CO ratio) is a function of the three key molar feed ratios previously described (steam, oxygen and CO_2). The ratio of hydrogen to carbon monoxide in the SCR outlet is currently targeted to be 2.0 to provide the correct H_2:CO ratio in the mixed feed into the FT synthesis reactor. The advantage of SCR process is to enable the development of low-quality natural gas fields containing CO_2 or reuses efficiently CO_2 emitted from various processes (CO_2 removal and FT process, etc.). **Figure 10** shows the simplistic integrated GT schematic process.

http://dx.doi.org/10.5772/67316

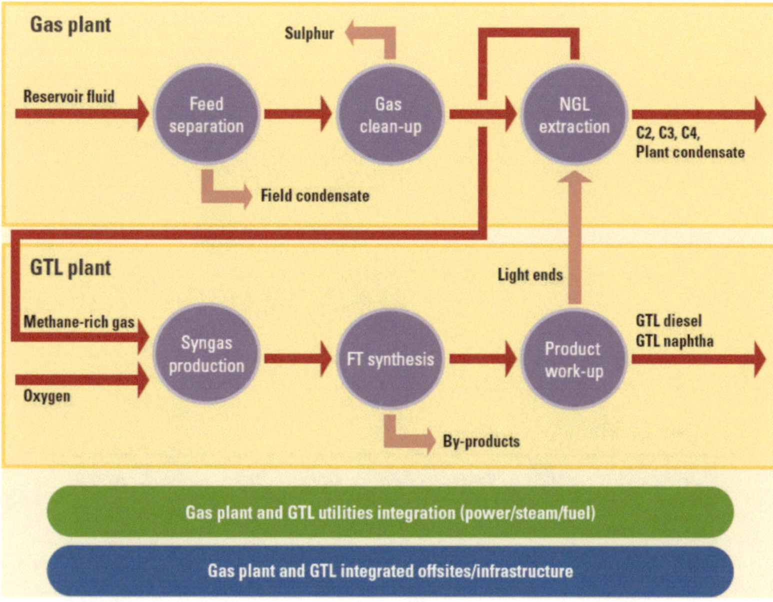

Figure 10. Simplistic integrated GTL schematic.

4.2.3. Step 3: FT synthesis and refining

FT synthesis is to convert synthetic gas into long-chain hydrocarbons. This conversion is made by catalyst. As for catalyst, cobalt is generally applied. For extracting substances for shipping as final products, fractional distillation is required.

4.3. Mini GTL

For successful mini GTL technology, it is required to develop compact and high-efficient GTL process and modularization techniques for being competitive even as small scaled. It will be efficient technology to be applied to small and medium gas field, associated gas of oilfield, and landfill gas on land and at sea. **Figure 11** shows the example of roadmap toward modular GTL plant.

GTL technology for developing small and medium gas field and associated gas requires following conditions: (1) minimization of plant construction cost for economic feasibility as small-scaled hundreds to thousands barrels/day, (2) compact and mobility for installation in places without infra such as frozen zone of Siberia, (3) easy installation in limited space for application to offshore, and (4) compactification and modularization of compressor and related equipments for simple process and high efficiency.

Further study is to be progressed for small and medium gas fields that are expected to have potential application by mini GTL technology. From long-term point of view, the development of gas-to-liquids-floating production storage and offloading (GTL-FPSO) linked to shipbuilding technology will be applied to small and medium-scaled offshore gas fields as well as the strategy of launching high value-added shipbuilding market is to be established.

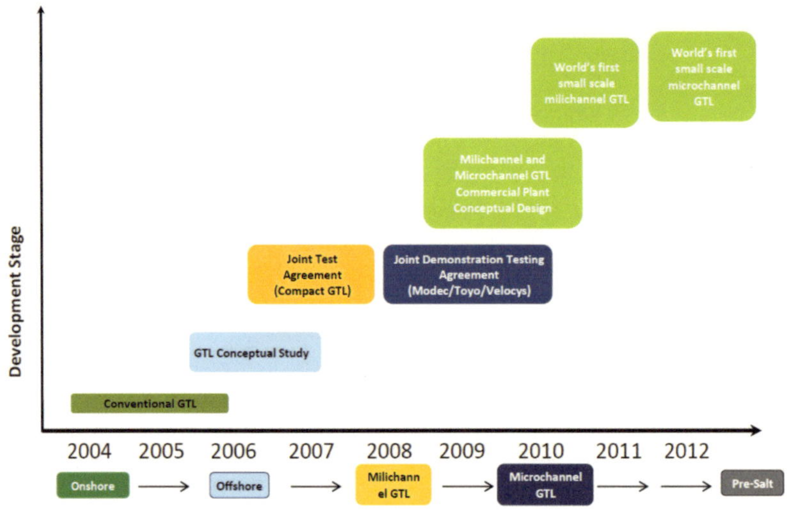

Figure 11. Example of roadmap toward modular GTL plant (source: S.A. Petrobras).

The micro reactor (synthetic gas + FT synthesis) was developed by CompactGTL Ltd., a manufacturer leading compact GTL technology, and its pilot operation was completed by applying to 20-barrels/day plant with Petrobras in 2011 [4–6]. **Figure 12** shows the example of compact GTL roadmap toward modular GTL plant.

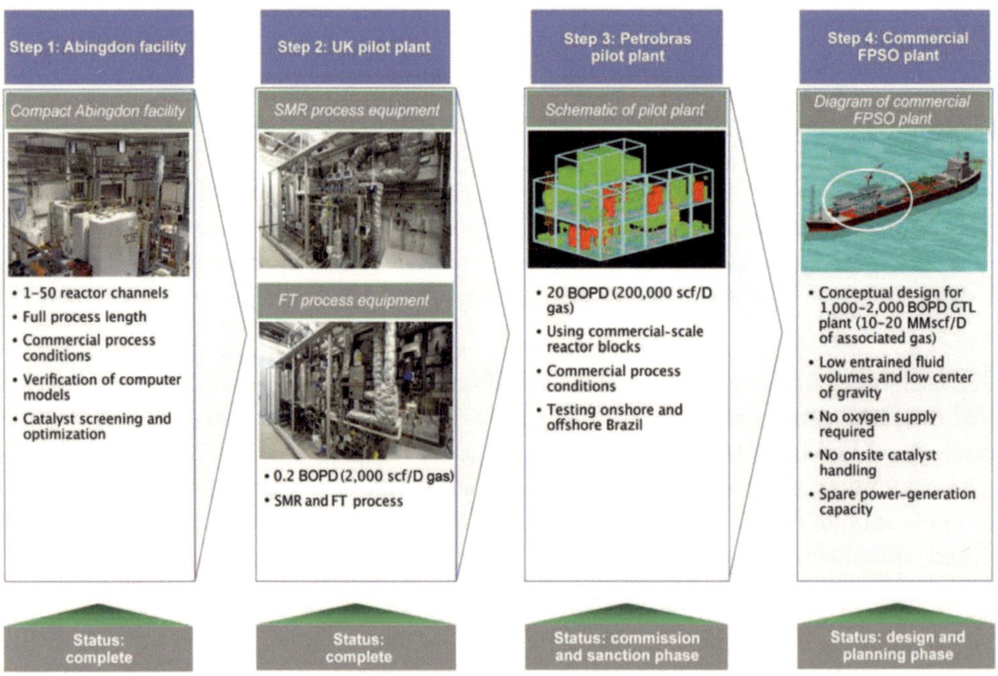

Figure 12. Example of compact GTL roadmap toward modular GTL plant (source: Compact GTL Limited).

http://dx.doi.org/10.5772/67316

Also by adopting micro channel technology, Velocys is developing mini GTL plants and reported that a pilot plant of 2.5 gallon per day has been developed. They constructed 6 BPD plant in Brazil under cooperation with Petrobras, MODEC and Toyo Engineering and plan the pilot operation in 2012.

Since micro reaction technology had advantages in its small volume, high heat transmission, and large reactive surface and control of exact reaction time, it will enhance high integration of chemical process, response selectivity, and stability.

5. Summary

The conversion of CO_2 to chemicals and energy products that is currently produced from fossil fuels is also promising due to the high potential market and promising benefits. Methanol is the key feedstock for C1 chemistry, as it is used for producing formaldehyde, acetic acid, chloromethane, and other chemicals for chemical industries. Also, industrial catalysts for methanol synthesis are available for gas containing H_2 and CO, which is applied with small quantity of CO_2 presented.

The utilization of CO_2 to produce chemicals like urea and cyclic carbonates is promising and can be a solution to reduce CO_2 emission. However, CO_2 still has certain disadvantages as a chemical reactant due to its inert, non-reactive, and low Gibbs free energy properties. DME is versatile and promising solution in the worldwide consideration of clean and low-carbon fuels. It has potential to solve forward problem of certain disadvantages as chemical reactant. Similar application of reaction schemes using CO_2 as applied to DME manufacturing is also possible in GTL fields, especially for low-value gas fields involving high CO_2 contents and for landfill gas fields.

Acknowledgements

This study was supported by "Ministry of Trade, Industry and Energy" and "Korea Evaluation Institute of Industrial Technology".

Author details

Wonjun Cho[1,*], Hyejin Yu[1] and Yonggi Mo[2]

*Address all correspondence to: williamcho86@gmail.com

1 R&D Division, Unisys International (& BF International), Daejeon City, Republic of Korea

2 Department of Chemical Engineering, Hanyang University, Seoul, Republic of Korea

References

[1] Wonjun Cho. Current Status of DME Technology Development in Korea. 2013. Available from: https://www.youtube.com/watch?v=mktzKbji58A.

[2] Wonjun Cho. KOGAS DME activities for commercialization. Asian DME Conference; 16 November 2011. Available from: http://aboutdme.org/aboutdme/files/cclibraryfiles/filename/000000001976/7asiandme_kogas_cho.pdf.

[3] Jumin Youn. The status of DME fuel in Korea. The 6th Korea-China-Japan Petroleum Technology Congress; 4–6 September 2013.

[4] Fabio Passarelli. Offshore GTL: modular solution for associated gas with variable CO_2 content. 25th World Gas Conference; 7 June 2012.

[5] Ana Paula Fonseca. Compact and other advanced GTL technologies. IGU Committee Meeting (WOC1 and PGC A); 18–21 February 2013.

[6] Available from: http://www.compactgtl.com/about/resources.

Chapter 10

Challenges Associated with CO_2 Sequestration and Hydrocarbon Recovery

Rouzbeh Ghanbarnezhad Moghanloo, Xu Yan, Gregory Law, Soheil Roshani, Garrett Babb and Wesley Herron

Additional information is available at the end of the chapter

http://dx.doi.org/10.5772/67226

Abstract

In the near- and midterm future, carbon capture and storage (CCS), also called CO_2 geo-sequestration, is likely to play a significant role in the reduction of atmospheric greenhouse gas. By expanding the set of possible sequestration targets, it is expected that CCS will enable larger quantities of CO_2 to be sequestered, mitigating human activity-driven climate change. In general, oil and gas reservoirs are ideal geologic storage sites for CO_2 because they have successfully held hydrocarbon molecules for millions of years. In addition to the significant and reliable storage capacity of hydrocarbon reservoirs, there is a considerable body of knowledge related to the behavior of hydrocarbon bearing reservoirs, and significant amounts of data are often acquired during their exploitation, factors which improve the economics and safety of any CCS project. By making use of existing and future oil and gas projects, CCS can become a major contributor in the fight against global warming, as well as a sizeable contributor to energy production worldwide. The CCS sequestration targets discussed in this study are sandstones, coal beds, shales, and carbonates. The potential and challenges associated with each of them are discussed in detail, and suggested topics for future research work are provided.

Keywords: CO_2 EOR, CO_2 storage, sandstone, carbonate, shale, coalbed methane

1. Introduction

Global levels of CO_2 in the atmosphere have been steadily rising with the increase of hydrocarbon production and usage. It is estimated that CO_2 emissions in the United States were approximately 5.5 billion tons in 2015, the largest volume yet. Anthropogenic greenhouse gases, such as carbon dioxide (CO_2), are considered a major contributor to global warming [1]. Sequestration of power plant-generated CO_2 through injection into petroleum and gas reservoirs through a process called carbon capture and storage (CCS) or "carbon sequestration" has been proposed as a method for reducing greenhouse gas emissions. Research on the use of CO_2 for enhanced oil recovery (EOR) continues with growing interest; however, research concerning terrestrial sequestration of CO_2 for environmental purposes, such as CCS, is relatively recent. As a result, fundamental topics of interest in sequestration research are concerned with scientific and technical aspects, as well as practical concerns such as the economic feasibility, safety, and the maximum possible amount of CO_2 storage [1]. Therefore, fighting CO_2 emissions with EOR and CCS is a priority, leading to innovations within the petroleum industry.

The process of CCS involves pumping sizeable quantities of atmospheric CO_2 underground, where, under the right circumstances, it can remain safely sequestered for thousands or millions of years. The economics of CCS are often unfavorable, especially as CO_2 is generally an expense rather than a revenue stream, but by combining the end goal of CCS with enhanced oil recovery (EOR) techniques used in the oil industry, there is the potential that CCS can be made economical while also increasing the productivity and efficiency of existing oil resources.

CO_2 EOR generally involves the injection of CO_2 into an oil-bearing reservoir in order to decrease oil viscosity, decrease the interfacial tension between oil and water, and increase the elastic energy of the formation, generally resulting in improved oil production. In the case of methane-bearing formations, most notably coal beds, injected CO_2 has a far stronger affinity to the formation than methane, resulting in the replacement of adsorbed methane with adsorbed CO_2, both increasing methane production and resulting in the sequestering of large volumes of CO_2.

EOR and CCS projects are both complicated tasks that require a vast understanding of the target reservoir in order to enhance storage capacity and storage time of CO_2, as well as hydrocarbon production. These topics will be discussed in greater detail throughout this paper.

1.1. Trapping mechanisms

One of the primary considerations when approaching a CCS project is the different mechanisms by which CO_2 can become safely sequestered underground. Generally, there are four different trapping mechanisms employed in the sequestration of CO_2, each of which contributes differently to the duration and volume of CO_2 trapping (**Figure 1**). In the different time stage, those four trapping mechanisms will work together.

- Structural/stratigraphic trapping: These types of traps are formed from tectonic forces and generally involve physical barriers to flow. An example of this is a thick layer of low permeability rock (caprock), such as shale, where, assuming a favorable structure, rising CO_2 will become trapped and begin accumulating.
- Residual trapping: This phase of trapping starts as soon as the CO_2 is injected. While the CO_2 is being injected, it is displacing the fluids that are inside the pores of the formation. As the primary CO_2 volume migrates upward, small volumes of CO_2 remain inside these tiny

http://dx.doi.org/10.5772/67226

pores due to capillary forces. This mechanism immobilizes the CO_2, potentially storing it in the formation for millions of years, just like the fluids it displaced.

- Solubility trapping: Solubility trapping refers to CO_2 being absorbed or adsorbed within the formation. Absorption occurs as CO_2 dissolves in the formation fluid, while adsorption occurs as CO_2 binds to the surface of the formation, like a piece of metal attaching to a magnet. After CO_2 has been adsorbed in the fluid, it will exist as a mixture, which will not be as buoyant as its gaseous form and will not migrate upward through the formation. This mixture will be denser than the surrounding fluids and will migrate downward over time.
- Mineral trapping: At the time of injection, this type of trapping is insignificant. Over a long period, after CO_2 has been dissolved in the formation fluids, it will begin reacting with the minerals in the surrounding formation and create solid carbonate minerals. These solid carbonate minerals will be attached to the rocks it reacted with and can be stored in the formation for millions of years.

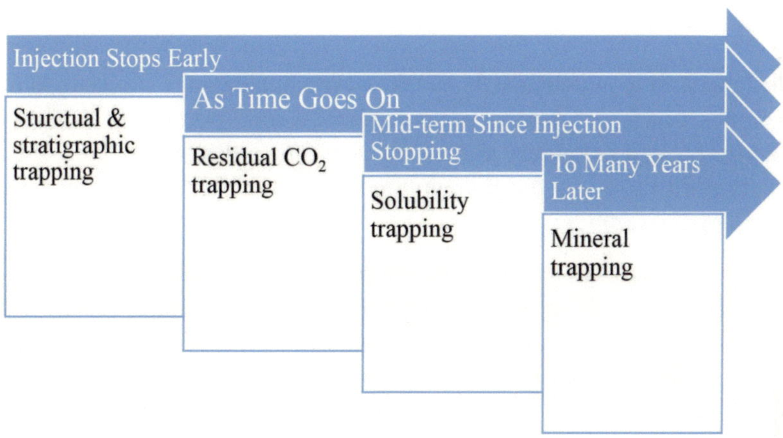

Figure 1. The four different mechanisms of CO_2 trapping.

Overall, these trapping mechanisms prevent carbon dioxide's upward travel and leakage while increasing the CO_2 storage potential and security of the desired formation. Assuming an ideal trapping mechanism, the temperature-related properties of a reservoir must be considered as well. The required temperature to store CO_2 underground should be less than the critical temperature of CO_2, making reservoirs such as those in Illinois Basin prime candidates. The critical temperature of CO_2 is 87.7°F; naturally, most geological formations exceed this temperature due to geothermal gradient [2].

1.2. Sandstone reservoirs

Sandstone reservoirs were the primary source of oil production during the early life of the oil industry. Many wells were produced and then abandoned long before the introduction of enhanced oil recovery (EOR) and other modern techniques that have enabled production from formations once thought of as nothing more than barriers to flow or geologic curiosities. In this modern day, sandstone reservoirs, once the workhorse of the oil industry but long since abandoned due to declining production, can be made to once again flow in economic quantities through the use of EOR techniques, such as CO_2 injection. While ensuring a renewed flow of oil to an energy-hungry world, CO_2 EOR in these old sandstone reservoirs may also play a major role in the preservation of our environment as injected CO_2 can be sequestered in subsurface formations for thousands of years. With these unique opportunities come unique challenges, ranging from the significant reservoir analysis required to ensure a safe sequestration to the infrastructure required to deliver such sizeable quantities of CO_2.

Sandstone reservoirs are particularly notable due to the sheer number of wells drilled in such formations that have been produced throughout the history of the oil industry and have since been abandoned. Due to their number, as well as how much time we have had to accumulate knowledge about their behaviors and the petrophysics involved in their production, sandstone reservoirs are likely to play a major role in any large-scale CCS program.

1.3. Coalbed reservoirs

Another ideal medium in which to store CO_2 is coal beds. Generally used to produce coalbed methane or coal at shallower depths, coal beds have a dual porosity system, which can be classified as primary and secondary porosity system. The pores within the coal matrix make up the primary porosity, while the pore volume of the numerous fractures permeating the coal bed makes up the secondary porosity.

The methane that is the primary target of coalbed drilling is stored in the coal matrix via adsorption. Because CO_2 has a greater affinity for coal than methane does, CO_2 is the desired choice to enhance methane recovery, and coal beds are a good place to store CO_2. Coal beds are distinctively different from the conventional hydrocarbon reservoirs in production as well as gas storage mechanisms. In conventional oil reservoirs, CO_2 is dissolved in the oil to decrease the viscosity of the oil, resulting in a great deal of CO_2 being recovered at the surface along with the produced oil. The "sequestered" CO_2 is then only that which dissolves into residual oil or is trapped due to one of the other trapping mechanisms [3]. In the case of coal beds, the majority of CO_2 adsorbs directly to the surface of the coal bed, providing a more efficient mechanism of sequestration while also forcing methane off the coalbed surface, helping to release any residual gas production. For example, methane recovery was improved from 77 to 95% of original gas in place at the Allison Unit CO_2-ECBM pilot in the San Juan Basin [4]. Coal beds can act as a significant contributor to CCS through the excellent economics of CO_2-based EOR, as well as the quality of their sequestration.

1.4. Shale reservoirs

As technology has advanced throughout the years, oil and gas exploration in unconventional shale reservoirs has become the main focus of the oil industry. Horizontal drilling and the hydraulic fracturing of shale formations have allowed us to unlock vast reserves of oil and gas production. Considering shale formations have extremely low permeability (of nanoDarcy in some cases), primary production does not produce the maximum amount of oil possible out of the formation. In most cases, tertiary production or EOR will begin with gas injection, such as carbon dioxide (CO_2), instead of water flooding because of shale's low permeability and the risk of reactions between the clays and injected water. During this production enhancement, some of the injected carbon dioxide will be permanently stored in the formation with different storage mechanisms, while some will be produced with the oil stream and get recycled back into the formation. CCS in shale reservoirs is often more difficult as less is known about their geology and long-term behaviors.

Shale reservoirs will likely play an important part in future CCS projects due to the scale of many shale reservoirs, their quality as a seal, and the importance of EOR techniques in existing shale plays.

1.5. Carbonate reservoirs

CO_2 injection into carbonate reservoirs was first considered in the 1930s but did not become a reality until 1964 in the Mead Strawn field located in Texas. CO_2 injection has since been established as a reliable form of EOR, with results regularly matching or surpassing those of other EOR techniques. In the 1964 example with the Mead Strawn field, oil production was increased by up to 82% beyond the results of a standard water flood [5]. Like many sandstone reservoirs, carbonate reservoirs have a long history and will likely play a significant role in future CO_2 EOR and CCS projects.

Hill et al. [6] state carbonate CO_2 EOR now produces approximately 305,000 bbls worldwide with an accelerating growth rate. The areas targeted for carbonate CO_2 projects in the United States are as follows: Louisiana, Michigan, Mississippi, New Mexico, Oklahoma, Texas, and Wyoming. CO_2 production wells provide immense amounts of data on the reservoir response to a CO_2 flood compared to saline projects. Azzolina et al. [7] discuss how CO_2 EOR is an established method for extending the life of a hydrocarbon sustaining carbonate reservoir.

Dissolved CO_2 injection into carbonate subsurface formation increases geologic carbon storage integrity by avoiding dependence on trapping mechanisms. As a result, solubility trapping will dominate until mineral trapping occurs, which is dependent on the formation rock [56]. Izgec et al. observed that solubility storage of CO_2 is larger than mineral trapping [1]. Eke et al. [8] state geological CO_2 storage in carbonate formations for long timescales (sequestration) relies on the contribution of several CO_2 trapping mechanisms: physical trapping in a subsurface formation, solubility trapping, hydrodynamic trapping, and mineral trapping.

2. Existing field applications

With the need for the prompt reduction in CO_2 emissions, the development of CCS must be taken seriously, as it has the potential to make a major difference in the levels of atmospheric CO_2. At one time, it was believed that oilfield reservoirs did not have sufficient pore volumes to have a significant impact on CO_2 emissions, but it is now understood that not only are there massive pore volumes available for CCS in depleted major pay zones (MPZs) of reservoirs, but there also exist residual oil zones (ROZs) and transition zones (TZs) in hydrocarbon fields that can be depleted and used for sequestration through quaternary production.

Traditionally, residual oil zones (ROZs) are considered to be uneconomic by the end of their primary or secondary recovery phase due to their extremely low oil saturation. However, Advanced Resources International [9] analyzed the feasibility of using CO_2 EOR to recover hydrocarbons from the ROZ and determined that a total of 55 fields in the Permian Basin have the potential to become economic ROZ resources. Simulations using CO_2 PROPHET, a water and CO_2 flood prediction software available through the US Department of Energy (DOE) website, estimated the recoverable ROZ at 11.9 billion bbls of the 30.7 billion bbls of TZ/ROZ oil in place in these five Permian Basin oil plays [9].

Usage of CO_2 injection as a form of EOR has not been limited to pilot and research tests. Kinder Morgan estimated that in the past 37 years, 655 million tons of CO_2 have been injected, produced, and recycled back into EOR. This is an average of 17.7 million tons per year, which is enough to negate the yearly emissions of six 500 MW coal-fired electric power plants [10]. Examples of some of these different field applications are given below.

2.1. Permian Basin

The Permian Basin in West Texas is one of the largest areas employing CCS techniques in ROZ and TZ and is currently undergoing the largest CO_2-enhanced oil recovery (EOR) operation in the world. Most of the ROZs created in this area are duc to lateral sweep by hydrodynamics and have thicknesses in excess of 300 feet [11]. While implementation of CO_2 floods is not particularly widespread due to the limited availability of CO_2, the Permian Basin has ready access to a pipeline of CO_2 originating in natural supplies in Colorado and New Mexico. Of the six CO_2 EOR projects in which recovery response has been published for Gulf Coast sandstone reservoirs, recovery factors are from 15 to 23% of original oil in place (OOIP) [12].

2.2. Port Neches

A CO_2 injection project in Port Neches, in a Texas Gulf sandstone, started in September 1993. The field had previously undergone water flooding, leaving a residual oil saturation of 30%. The goal of the project was to recover an additional 10% original oil in place (OOIP) [13]. A follow-up paper recorded that the production peaked at 500 barrels of oil per day (Bopd) (**Figure 2**) and later at 800 Bopd with CO_2 injection. The OOIP reduced from 12 to 7 million stock tank barrels (MMSTB) in the main fault block of the reservoir [14].

2.3. Bati Raman field

In 1986, the Turkish Petroleum Corporation started a large immiscible CO_2 injection project; the trend can be seen in **Figure 3**.

2.4. Ordos' Basin

The evaluation of Changqing oil field, Ordos' Basin, Northwest China, concluded that conducting a CO_2 flood after water flooding could produce 119 million tons of oil and sequestrate 273 million tons of CO_2 [16]. In 2000, the International Energy Agency Weyburn CO_2 Monitoring and Storage Project did a study on CO_2 storage in a partially depleted oil reservoir and found that a $1.5 billion, 30-year commercial CO_2 EOR produced an additional 130 million barrels of oil.

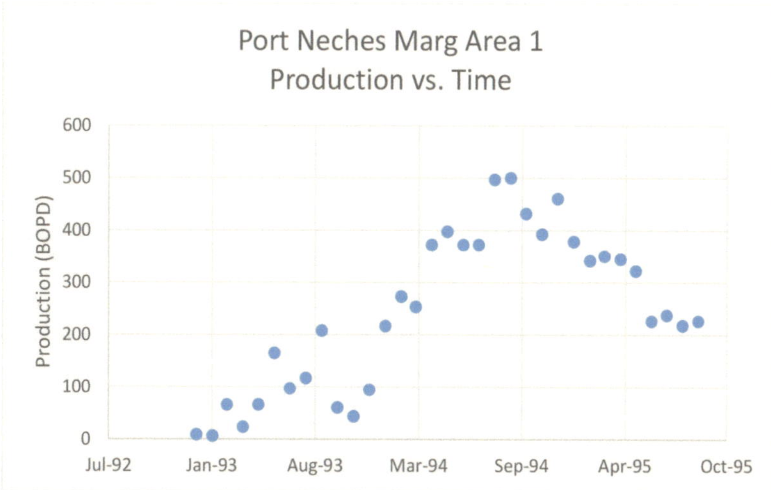

Figure 2. Production vs. time plot.

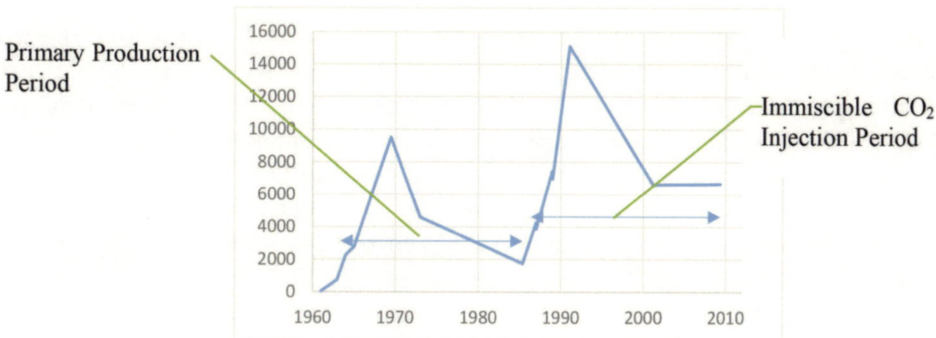

Figure 3. The Bati Raman field's production trend [15].

2.5. SECARB

The Southeast Regional Carbon Sequestration Partnership (SECARB) [17] operated a test for CO_2 sequestration at Black Warrior Basin in Alabama from 2006 to 2009 and determined that more than 360 million tons could be sequestered while increasing coalbed methane reserves by more than 20%. The SECARB set up monitoring systems in shallow boreholes and continues to monitor the local soil profile to determine if seepages of their test injection of 1000 tons of CO_2-injected gas occur and to facilitate the development of monitoring protocols that will ensure the safe conduct of CO_2 injection activities.

2.6. SWP projects

The Southwest Regional Partnership for Carbon Sequestration (SWP) indicates that over 2 million metric tons out of a total of 7 million metric tons retained CO_2 in the Scurry Area Canyon Reef Operators (SACROC) project were dissolved in the aqueous phase. That report does not include nor report CO_2 dissolution in oil, and therefore the numbers for CO_2 dissolution in the aqueous phase may be compromised. In addition to CO_2 dissolution in oil, the presence of a hydrocarbon phase can limit the contact between injected CO_2 and the aqueous phase even in depleted carbonate reservoirs. This work will, therefore, enhance estimates of predicted storage capacity both in depleted and producing oil reservoirs by revisiting and considering CO_2 solubility in the oil phase.

Additionally, rock wettability determines whether hydrolyzed CO_2 and the resulting acid in the aqueous phase can come into contact with the rock surface. When the rock is strongly oil wet such as in most carbonates, carbonate dissolution cannot take place; therefore, requirements for the mineralization trapping mechanism will not be met. In that case, the current estimation of CO_2 storage capacity in oil reservoirs because of the mineralization mechanism should be revisited. There is no indication of wettability measurement in the SACROC project. The SACROC project seeks to develop a subsurface geochemical-compositional flow model that incorporates the physics learned from lab-based measurements conducted throughout the course of its work, which will add considerably to the body of knowledge for carbonate reservoirs.

2.7. Existing exploited CO_2 sources

The majority of CO_2 injected into formation during operations is from natural reservoirs; however, problems arise such as climate change, diminished supply, and large demand. Innovation provides the solution by capturing CO_2 previously released to the atmosphere and using it for CO_2 EOR. During the production process, produced CO_2 is captured at the surface and reinjected, thus trapping the majority of injected CO_2 in formation. In Wyoming, natural gas processing plants produce approximately 716 Tcf of CO_2 while injecting 705 Tcf [7] in carbonate formations. In Michigan, an existing source of CO_2 provides the opportunity for carbonate CO_2 EOR in the NPRT; thus, Core Energy is using CO_2 emissions for EOR operations exploiting carbonate reef deposits [18]. These examples are helping reduce the emissions that would otherwise be vented to the atmosphere.

2.8. ECBM studies

Due to the effectiveness of CO_2 EOR and sequestration in coal beds, numerous studies have examined the usage of CO_2 sequestration in enhanced coalbed methane (ECBM) fields, and there are many field cases.

Mastalerz et al. [19] studied CO_2 sequestration and ECBM in unminable coal seams of the Illinois Basin. They found that approximately 271 billion tons of CO_2 could potentially be sequestered in the basin. Moreover, they found that potentially 1.6–4.6 billion tons of CO_2 could be sequestered in Illinois Basin coals and 70–280 billion m^3 (2.4–9.8 Tcf) of CH_4 is potentially recoverable as a result of CO_2 ECBM practices. The paper does suggest that volumetric strain due and coal swelling, which causes permeability damage, should be considered in any CCS or CO_2 EOR project.

Yu et al. [20] predicted in 2007 that the CO_2 sequestration throughout all ECBM projects (existing and potential) in China could result in over 3.751 Tm^3 of additionally recoverable methane, with a CO_2 sequestration capacity of around 142.67 billion tons.

2.9. Shale storage capacity (the United States and Canada)

The amount of available storage for CO_2 in oil and gas shale is currently unknown, but the vast volumes of shale formations indicate that the storage capacity is significant. A recent report has estimated between 1.85 trillion and 20.5 trillion tons of carbon dioxide storage capacity is available in oil and gas reservoirs just in the Unites States and Canada. These estimates suggest the availability for storing centuries worth of CO_2.

2.10. Additional possible locations and projects

Depleted oil and gas fields in the SECARB region could provide 29.7–34.7 billion tons of CO_2 storage with 24 million recovered oil barrels [21]. Almost 60% of the estimated volume relate to offshore fields. Coal and organic-rich shale formations can also offer a significant place for storage due to high absorption capacity of CO_2 in addition to potential EOR applications. A tertiary coal in the Gulf of Mexico is estimated to have 20–28 billion tons of CO_2 storage [18].

The potential storage capacity of the Barnett Shale is estimated to be 19–27 Gton, while other shale formation, Fayetteville Shale, is estimated to be capable of sequestrating 14–20 Gton of CO_2 [18]. There are still a lot more fields in the SECARB region to be evaluated on a possibility of a potential CO_2 storage and sequestration site. The SECARB region has a large annual CO_2 emission from coal-fired power generation and other fossil-fueled plants. In 2008 it was estimated to emit almost 2.9 [22] billion metric tons of CO_2.

An estimation of possible CO_2 sequestration volume was done by a "production replacement" principle, where for every volume of hydrocarbon, a 1:1 replacement ratio of CO_2 volume takes place. For the 2008 rates of CO_2 emission, SECARB region was capable of providing at least 28 years of CO_2 storage [18]. A case with a CO_2 EOR and sequestration in the Bell Creek oil field has a promising estimation of a recovery of additional 35 [23] million bbl of incremental oil through CO_2 flooding. Current plans exist to build a 232-mile pipeline from ConocoPhillips Lost Cabin gas producing plant to the Bell Creek field. This will help to integrate the large-scale storage of over 1 million tons of CO_2 per year.

3. Upcoming improvements to field applications

Kuuskraa, Godec and Dipeitro [24] analyzed primary and enabling next-generation technologies with applications in CO_2 sequestration, as shown in **Table 1**, and approximated the benefits of these technologies on a sample field area, as shown in **Table 2**. Notably, using their sample and estimates, they predict an increase in economically recoverable resource from 21.4 to 63.3 billion bbls.

Technologies		Technology implementation	The use of enabling technologies
I. Primary technologies			
1.	Improved reservoir conformance	Divert CO_2 from high permeability reservoir channels	Reservoir characterization and MDC
2.	Advance CO_2 flood design	Realign CO_2 flood pattern; drill additional wells to flood poorly swept zone(s)	Reservoir characterization and MDC
3.	Enhanced mobility control	Increase viscosity of drive water (WAG) to 2 cp	Enhanced fluid injectivity
4.	Increased volumes of efficiently used CO_2	Increase CO_2 injection from 1 HCPV to 1.5 HCPV; reduce sorm from 0.1 to 0.08	MDC and enhanced fluid injectivity
5.	Near-miscible CO_2 EOR	Apply CO_2 EOR to oil reservoirs with max pressure within 80% of MMP; reduce sorm based on reservoir pressure	–
II. Enabling technologies			
1.	Robust reservoir characterization	Advanced logging, seismic monitoring and core analysis	Essential for technologies 1 and 2
2.	Enhanced fluid injectivity	Effective near-wellbore stimulation methods	Essential for technologies 3 and 4
3.	Monitoring, diagnostics and control (MDC)	Downhole monitoring systems, real-time diagnostics, smart wells, etc.	Essential for technologies 1, 2, and 4

Table 1. Technologies used in next-generation CO_2 EOR [22].

Resource area	Economic oil recovery (billion bbls) *		Demand for CO_2 (billion metric tons)		Average CO_2 utilization (bbls/ $mtCO_2$)	
	SOA	Next generation	SOA	Next generation	SOA	Next generation
Miscible	19.6	60.8	8.4	15.4	2.3	3.9
Near miscible	1.8	2.6	0.5	0.8	3.9	3.3
Total	**21.4**	**63.3**	**8.9**	**16.2**	**2.4**	**3.9**

*At $90 per barrel oil price and $40 per metric ton CO_2 price, with 20% rate of return (before tax). Results compiled from simulations of CO_2 EOR floods at 1800 oil-bearing formations in the onshore continental United States. Reservoir characterization data drawn from the Big Oil Fields database, simulations conducted using the PROPHET stream tube model.

Table 2. Results from next-generation CO_2 EOR [22].

3.1. Simultaneous injection into pay zones and aquifers for ECBM

Ahmadi et al. [25] performed numerical modeling to investigate reasonable CO_2 injection scenarios, which were applied to CO_2 sequestration and ECBM. In their study, the main goal was to study different CO_2 injection methods and the effect of operational factors on the performance of each method by a numerical simulation model. There were three different strategies

http://dx.doi.org/10.5772/67226

concentrated, which were soluble and insoluble CO_2 injection into the bottom aquifer, CO_2 injection into pay zone, and simultaneous CO_2 injection into aquifer and pay zone. The result was that simultaneous injection into aquifer and pay zone leads to higher final oil recovery in EOR schemes.

3.2. Modifications to shale CO_2 processes

Due to the low porosity of shale, capillary forces are not negligible. Furthermore, adsorption has to be carefully considered due to a large specific area in shale. Pu and Li [26] gave a new formulation that includes the capillary force and adsorption through pore size distribution. A local density optimization algorithm was used to the adsorption model. In the Bakken field, the results of their investigations reduced the soaking time of the CO_2 huff "n" puff process and increased the 18% OOIP ultimate recovery.

3.2.1. Shale heterogeneity needs to be considered

Most of the unconventional reservoirs are heterogeneous, which influences the application of the huff "n" puff method. Chen et al. [27] studied the relationship between the reservoir heterogeneity and CO_2 huff "n" puff recovery through running simulations in the Elm Coulee Field of the Bakken. Shale heterogeneity had a significant negative impact, reducing the final recovery rate of the well.

3.2.2. There have not been large-scale CO_2 sequestration projects with shale

Large-scale demonstrations to prove CO_2 storage capability and capacity for very long periods of time in shale have not yet occurred [28]. According to Global CCS Institute [29], only 15 large-scale projects on CO_2 storage are taking place around the world with CO_2 capture capacity volumes ranging from 0.7 to 7 million tons per annum (Mtpa) in countries such as Norway, Algeria, Canada, and the United States. These do not include smaller projects that use CO_2 injection and end up sequestering smaller volumes, i.e., CO_2 EOR projects.

3.2.3. Improving CO_2 sweep efficiency

To maximize the effectiveness of CO_2 sequestration and adsorption in shale, it is important for the injected carbon dioxide to come in contact with as much reservoir volume as possible, a phenomenon known as sweep efficiency. Again, not enough CO_2 sequestration projects in shale formations have taken place and been monitored to show what the most effective conditions are to keep carbon dioxide sequestered. An increase in recovery rate from CO_2 injection under specified conditions can be used to estimate the optimum requirements to achieve utmost levels of sweep efficiency, but this is not necessarily the ideal condition for sequestration.

The available knowledge suggests recovery factors increase drastically when carbon dioxide is injected around minimum miscible pressure (MMP) that is around 1500 psi [30]. MMP can change by a few percentages depending on reservoir pressure, permeability, heterogeneity, and pore geometry.

One of the advantages of carbon dioxide is that its MMP is much lower than other gases; therefore, CO_2 MMP injection is possible under a wide range of reservoir pressures [31]. At around MMP, carbon dioxide and oil are miscible which leads to a zero entry capillary pressure. This allows carbon dioxide to enter the oil filled tight pores of shale and increase sweep efficiency and storage with high displacement efficiency. Also, the required soaking time, the time needed for injected gas to pierce and spread throughout the formation, appears to have a significant effect on sweep efficiency because of exceedingly low permeability of shale formations. Longer shut-in periods after CO_2 injection show higher oil recoveries that indicate a greater sweep efficiency [28].

3.3. Carbonate potential

In the South Sumatera Basin, 98 carbonate oil fields represent 59% of total original oil in place (OOIP) [32]. A study ranked these reservoirs based on CO_2 EOR and sequestration.

3.3.1. *Challenges in carbonates present opportunities in CCS*

Carbonate hydrocarbon reservoirs remain poorly understood; opposed to other storage sites, carbonates are likely to be hydrophobic (2/3rd of the world's carbonate reservoirs are oil wetting). CO_2 dissolution in the oil phase is orders of magnitude higher than its solubility in brine as seen in **Figures 4** and **5**. In the context of CO_2 sequestration in carbonate hydrophobic storage sites, dissolution of CO_2 in the oil phase is favorable for the long-term CO_2 storage in comparison with free supercritical CO_2 storage or CO_2 dissolution in brines.

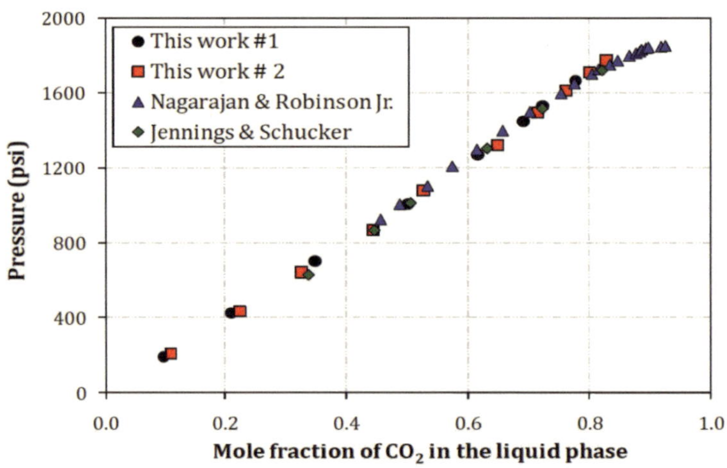

Figure 4. Pressure dependence of CO_2 dissolution in an oleic phase in 71°C [33].

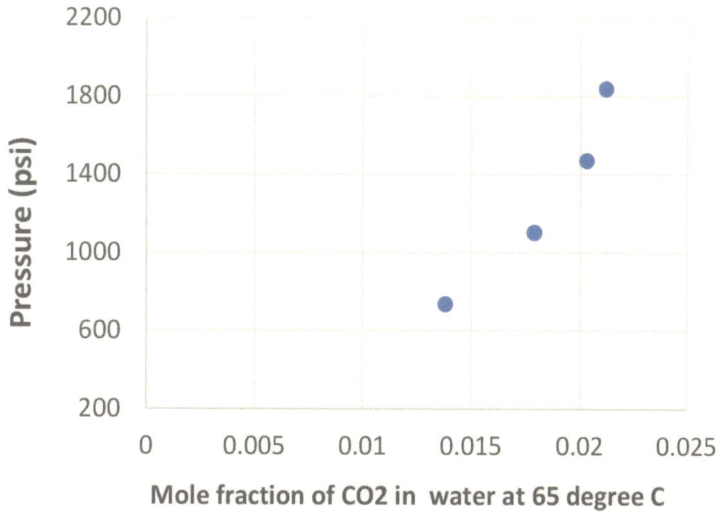

Figure 5. Pressure dependence of CO_2 dissolution in water at 65° [34].

4. Economics

One of the primary challenges facing CCS and CO_2 EOR is the cost of trapping and delivering CO_2. Large-scale injection of CO_2, for any purpose, can only reach its full potential when a supply chain and infrastructure are established, and most locations do not have access to a preexisting CO_2 infrastructure [16, 35].

For instance, while the impermeable shale barriers in an Illinois Basin are a perfect seal for a long-term sequestration of CO_2, the absence of a CO_2 delivery infrastructure, despite local electrical power facilities emitting over 255 [20] metric tons of CO_2 annually, still overcomes all the scientific potential in the area. The same scientific potential could allow low-temperature oil reservoirs to become sequestration targets, and to increase the local CO_2 storage capacity 20 times, at the same time, to enhance the oil recovery by another 6–18% (360–1100 MMSTB) [21].

In one case in the Gazran field, the costs to acquire CO_2 were approximately 11$/metric ton, with recycling costs of approximately 8$/metric ton [16]. In other areas, such as West Texas, prices can be as high as $40/ton with 18 billion tons of CO_2 required, making it very difficult to initiate large-scale CO_2 projects without a proper supply chain. Ghomian et al. [36] estimated that the total costs of CO_2 sequestration are in the range of $40–$60 per ton of CO_2 stored, primarily due to the costs of CO_2 capture and compression. In cases where a proper CO_2 infrastructure can be created, CO_2 transported via a pipeline with rates above 10 million tons of CO_2 per year often cost less than $1/metric ton of CO_2 per 100 km, with lower flow rates costing as much as double that amount [34]. This suggests that once a basic infrastructure has been created, the capture cost of CO_2 will become the limiting factor in CCS and CO_2 EOR projects.

4.1. Coal bed

No matter how efficiently CO_2 ECBM and CO_2 sequestration works when CO_2 is readily available, economic problems cannot be ignored. Robertson [37] provided the economic analysis of CO_2 sequestration and CO_2 ECBM of the Powder River Basin in Wyoming. He evaluated three production scenarios (no gas injection, flue gas injection, CO_2 injection). Strategies were analyzed using a discount rate of 10% and the rate of return on investment. A Monte Carlo model was used to analyze the CO_2 injection method and the mean value of the CO_2 injection scenario (**Figure 6**). It was found that for the mean case, a cost of CO_2 of approximately $4.81/Mg (or $4.81/metric ton) is required to maintain the economic viability [35].

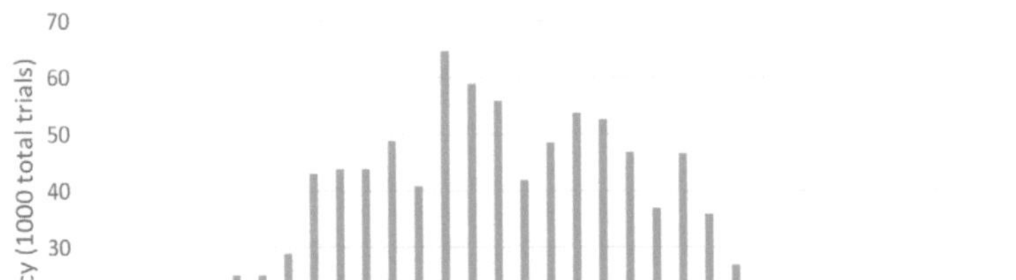

Figure 6. Distribution and mean value of the cost of CO_2 separation/capture required to yield a 10% rate of return [35].

Robertson also suggested that separating CO_2 from flue gas and injecting it into the unminable coal zones of the Powder River Basin seam, while currently uneconomical, can increase recovery of methane by 17% and could sequester over 86,000 tons CO_2/ac [35].

A 2009 economic analysis by Gonzalez et al. investigated the effectiveness of CO_2 EOR and sequestration on coal beds of different initial permeability values and determined that CO_2 storage was often quite economical on wells of moderate permeability (10 milliDarcy) and high permeability (100 milliDarcy). In their study, none of the low permeability cases were economical. It is worth to mention that high-rank coals (those containing higher levels of carbon) showed the strongest economics [38].

5. Injection and sequestration

Unlike in the oil industry where the inability to recover injected resources is often a cause for concern and additional economic strain, CCS inherently requires the permanent sequestration of CO_2 in the given reservoir. These conflicting intentions will need to be overcome

http://dx.doi.org/10.5772/67226

for economic purposes if CCS and CO_2 EOR are to become major players in the fight against climate change. Once these challenges have been overcome, the effectiveness with which CO_2 can be sequestered into different formations becomes a major point of importance.

Examples of the effectiveness of CO_2 sequestration are fairly common. Yamaguchi et al. [39] investigated the Ishikari Coalfield in Japan, where a multi-well test was able to inject 600 tons of CO_2 with an estimated 96% of the CO_2 being successfully adsorbed into the coal bed. Mavor et al. [40] analyzed a project by the Alberta Research Council which operated a two-well pilot test, where they determined that the increase in CO_2 injectivity (owing ballooning and water saturation reduction) was able to overcome injectivity losses due to swelling. Results were greatly improved by reducing injection periods, which allowed for adsorbed gas in the coal bed to finish swelling and for CO_2 to diffuse throughout the reservoir. These results were mirrored by Wan and Sheng [41], who determined that in fractured reservoirs, cyclic gas injection could increase oil recovery to 29%, while primary production only produced about 6.5% of OOIP [39].

Sheng and Chen [42] compared CO_2 and water flooding and were able to achieve superior results for CO_2 injection both in the case of flooding and huff "n" puff scenarios, with the best results (production of 32.46% OOIP) occurring using the huff "n" puff method.

5.1. CO_2 EOR in gas condensate wells

Higher densities of CO_2 relative to the native gas condensate cause CO_2 to migrate downward; with an increase of viscosity, CO_2 will displace the hydrocarbon gas phase. CO_2 EOR is very effective in light and medium gravity reservoir oils, in addition to being effective at recovery of gas condensates [43]. The dissolution of CO_2 into the oil decreases its interfacial tension; this creates a chance for the capillary force to enhance the recovery of the residual oil. This aspect heavily depends on the pressure and thus the depth. The properties of depleted gas/condensate reservoirs make them favorable for repressurization and enhanced gas recovery using CO_2 [41].

6. Possible geomechanical problems

EOR through CO_2 sequestration provides great opportunities for improving hydrocarbon recovery and the reduction of the greenhouse effect. Yet there are still problems about CCS that need to be addressed. A study on a pressure-depleted gas reservoir in the southern North Sea provided insight on CO_2 sequestration in depleted hydrocarbon reservoirs [60]. Their sequestration led to multiple geomechanical problems during drilling, completion, and CO_2 injection.

These depleted reservoirs have a narrow window of drilling mud weights that will not result in reservoir problems, and well completions can be affected by potential solid flow back when the injection of CO_2 is interrupted, while the temperature changes near the wellbore can lead to thermal fracturing and reactivation of faults. CO_2 sequestration can sometimes require drilling additional injection wells, which can be a problem with a narrow mud weight window because of the increased chance of a wellbore collapse.

The narrow mud weight window can make it nearly impossible to avoid falling out of the ideal range of mud weights, leading to a number of risks and an increase of nonproductive time and additional costs. During the injection stage, if there are problems with CO_2 supply, resulting in an interruption of CO_2 injection, solids will flow back into the well, resulting in a risk of rock failure or erosion of a pipeline.

Well integrity is the achievement of fluid containment and pressure containment within the well throughout its whole life cycle. The CO_2 injection can lead to the corrosion and degradation of the injection tubing, injection casing, and cement and packer material. The trickiest part is keeping the well leak-free. A CO_2 sequestration well has to be designed for over 40 years of continued well integrity. Some potential methods of protecting well integrity include the injection of supercritical CO_2 fluid, as it is dry and noncorrosive, protecting a well for a much longer period [44]. Usage of supercritical CO_2, unfortunately, increases costs and can increase issues with temperature changes, which can hydraulically and thermally fracture a rock in a near-wellbore region. This risk can be mitigated by keeping the fluid pressure that acts on a caprock outside of its fracturing pressure. Most other well failure problems can be reduced by keeping a well straight instead of inclined [60].

6.1. Offshore leak issues

Offshore injections of CO_2 for EOR and sequestration lead to alterations and deformations of caprock, affecting seal integrity. A break in a cap rock can result in a large burst of CO_2 from a reservoir and ultimately the seabed. When evaluating long-term caprock integrity, it is important to note the intrinsic caprock properties, chemical conditions at reservoir/caprock interface, and injection-induced pressure perturbation [61].

The caprock properties to look for are fracture normal stiffness, bulk concentration, and carbonate-forming cations. The enhancement or degradation of a caprock is related to the reduction and widening of microfracture apertures. During an injection process, initial mineral trapping takes place, which can have a significant impact on maintaining initial CO_2 injectivity and can delineate and partially self-seal plume boundaries while also reducing caprock permeability. Many CO_2 migration and sequestration processes in saline aquifers are equally applicable to CO_2 flood EOR in shale-capped water-wet oil reservoirs [21]. The CO_2 storage capacity is inverse proportional on reservoir permeability, which, in pure sequestration scenarios with high injection pressure, benefits from an increased storage and delayed migration, providing a noncompromised caprock performance.

Injection could also lead to pressures exceeding the formations natural fracturing pressure, resulting in the reactivation of a fault or the reservoir rock becoming hydraulically and thermally fractured. This creates a potential breach in the caprock that prevents CO_2 migration to the surface or flow into an adjacent formation [60]. An injection is followed by a change of reservoir temperatures that result in expansion and contraction of materials and ultimately result in changes of the field stresses, which creates a risk of breaching the caprock over time.

All geomechanical problems impose a great risk on CO_2 storage, which means the caprock integrity must be addressed when selecting a storage well site. The sealing efficiency is dependent

http://dx.doi.org/10.5772/67226

on many factors, including caprock, well cement, capillary threshold pressure, and chemical reactivity to CO_2. A proper geological evaluation is required to investigate the possible paths for CO_2 migration to the surface through the faults and fractures. Well sites with microseismic activity are generally poor candidates for the long-term containment of CO_2. Topography has to be addressed in the same manner, in the case of CO_2 leakage; the surface has to be well ventilated to prevent an accumulation of CO_2 cloud.

6.2. Risks and examples of CO_2 leakage

Equipment degradation is a big problem in abandoned wells, as well as currently operating wells. Individual wells have to be monitored in order to spot a leakage of CO_2 through the annulus of a wellbore. Leakage can result in not only a migration of CO_2 to the surface but also a contamination of surrounding reservoirs and aquifers [43]. This can happen because of wellbore expansion and contraction due to temperature and pressure changes.

Nygaard et al. [45] wrote a paper regarding wellbore well leakage and found that 95 out of 1000 wells near Wabamun Lake in Alberta identified as potential leakage pathways caused by an immediate caprock penetration. This sort of issue is common in poorly plugged wells which can leak CO_2 at the cement-rock interface or through a cement plug. Any mechanical load during a completion or stimulation can affect the integrity of the cement, in addition to corrosion and chemical reactions near the wellbore. Issues such as those found in this study must be considered during the life and abandonment of well to ensure a reliable seal.

6.3. Actions to prevent future CO_2 leakage

There are multiple options for sealing abandoned wells, but all of them require at least 8 m of cement inside the casing. Most abandoned wells after 1995 have sufficient integrity. In order to improve the seal integrity, it is suggested to remove the casing steel from abandoned wells before the final cement plug, and an injection of the CO_2-resistant polymer is executed [44]. The cement samples from 30- to 50-year-old wells kept a good sealing integrity and prevented leakage of CO_2, even though they contained a degree of carbonation [10]. It is not suggested to squeeze the cement into an opening in the casing, but a melted alloy can fill most openings, and its expansion will mitigate microfissures [44].

CO_2 injection affects the mineralogy and structural heterogeneity of the reservoir, which will have an impact on the porosity, permeability, and storage stability of the well. Better predictions of reservoir response to CO_2 injection are a necessary step in the evaluation of possible long-term sequestration of CO_2. A well-proven method for CO_2 testing is Hassler cell core testing, but unfortunately, there is no standard protocol for CO_2 testing, which can lead to errors in results [10].

6.4. Selection of readings on stress and possible leakage in ECBM wells

As ECBM reservoirs often do not have a standard caprock to prevent leakage, their long-term viability as CO2 sequestration targets must be carefully considered. Numerous papers have explored this question and are briefly listed and summarized below:

- In the report of Myer [46], geomechanical factors that risk CO_2 leakage in sequestration, as well as the risk of CO_2 leakage from drilling and completion, production, and repressurization, are discussed.
- Mitra and Harpalani [47] investigated the matrix strain resulting from a CO_2 injection.
- Chen et al. [48] investigated how the effective stress factor and methane CO_2 counter-diffusion work on the CO_2 recovery using a finite element model that coupled coal deformation, gas flow, and methane-CO_2 counter-diffusion. Through their study, it was found that permeability loss/gain is influenced by effective stress and methane CO_2 counter-diffusion and that the gas pressure distribution is related to gas composition [47].
- Fathi and Akkutlu [49] investigated counter-diffusion and competitive adsorption effects according to the new one-dimensional theoretical model they created. Compared with the conventional model, they created a new triple porosity dual permeability multi-continuum model to describe the gas release from macro-pores and micro-pores to the fracture.

6.5. CO_2 monitoring

Due to the quantities of CO_2 being sequestered in large CCS projects and the importance of keeping that CO_2 permanently underground, monitoring is a very important part of any CCS project. 3D seismic survey has proven to be effective at monitoring CO_2 storage but is prohibitively expensive. Gasperikova and Hoversten [50] investigated using a combination of gravity inversion, electromagnetic (EM), and amplitude vs. angle (AVA) monitoring analysis to detect changes in CO_2 saturation. Gravity inversion detects density changes in the injected layer. EM and AVA can be used to estimate CO_2 saturation changes. Macdonald [51] provided field and lab measurements of CO_2 using Raman spectroscopy, which improved monitoring of the prevised amounts of CO_2 dissolved in reservoir brine. Through the Saptharishi and Makwana [52] study, various monitoring techniques are summarized, which include but are not limited to techniques for coal beds.

6.6. Risks of CO_2 injection, possible failure modes

Carbon dioxide storage is not a risk-free task. As years go by after CO_2 has been injected into a formation, it is possible for the CO_2 to begin migrating upward and leak out of the ground back into the atmosphere through openings in the caprock or fractures, faults, and poorly completed preexisting wells [53]. This problem can be prevented or reduced if the formation of interest for CO_2 storage has a caprock with ideal qualities.

An ideal caprock is a layer of the formation with very low permeability that can prevent oil and gas from migrating upward and out of the reservoir formation. In any case of CO_2 storage, a thick shale layer is the most desirable type of caprock. Due to shale's extremely low permeability, causing a more tortuous flow path, CO_2 migration vertically is tremendously limited [54]. The degradation of cement and metal casing with a presence of CO_2 is currently a topic that needs extensive investigation [52]. As the **Figure 7** shows, there are five possible leakage pathways in an abandoned well. Label a and label b are the pathway between casing

http://dx.doi.org/10.5772/67226

and cement wall and plug, respectively. Label c shows leakage through cement plugs. Label d represents leakage through casing. Label e shows leakage through the cement wall. Label f represents leakage pathways between the cement wall and the formation [52].

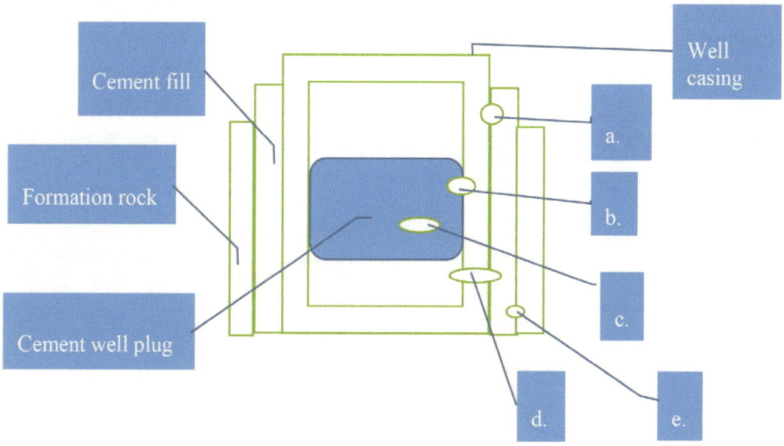

Figure 7. CO_2 leakage pathways.

The Sleipner Project, located in Norway, is currently storing more than 2700 tons of CO_2 per day below an extensive and thick shale layer [27]. Monitoring the injected CO_2 during the past 13 years is showing that the gas has spread out nearly two square miles below the shale layer without moving upward or leaking out of the reservoir storage [55]. This is one of the most significant evidence that proves how effective shale formations can be as CO_2 storage reservoir and caprock, where carbon dioxide will be trapped and immobile. In short, the ultimate geological storage reservoir should have sufficient capacity, be a thick shale layer acting as a caprock simultaneously, and be a stable storage environment maintaining the original characteristics of the reservoir.

6.6.1. Overcoming the high risk of CO_2 leakage in carbonate reservoirs

Carbonate reservoirs do not generally possess an impermeable boundary or caprock, and therefore permanent trapping of CO_2 through geomechanical means is unrealistic [57]. Solubility storage decreases potential leakage in carbonate formations, as the dissolution of CO_2 into water promotes mineralization, but this will need to be studied further before carbonate reservoirs can be relied upon to properly sequester large volumes of CO_2 [56].

Agada et al. [57] did extensive research on how fracture network geometry affected oil recovery and CO_2 storage in carbonate reservoirs. They noted that many of the problems associated with high fracture-matrix connectivity, such as bypassing of oil, early water breakthrough, and rapid CO_2 migration, could be mitigated by foam flooding. Sehbi et al. [58] proposed a low injection rate, longer in-reservoir CO_2 retention time, and good pore structure to improve the efficiency in carbonate reservoirs. Carbonate formations showed an increase in effective permeability resulting from chemical dissolution in the matrix, thus enhancing pore connectivity [59].

6.6.2. Examples of natural carbonate sequestration

The Colorado Plateau-Southern Rocky Mountains region contains natural CO_2, which has been discovered during exploration for oil and gas fields (**Figure 8**), thus providing a natural laboratory for studying the effects of long-term, subsurface CO_2 storage in carbonate reservoirs. These laboratories yield information such as that injecting CO_2 separated from flue gases ensures the subsurface migration path is long, thereby yielding optimal sequestration. Despite the number of carbonate CO_2 reservoirs in the region and active flux of CO_2 to the surface, no hazards from CO_2 surface accumulations are known. The nature and rate of CO_2 surface leakage in carbonate formations are still unknown [60].

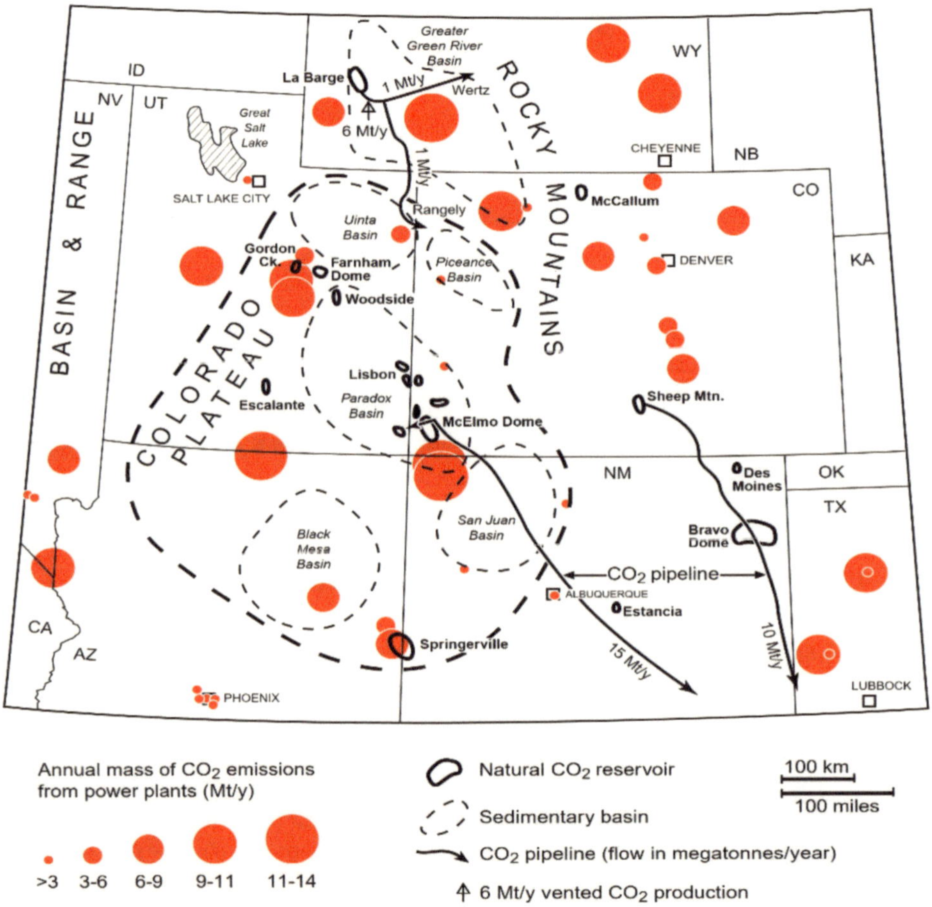

Figure 8. Synthesis of data relating to CO_2 fluxes and concentrations around the Colorado Plateau [8].

6.7. Additional potential concerns

Despite the many benefits of CO_2 EOR and CCS programs, it must be remembered that these are complicated projects being undertaken in complex geological environments. A 2004–2008 project in Algeria stored over 2.5 [61] million tons of CO_2 in a carboniferous sandstone reservoir. During the shut-in process, the CO_2 injection was unexpectedly interrupted, and the wellbore pressure went lower than the reservoir pressure, risking rock failure, sand production, and possible blowout.

Potential concerns also include preventing potentially catastrophic failure of the reservoir seal. For instance, if injection pressures exceed the breakthrough pressure of the sealing caprock, the CO_2 would break through and risk flowing back to the surface [60]. Reactive transport modeling shows that for a typical shale caprock, geochemical processes continuously improve isolation performance, and geomechanical processes first rapidly degrade and then improve isolation performance over time. There is a possibility for a counterbalancing of geomechanical and geochemical effects, but they must be carefully monitored [62].

Some issues are more minor, not directly threatening the safety of the operation, but nonetheless affecting the economics of a combined CO_2 EOR and CCS project. Bou-Mikael [14] wrote about the performance of a CO_2 flood at Port Neches in the Gulf of Mexico, with a partnership of Department of Energy and Texaco E&P. The CO_2 flood underperformed [13], with 500 bbl/day instead of 800 bbl/day; with this underperformance was attributed to the following reasons: reservoir characterization, oil saturation, water blockage, and wellbore mechanical problems. After a careful evaluation of the project, it was determined that in ideal circumstances and if related criteria are met, CO_2 injection can accelerate production two to three times compared to unassisted primary production [13].

6.7.1. *Potential coalbed problems*

Coal beds, despite offering unique opportunities, also offer unique challenges. In particular the coal matrix swells during CO_2 adsorption. Coal matrix swelling can cause reductions in permeability. Bustin et al. [63] experimented on the volumetric strain from three western Canadian coals and found that a mixture of N_2 and CO_2 injection would improve CO_2 injection rates greatly but that CO_2 sequestration capacity decreased wildly. However, pure CO_2 injection could cause the reduction of permeability by two orders of magnitude. The applicability of the CO_2-ECBM process in any coal seam is mainly governed by the seam's permeability and its adsorption process [62]; therefore, these concerns must be explored further.

7. Conclusion

Numerous studies support the potential of major sequestration projects, and due to the negative impacts of atmospheric CO_2, CCS will continue to be an important part of protecting our environment. While EOR through CO_2 sequestration has proved to be valuable, there are still challenges that need to be addressed in the future. Reservoir properties must continue to be carefully considered for all CCS projects due to their impact on successful EOR and CO_2 sequestration.

The major challenges currently facing CCS projects are primarily those of economics and transportation. Limited CO_2 transportation supply chains act as a barrier for CO_2 EOR utilization in the oil industry. When this barrier has been removed and a large network of CO_2 capturing mechanisms have been created, it will open the petroleum industry to a breadth of new possibilities both in terms of improved recovery and environmental sustainability.

For the purposes of having a significant impact on atmospheric CO_2 levels, the simple merging of CO_2 EOR and CCS may not be enough. In every reservoir type, in every circumstance, there are diminishing returns as far as incremental production as additional CO_2 is injected into a well. As such, as long as CO_2 remains an expense, rather than a revenue stream, the full potential of CCS will not be realized. In the meantime, however, there are numerous projects that hold a good deal of promise and are economical under current conditions due to the benefits of CO_2 injection on ultimate hydrocarbon recovery.

7.1. Closing notes on shale reservoirs

Shale reservoirs still hold a great deal of promise for CCS and CO_2 EOR, as the benefits for production are significant, and the formations themselves provide excellent seals against any risk of CO_2 migration. Unfortunately, a great deal of research and monitoring is still required in order to ensure that shale beds maintain the quality of their seals over time and to maximize CO_2 sequestration. Knowledge gaps such as lack of information on available storage capacity, lack of formation and reservoir data that specifies favorable sequestration settings, understanding long-term CO_2 interaction in shale, and testing different strategies for CO_2 injection and well patterns to achieve efficient carbon dioxide sequestration and EOR still exist [52]. Many questions regarding this topic will remain unanswered until additional, large, in situ field tests take places.

7.2. Closing notes on carbonate reservoirs

The future of CO_2 EOR and sequestration in carbonate reservoirs will steadily improve due to the statistical data being acquired from existing field tests. The United States' carbonate formations provide the foundation for CO_2 injection in carbonate reservoirs [64]. The Bati Raman reservoir provides a significant opportunity to further carbonate CO_2 EOR operations. Sahin et al. [65] state this reservoir could easily yield a billion dollars in revenue as a CO_2 EOR project. Hydrocarbon fuels can supply relatively pure CO_2 for EOR allowing the byproducts of the industry's previous production to add in new production while also creating a more environmentally friendly outcome. CO_2 that cannot be used for EOR can be stored in depleted carbonate formations, thus furthering the climate-friendly initiative [66]. Recent estimates of future CO_2 demand suggest that large volumes will be required to meet the promise of next-generation EOR including the development of residual oil zones [7].

7.2.1. Specific challenges in carbonate reservoirs

As previously discussed, dissolved CO_2 injection is recommended for reactive fractured formations and formations with uncertain caprock integrity [7]. The challenges of the carbonate pinnacle reef data analysis are as follows: an increase in pressure with CO_2 injection, the

http://dx.doi.org/10.5772/67226

presence of multiple reservoir fluids, and unique CO_2 phase behavior due to changing pressure and temperature.

Izgec et al. [1] discuss challenges of mineral trapping, the effects of changing rock properties, and the residual impact on CO_2 CCS in carbonate reservoirs. Puon et al. [67] state that other challenges in carbonate formations include CO_2 tendency to bypass a large percentage of pore volume, yielding an early breakthrough, and reductions in recovery efficiency. As a result, CO_2 flooding is not economically feasible without improved mobility control. Several mobility control methods have been attempted with limited success; therefore, concepts for CO_2 mobility control are required to increase the overall recovery efficiency and economics in carbonate reservoirs [66]. Eke et al. [8] suggested the injection of denser CO_2 saturated brine in carbonate formations, which should be capable of eliminating much of buoyancy force. Thus, CO_2 brine surface mixing strategy is recommended due to the enhancement and secure storage of CO_2 in subsurface carbonate formations.

8. Suggestions for future study

For sandstone and coalbed reservoirs, the last major remaining barrier to large-scale implementation of CO_2 EOR and CCS is the economic burden of CO_2 capture and transportation. Research into improving capture and transport techniques, as well as how to structure intelligent government incentives, will go a long way in increasing CO_2 sequestration rates.

Unlike with sandstone and coalbed reservoirs, the primary barrier to CO_2 sequestration in shale reservoirs is a lack of research and monitoring work after CO_2 injection. The lack of research is in fact only aggravated by the lack of monitoring and in situ data.

At last, the study of CO_2 sequestration in carbonate reservoirs needs to expand to include the effects of CO_2 on carbonate rock properties. Issues ranging from an early breakthrough, to low sweep efficiency, to structural problems within the formation, all, limit the viability of large-scale carbonate projects. Future experiments will need to be performed using a high-pressure carbonate core flooding system optimized for use within different lab apparatus so that experiments can be conducted to better understand the complications and benefits of supercritical CO_2 [57].

There is no doubt that CCS and CO_2 EOR/ECBM will play a major role in the future of the energy industry. However, besides the economic issues with many CO_2 implementations, legal risks must also be considered as well. Unitization is important for CO_2 EOR in order to avoid the trespass claim [68]. In addition, different states and regulators treat CO_2 differently, sometimes as a pollutant, other times as a natural gas [67]. Despite not being research-based concerns, the legal climate of the United States must also change for a truly successful CCS program to take hold and to make the best use of what could become a CO_2 revolution.

Nomenclature

AVA	Amplitude vs. angle
bbl	barrel
CCS	Carbon capture and storage
CO_2	Carbon dioxide
DOE	Department of Energy
ECBM	Enhanced coalbed methane
EM	Electromagnetic
EOR	Enhanced oil recovery
MMP	Minimum miscible pressure
MPZ	Major pay zone
ROZ	Residual oil zone
SACROC	Scurry Area Canyon Reef Operators
SECARB	Southeast Regional Carbon Sequestration Partnership
SWP	Southwest Regional Partnership for Carbon Sequestration
TZ	Transition zones

Author details

Rouzbeh Ghanbarnezhad Moghanloo*, Xu Yan, Gregory Law, Soheil Roshani, Garrett Babb and Wesley Herron

*Address all correspondence to: rouzbeh.gm@ou.edu

The University of Oklahoma, MPGE, Norman, United States

References

[1] Izgec, O., B. Demiral, H. J. Bertin, and S. Akin, 2005, CO_2 Injection in Carbonates: SPE Western Regional Meeting, doi:10.2118/93773–ms.

[2] Frailey, S., J. Grube, B. Seyler, and R. Finley, 2004, Investigation of Liquid CO_2 Sequestration and EOR in Low Temperature Oil Reservoirs in the Illinois Basin: Proceedings of SPE/DOE Symposium on Improved Oil Recovery, doi:10.2523/89342–ms.

[3] Shi, J. Q., and S. Durucan, 2005, CO_2 Storage in Deep Unminable Coal Seams: Oil & Gas Science and Technology–Rev: IFP Oil & Gas Science and Technology, v. 60, no. 3, pp. 547–558, doi:10.2516/ogst:2005037.

[4] Reeves, S., A. Taillefert, L. Pekot, and C. Clarkson, 2003, The Allison unit CO_2-Ecbm Pilot: A Reservoir Modeling Study, doi:10.2172/825083.

[5] Holm, L., and L. O'brien, 1971, Carbon Dioxide Test at the Mead-Strawn Field: Journal of Petroleum Technology, v. 23, no. 04, pp. 431–442, doi:10.2118/3103–pa.

[6] Hill, B., S. Hovorka, and S. Melzer, 2013, Geologic Carbon Storage Through Enhanced Oil Recovery: Energy Procedia, v. 37, pp. 6808–6830, doi:10.1016/j.egypro.2013.06.614.

[7] Azzolina, N. A., D. V. Nakles, and C. D. Gorecki, 2015, CO_2 Storage Associated with CO_2 Enhanced Oil Recovery: A Statistical Analysis of Historical Operations: International Journal of Greenhouse Gas Control, v. 37, pp. 384–397, doi:10.1016/j.ijggc.2015.03.037.

[8] Eke, P. E., Naylor, M., Haszeldine, S., & Curtis, A. (2011, March 1). CO_2/Brine Surface Dissolution and Injection: CO_2 Storage Enhancement. Society of Petroleum Engineers. doi:10.2118/124711-PA

[9] Koperna, G. J., Kuuskraa, V. A., 2006, Technical Oil Recovery Potential from Residual Oil Zones: Permian Basin: Prepared for U.S. Department of Energy, Office of Fossil Energy – Office of Oil and Natural Gas. February.

[10] Sweatman, R. E., A. K. Santra, D. S. Kulakofsky, and D. G. Calvert, 2009, Effective Zonal Isolation for CO_2 Sequestration Wells: SPE International Conference on CO_2 Capture, Storage, and Utilization, doi:10.2118/126226–ms.

[11] Ming, C. M., and L. S. Melzer, 2010, Symposium on the Role of EOR in Accelerating the Deployment of CCS.

[12] Holtz, M. H., 2008, Summary of Sandstone Gulf Coast CO_2 EOR Flooding Application and Response: SPE Symposium on Improved Oil Recovery, doi:10.2118/113368–ms.

[13] Davis, D., 1994, Project Design of a CO_2 Miscible Flood in a Waterflooded Sandstone Reservoir: Proceedings of SPE/DOE Improved Oil Recovery Symposium, doi:10.2523/27758–ms.

[14] Bou-Mikael, S., 1996, A New Analytical Method to Evaluate, Predict, and Improve CO_2 Flood Performance in Sandstone Reservoirs: Proceedings of SPE/DOE Improved Oil Recovery Symposium, doi:10.2523/35362–ms.

[15] Sahin, S., U. Kalfa, and D. Celebioglu, 2010, Unique CO_2-Injection Experience in the Bati Raman Field May Lead to a Proposal of EOR/Sequestration CO_2 Network in the Middle East: SPE International Conference on CO_2 Capture, Storage, and Utilization, doi:10.2118/139616–ms.

[16] Ran, X., Y. Zhao, and X. Liao, 2012, An Assessment of a CO_2 Flood for EOR and Sequestration Benefits in the Ordos Basin, Northwest China: Carbon Management Technology Conference, doi:10.7122/150272–ms.

[17] Pashin, J. C., P. E. Clark, M. R. Mcintyre-Redden, R. E. Carroll, R. A. Esposito, A. Y. Oudinot, and G. J. Koperna, 2015, SECARB CO_2 Injection Test in Mature Coalbed Methane Reservoirs of the Black Warrior Basin, Blue Creek Field, Alabama: International Journal of Coal Geology, v. 144–145, pp. 71–87, doi:10.1016/j.coal.2015.04.003.

[18] Barnes, D., B. Harrison, and G. M. Grammer, 2013, CO_2/EOR and Geological Carbon Storage Resource Potential in the Niagaran Pinnacle Reef Trend, Lower Michigan, USA: Energy Procedia, v. 37, pp. 6786–6799, doi:10.1016/j.egypro.2013.06.612.

[19] Mastalerz, M., J. Rupp, A. Drobniak, S. Harpalani, A. Anderson, C. Korose, S. Frailey, and D. Morse, n.d., Assessment of CO_2 Sequestration and Enhanced Coalbed Methane Potential in Unminable Coal Seams of the Illinois Basin, essay, in Carbon dioxide Sequestration in Geological Media—State of the Science: AAPG Studies in Geology, pp. 149–171.

[20] Yu, H., G. Zhou, W. Fan, and J. Ye, 2007, Predicted CO_2 Enhanced Coalbed Methane Recovery and CO_2 Sequestration in China: International Journal of Coal Geology, v. 71, no. 2–3, pp. 345–357, doi:10.1016/j.coal.2006.10.002.

[21] Petrusak, R. L., D. E. Riestenberg, P. L. Goad, K. C. Schepers, J. Pashin, R. A. Esposito, and R. C. Trautz, 2009, World Class CO_2 Sequestration Potential in Saline Formations, Oil and Gas Fields, Coal, and Shale: The US Southeast Regional Carbon Sequestration Partnership Has It All: SPE International Conference on CO_2 Capture, Storage, and Utilization, doi:10.2118/126619–ms.

[22] United States Department of Energy, National Energy Technology Laboratory, 2008, Carbon Sequestration Atlas of the United States and Canada. http://www.netl.doe.gov/technologies/carbon_seq/refshelf/atlas/.

[23] Gorecki, C. D., J. A. Hamling, J. Ensrud, E. N. Steadman, and J. A. Harju, 2012, Integrating CO_2 EOR and CO_2 Storage in the Bell Creek Oil Field: Carbon Management Technology Conference, doi:10.7122/151476–ms.

[24] Kuuskraa, V. A., M. L. Godec, and P. Dipietro, 2013, CO_2 Utilization from "Next Generation" CO_2 Enhanced Oil Recovery Technology: Energy Procedia, v. 37, pp. 6854–6866, doi:10.1016/j.egypro.2013.06.618.

[25] Ahmadi, M. A., B. Pouladi, and T. Barghi, 2016, Numerical Modeling of CO_2 Injection Scenarios in Petroleum Reservoirs: Application to CO_2 Sequestration and EOR: Journal of Natural Gas Science and Engineering, v. 30, pp. 38–49, doi:10.1016/j.jngse.2016.01.038.

[26] Pu, H., and Y. Li, 2015, CO_2 EOR Mechanisms in Bakken Shale Oil Reservoirs: Carbon Management Technology Conference, doi:10.7122/439769–ms.

[27] Chen, C., M. T. Balhoff, and K. K. Mohanty, 2014, Effect of Reservoir Heterogeneity on Primary Recovery and CO_2 Huff 'n' Puff Recovery in Shale-Oil Reservoirs: SPE Reservoir Evaluation & Engineering, v. 17, no. 03, pp. 404–413, doi:10.2118/164553–pa.

[28] IEAGHG, 2013, Potential Implications on Gas Production from Shales and Coals for Geological Storage of CO_2, 2013/10, September.

[29] Large Scale CCS Projects, n.d., https://www.globalccsinstitute.com/projects/large-scale-ccs-projects (accessed August 16, 2013).

[30] Gamadi, T., J. Sheng, M. Soliman, H. Menouar, M. Watson, and H. Emadibaladehi, 2014, An Experimental Study of Cyclic CO_2 Injection to Improve Shale Oil Recovery: SPE Improved Oil Recovery Symposium, doi:10.2118/169142–ms.

[31] Vega, B., W. J. O'brien, and A. R. Kovscek, 2010, Experimental Investigation of Oil Recovery From Siliceous Shale by Miscible CO_2 Injection: SPE Annual Technical Conference and Exhibition, doi:10.2118/135627–ms.

[32] Iskandar, U. P. and S. H. Lastiadi, 2014, A Systematic Approach to Source-sink Matching for CO_2 EOR and Sequestration in South Sumatera: Energy Procedia, v. 63, pp. 7750–7760, doi:10.1016/j.egypro.2014.11.809.

[33] Tsau, J. S., 2010, Near Miscible CO_2 Application to Improve Oil Recovery for Small Producers.

[34] Moghanloo, R. G., 2012, Modeling the Fluid Flow of Carbon Dioxide through Permeable Media, dissertation.

[35] Gray, L., and S. G. Goodyear, 2014, Overcoming the CO_2 Supply Challenge for CO_2 EOR: Abu Dhabi International Petroleum Exhibition and Conference, doi:10.2118/172105–ms.

[36] Ghomian, Y., M. B. Urun, G. A. Pope, and K. Sepehrnoori, 2008, Investigation of Economic Incentives for CO_2 Sequestration: SPE Annual Technical Conference and Exhibition, doi:10.2118/116717–ms.

[37] Robertson, E. P., 2010, Enhanced Coal Bed Methane Recovery and CO_2 Sequestration in the Powder River Basin, doi:10.2172/983351.

[38] Gonzalez, R. J., K. C. Schepers, G. J. Koperna, and A. Y. Oudinot, 2009, Assessment of the Potential and Economic Performance for ECBM Recovery and CO_2 Sequestration: Latin American and Caribbean Petroleum Engineering Conference, doi:10.2118/121157–ms.

[39] Yamaguchi, S., K. Ohga, M. Fujioka, and S. Muto, 2005, Prospect of CO_2 Sequestration in the Ishikari Coal Field, Japan: Greenhouse Gas Control Technologies 7, pp. 423–430, doi:10.1016/b978-008044704–9/50043–4.

[40] Mavor, M., W. Gunter, and J. Robinson, 2004, Alberta Multiwell Micro-Pilot Testing for CBM Properties, Enhanced Methane Recovery and CO_2 Storage Potential: Proceedings of SPE Annual Technical Conference and Exhibition, doi:10.2523/90256–ms.

[41] Wan, T., and J. J. Sheng, 2015, Evaluation of the EOR Potential in Hydraulically Fractured Shale Oil Reservoirs by Cyclic Gas Injection: Petroleum Science and Technology, v. 33, no. 7, pp. 812–818, doi:10.1080/10916466.2015.1010041.

[42] Sheng, J. J., and K. Chen, 2014, Evaluation of the EOR Potential of Gas and Water Injection in Shale Oil Reservoirs: Journal of Unconventional Oil and Gas Resources, v. 5, pp. 1–9, doi:10.1016/j.juogr.2013.12.001

[43] Shtepani, E., 2006, CO_2 Sequestration in Depleted Gas/Condensate Reservoirs: Proceedings of SPE Annual Technical Conference and Exhibition, doi:10.2523/102284–ms.

[44] Bilardo, U., and F. Panvini, Carbon Sequestration: Key Features and Issues: Presented at Offshore Mediterranean Conference and Exhibition, Ravenna, Italy, 28–30 March 2007.

[45] Nygaard, R., S. Salehi, and R. G. Lavoie, 2011, Effect of Dynamic Loading on Wellbore Leakage for the Wabamun Area CO_2 Sequestration Project: Canadian Unconventional Resources Conference, doi:10.2118/146640–ms.

[46] Myer, L. R., 2003, Geomechanical Risks in Coal Bed Carbon Dioxide Sequestration, doi:10.2172/815530.

[47] Mitra, A., and S. Harpalani, 2007, Modeling Incremental Swelling of Coal Matrix with CO_2 Injection in Coalbed Methane Reservoirs: Proceedings of Eastern Regional Meeting, doi:10.2523/111184–ms.

[48] Chen, Z., Liu, J., Connell, L., Pan, Z., 2008, Impact of Effective Stress and CH4-CO_2 Counter-Diffusion on CO_2 Enhanced Methane Recovery: Proceedings of SPE Asia Pacific Oil & Gas Conference

[49] Fathi, E., and I. Y. Akkutlu, 2008, Counter Diffusion and Competitive Adsorption Effects During CO_2 Injection and Coalbed Methane Production: SPE Annual Technical Conference and Exhibition, doi:10.2118/115482–ms.

[50] Gasperikova, E., and G. Hoversten, 2008, Modeling the Resolution of Inexpensive, Novel non-Seismic Geophysical Monitoring Tools to Monitor CO_2 Injection into Coal Beds, doi:10.2172/974262.

[51] Macdonald, S., 2007, Measuring CO_2 in Coalbed Reservoirs: Proceedings of Rocky Mountain Oil & Gas Technology Symposium, doi:10.2523/107742–ms.

[52] Saptharishi, P., and M. Makwana, 2011, Technical and Geological Review of Carbon dioxide Geo Sequestration Along with Analysis and Study of Various Monitoring Techniques: International Petroleum Technology Conference, doi:10.2523/iptc-15402–ms.

[53] IPCC, 2005, IPCC Special Report on Carbon Dioxide Capture and Storage: In Prepared by Working Group III of the Intergovernmental Panel on Climate Change: In Metz, B., O. Davidson, H. C. de Coninck, M. Loos, and L. A. Meyer (eds.). Cambridge University Press, Cambridge, United Kingdom and New York, NY, USA, 442 pp.

[54] Bachu, S., 2015, Review of CO_2 Storage Efficiency in Deep Saline Aquifers: International Journal of Greenhouse Gas Control, v. 40, pp. 188–202, doi:10.1016/j.ijggc.2015.01.007.

[55] Carbon Sequestration Leadership Forum, 2011. Underground CO_2 Storage: A Reality?. pp. 1–5.

[56] Sigfusson, B. et al., 2015, Solving the Carbon-Dioxide Buoyancy Challenge: The Design and Field Testing of a Dissolved CO_2 Injection System: International Journal of Greenhouse Gas Control, v. 37, pp. 213–219, doi:10.1016/j.ijggc.2015.02.022.

[57] Agada, S., S. Geiger, and F. Doster, 2016, Wettability, Hysteresis and Fracture–Matrix Interaction During CO_2 EOR and Storage in Fractured Carbonate Reservoirs: International Journal of Greenhouse Gas Control, v. 46, pp. 57–75, doi:10.1016/j.ijggc.2015.12.035.

[58] Sehbi, B. S., S. M. Frailey, and A. S. Lawal, 2001, Analysis of Factors Affecting Microscopic Displacement Efficiency in CO_2 Floods: SPE Permian Basin Oil and Gas Recovery Conference, doi:10.2118/70022–ms.

[59] Vogt, S. J., C. A. Shaw, and J. E. Maneval, 2014, Magnetic Resonance Measurements of Flow-Path Enhancement During Supercritical CO_2 Injection in Sandstone and Carbonate Rock Cores: Journal of Petroleum Science and Engineering, v. 122, pp. 507–514, doi:10.1016/j.petrol.2014.08.013.

[60] White, S., R. Allis, and J. Moore, 2003, Natural CO_2 Reservoirs on the Colorado Plateau and Southern Rocky Mountains, USAA Numerical Model: Greenhouse Gas Control Technologies – 6th International Conference, pp. 423–428, doi:10.1016/b978-008044276–1/50068–4. (12 15)

[61] Fang, Z., A. Khaksar, and K. Gibbons, 2012, Geomechanical Risk Assessments for CO_2 Sequestration in Depleted Hydrocarbon Sandstone Reservoirs: SPE Drilling & Completion, v. 27, no. 03, pp. 368–382, doi:10.2118/133994–pa.

[62] Johnson, J., J. Nitao, J. Morris, and S. Blair, 2003, Reactive Transport Modeling of Geohazards Associated with CO_2 Injection for EOR and Geologic Sequestration: Offshore Technology Conference, doi:10.4043/15119–ms.

[63] Bustin, R. M., X. Cui, and L. Chikatamarla, 2008, Impacts of Volumetric Strain on CO_2 Sequestration in Coals and Enhanced CH4 Recovery: Bulletin AAPG Bulletin, v. 92, no. 1, pp. 15–29, doi:10.1306/08220706113.

[64] Manrique, E., V. Muci, and M. Gurfinkel, 2006, EOR Field Experiences in Carbonate Reservoirs in the United States: Proceedings of SPE/DOE Symposium on Improved Oil Recovery, doi:10.2523/100063–ms.

[65] Sahin, S., U. Kalfa, and D. Celebioglu, 2007, Bati Raman Field Immiscible CO_2 Application: Status Quo and Future Plans: Proceedings of Latin American & Caribbean Petroleum Engineering Conference, doi:10.2523/106575–ms.

[66] Herzog, H. J., 2001, Peer Reviewed: What Future for Carbon Capture and Sequestration?: Environmental Science & Technology, v. 35, no. 7, doi:10.1021/es012307j.

[67] Puon, P., S. Ameri, K. Aminian, D. Durham, J. Wasson, and R. Watts, 1988, CO_2 Mobility Control Carbonate Precipitation: Experimental Study: Proceedings of SPE Eastern Regional Meeting, doi:10.2523/18529–ms.

[68] Durrant, M. B., n.d., Legal Basics for CO_2 Enhanced Oil Recovery: the Rocky Mountain Mineral Law Foundation. https://www.oilgasandmining.com/volume1/issue2/84–v1n2–durrant (accessed September 7, 2016).

Section 5

Economics of CCS

Chapter 11

Economics of Carbon Capture and Storage

John C. Bergstrom and Dyna Ty

Additional information is available at the end of the chapter

http://dx.doi.org/10.5772/67000

Abstract

Human-engineered capture of CO_2 emissions at the point source and subsequent long-term storage of this CO_2 underground represent a potential mitigation strategy for global warming. The so-called carbon capture and storage (CCS) projects are technically feasible but have not been well established from an economic efficiency perspective. This chapter uses economic theory to describe the costs, benefits, and economically efficient level of CCS provision. Achieving the economically efficient level of CCS provision requires consideration of both the private and public costs and benefits of CCS and will also likely require some degree of government intervention in the form of economic incentives and/or direct regulation.

Keywords: CO_2, emissions, point source, capture, storage, economics, costs, benefits

1. Introduction

Since the late twentieth century, a newly developed technology has become one of the tools that can help mitigate the negative impacts on climate change from rising levels of greenhouse gases, especially CO_2. This technology is commonly known as the carbon capture and storage (CCS). CCS technology involves "capturing" CO_2 emissions, say from a coal-fired power plant, and then depositing the captured CO_2 gas in a storage site, such as an underground geological formation, where it will not enter the atmosphere. CCS projects are currently being tested and implemented throughout the world. However, economic feasibility of human-engineered CCS is not well established [1–4]. The purpose of this chapter is to discuss the economic benefits and costs of CCS projects from both private and public perspectives in order to shed light and provide insight on the potential for CCS technology to provide a viable mitigation strategy for helping to meet twenty-first century global CO_2 emission reduction goals, such as set forth in the 2015 United Nations Climate Change Conference in Paris, France.

2. Carbon and oxygen cycles[1]

Carbon (C) is the basic building block for plant, animal and human life—all are "carbon-based" organisms. Plants, animals, and humans also depend on oxygen (O_2) for survival. The cycling of carbon and oxygen in ecosystems is ultimately powered by solar energy. In photosynthesis, plants combine carbon dioxide (CO_2), water (H_2O), and solar energy to produce sugars, oxygen, and energy. In cellular respiration, animals and humans combine sugars and oxygen to produce carbon dioxide, water, and energy. Carbon-oxygen-hydrogen compounds (e.g., sugars) pass through the food chain or web in ecosystems via herbivores, carnivores, and omnivores. In the food chain, some of the carbon and oxygen stored in organic compounds are returned to the environment in the form of CO_2 and H_2O via cellular respiration. When a large organism such as a plant or an animal dies and is decomposed by microorganisms, more of the CO_2 and H_2O stored within the plant or animal is returned to the environment where it can be taken up again by plants to produce more carbon-oxygen-hydrogen compounds which can then be taken up again by animals and humans.

Not all carbon and oxygen are recycled in the relatively short-term cycle described above. Some carbon and oxygen from decomposing plants and animals are converted by relatively long-term geologic processes into rocks (e.g., carbonate rock formations such as limestone) and minerals (e.g., coal, oil, and natural gas) stored in the earth's crust. When coal, oil, and natural gas enter economic systems, they are termed fossil fuels. The "fossil" part of this term derives from the fact that they come from fossilized remains of plants and animals. The "fuel" part is derived from the fact that coal, oil and natural gas, and their processed derivatives (e.g., gasoline) are burned as fuel in engines and other machinery found throughout our economic system (e.g., planes, trains, automobiles, electricity power plants, and home furnaces). When fossil fuels are burned, CO_2 (and other emission gases—CH_4, N_2O) stored in these minerals is released back into the environment. The release of CO_2 from burning fossil fuels is the focus of recent concern and debate over global climate change.

As indicated in the discussion above, human activities affect global climate change through impacts on the carbon and oxygen cycle. Burning of fossil fuels is a major contributor to releasing more CO_2 into the atmosphere, primarily from terrestrial sources of stored carbon (e.g., coal deposits, oil deposits, and trees). Human activities can also help to remove CO_2 from the atmosphere, with one of the primary means being increasing the storage of carbon in terrestrial plants. For example, taking actions to protect "green space" including farmland from development (and managing forests in a sustainable manner following an optimal harvest and replanting schedule) helps to remove CO_2 in the atmosphere through carbon sequestration in plants via photosynthesis. Farms, forests, and other green space

[1]This section appears also in Ref. ([5], p. 16–18).

http://dx.doi.org/10.5772/67000

areas thus act as "carbon sinks" helping to counteract the greenhouse effect. Another means for storing carbon is through human-engineered carbon capture and storage projects.

3. CCS costs

3.1. Components of total fixed costs and total variable costs

CCS projects are not cheap. For example, in the United States, NRG Energy and JX Nippon Oil and Gas Exploration, Inc., are investing around $1 billion USD on the Petra Nova CCS project. This project when completed in late 2016 is projected to capture and store about 1.4 million tons of carbon per year from one of NRG's existing coal-fired power plants in the State of Texas, USA [6, 7]. In this section, we discuss the concepts and components of CCS costs.

First, we need to realize that CCS projects are actually two interconnected projects in one. The first project is "carbon capture" and the second project is "carbon storage." Each of these projects has various options with different costs. As indicated in the previous section, ecosystems via the carbon and oxygen cycle will naturally capture carbon dioxide from the air (e.g., through photosynthesis) and then store the captured carbon in plants, the soil, and rocks and minerals. While CCS through natural ecosystem processes and functions is a viable mitigation strategy in response to CO_2-induced global climate change concerns (e.g., planting trees), the focus of this chapter is on human-engineered CCS.

In the case of carbon capture, human-engineered means of capturing carbon focus on "end-of-pipe technologies" that remove CO_2 from industrial emissions, particularly fossil fuel-fired (e.g., coal) electricity power plants. The "best available technology" (BAT) in the current time period (2016) is chemical absorption of CO_2 from emissions at the point source (e.g. power plant smokestack). Once the CO_2 has been removed from emissions, say from a coal-fired power plant, the CO_2 can then be converted by pressurization to a liquid for transportation and storage [1, 2, 8].

Thus, one component of the costs of human-engineered carbon capture is the costs of the equipment (e.g., "scrubbers") and absorption chemicals used to remove CO_2 from emissions [4, 9]. From a neoclassical microeconomics theory perspective, the "scrubber" equipment costs are "fixed costs" and the absorption chemicals are "variable costs." Fixed costs are so-called because they are a sunk cost which does not vary with the level of production. For example, once purchased and installed, a coal-fired power plant owner must incur the costs of scrubber equipment whether they are producing electricity or not (e.g., they still have to pay off the equipment as a capital cost).

Variable costs are so-called because they vary with the level of production. For example, as more (less) electricity is produced from a coal-fired power plant, more (less) emissions are generated, and more (less) absorption chemicals must be purchased. The fixed costs of human-engineered carbon capture can be quantified by multiplying the units of equipment purchased by the market price of equipment per unit (plus loan fees and interest if the equipment is financed). The variable costs can be quantified by multiplying, say the units of absorption chemicals purchased by the market price of chemicals per unit.

In addition to the direct, out-of-pocket fixed and variable costs of carbon capture discussed above, there are also opportunity costs of human-engineered carbon capture. For example, from an energy use perspective, human-engineered carbon capture at an electricity power plant comes with an energy use cost in the form of electricity generation that must be given up in support of carbon capture at the plant. This so-called energy penalty can be quantified by multiplying the amount of electricity lost in order to support carbon capture times the market price of electricity [2, 9–11].

After carbon is captured at a point source such as a coal-fired electricity power plant, it must be transported to and stored at a long-term storage site. At the time this chapter is being written, the most practical long-term storage sites appear to be various forms of natural underground geologic cavities (NUGCs). One option under this category is NUGC which once held crude oil and/or natural gas deposits but has been depleted through mining (e.g., oil and gas wells). Oil and gas companies already inject CO_2 into operational oil and gas wells in order to squeeze more oil and gas out of the resource deposit. Thus, the technology for injecting CO_2 captured from point source emissions into NUGCs where oil and gas deposits have been depleted through mining is well proven [9, 12, 13].

Because natural deposits of oil and gas have been stored by the carbon and oxygen cycle (see above) in NUGCs for thousands and millions of years, NUGCs have displayed the ability to store new CO_2 injected into these formations for long periods of time with minimal leakage of CO_2 back into the atmosphere. In addition to NUGCs where oil and gas deposits have been depleted, geologists and engineers can locate new NUGCs capable of storing large quantities of CO_2 with minimal leakage for long time periods [9, 13].[2]

In order for carbon captured at the point source to be stored at long-term storage site, it must be transported from the point source to the storage site. The process for transport is generally to convert CO_2 captured at the point source to a liquid through pressurization, and then move this liquid to the storage facility by truck, train, or pipeline. Assuming that NUGCs are used for long-term storage, the costs of carbon storage will mostly be the fixed and variable costs of converting CO_2 to a liquid, transporting it to the storage site, and then injecting it into the NUGCs [9, 13]. After injecting the CO_2 into an NUGC, the ongoing costs of storage should be minimal (e.g., limited to costs of monitoring for leakages).

The fixed costs of carbon storage (including transportation) include the costs of pressurized transport trucks and train cars, and the costs of installing a pipeline. Fixed costs also include the costs of any equipment needed to remove captured CO_2 from a truck, train car, or pipeline and inject it into NUGCs. These fixed costs can be quantified by multiplying the units of equipment (e.g., transport truck or rail car) purchased by its market price per unit. The variable costs of carbon storage include payments to labor (e.g., workers who operate and main-

[2]As discussed in Section 2, natural chemical cycles covert carbon to hard rock and mineral deposits which further enhances long-term storage with minimal leakage.

http://dx.doi.org/10.5772/67000

tain trucks, trains, pipelines, and injection equipment), purchase of replacement parts, and the costs of fuel and power needed to operate and maintain trucks, trains, pipelines, and injection equipment. These variable costs can be quantified by multiplying the units employed (e.g., number of workers) or purchased (e.g., number of replacement parts) by the market wage rate for labor or the market price for replacements parts [8, 9, 13].

We can now define the total costs of carbon capture and storage (TC_{ccs}) as

$$TC_{CCS} = \left(TFC^{c}_{CCS} + TVC^{c}_{CCS}\right) + \left(TFC^{T}_{CCS} + TVC^{T}_{CCS}\right) + \left(TFC^{s}_{CCS} + TVC^{s}_{CCS}\right) \tag{1}$$

where

TFC^{c}_{ccs} = is the total fixed costs of carbon capture at the point source;

TVC^{c}_{ccs} = the total variable costs of carbon capture at the point source;

TFC^{T}_{ccs} the total fixed costs of captured carbon transportation to storage site;

TVC^{T}_{ccs} the total variable costs of captured carbon transportation to storage site;

TFC^{s}_{ccs} the total variable costs of carbon storage at storage site;

TVC^{s}_{ccs} is the total variable costs of carbon storage at storage site.

With respect to economic efficiency, it is imperative we measure the marginal costs of human-engineered CCS. The short-run marginal costs (MC_{CCS}) of human-engineered CCS are defined as

$$MC_{CCS} = \frac{\partial TC_{CCS}}{\partial Q_{CO_2}} \tag{2}$$

where Q_{CO_2}= is the quantity of CO_2 captured and stored.

3.2. Measures of total marginal fixed costs and marginal variable costs

In practice, there are two common measures used in the cost-benefit analysis to make per-unit CSS costs and benefits comparably equivalent for any given potential level of optimal quantities of carbon dioxide (CO_2) being captured and stored. These units are measured through time and space either in millions of tons of carbon (MtC) or of CO_2 ($MtCO_2$) avoided per year, that is, MtC/year or $MtCO_2$/year.

As described above in the definition of the total costs of carbon capture and storage (TC_{CCS}), TC_{CCS} consists of total fixed costs and total variable costs of carbon capture at the point of source, captured carbon transportation to storage, and carbon storage at the storage site. With respect to economic efficiency, the marginal cost (MCccs) is the imperative measure of the costs of human-engineered carbon capture and storage technology. In this chapter, marginal costs of employed CCS technology (MCccs), as well as marginal benefits received from employed CCS technology (MBccs), are quantified as US dollar per ton carbon ($/tC) or US dollar per ton carbon dioxide ($/t$CO_2$),[3] where one ton of carbon equals 3.67 tons of carbon dioxide.[4]

According to recent literature, an estimated avoided total cost of CCS per unit (MCccs) is between US $225/tC and $315/tC (or US $61/$CO_2$ and $86/t$CO_2$), but a considerable reduction in MCccs can arise in the near future because of continuously technological improvements in CCS [8]. To give a breadth of findings, estimates of marginal cost avoided can be shared in three cost components: (1) marginal costs of carbon captured at the point of source, which range from US $200/tC to $250/tC [8]; (2) marginal costs of captured carbon transportation to storage, which range from US $5/tC to $10/tC per 100 km [8][5]; and (3) marginal costs of carbon stored at the storage site, which range from US $20/tC to $55/tC [14].[6]

4. CCS benefits

4.1. Private benefits and public benefits from employed CCS technologies

The private benefits of carbon capture and storage include the proven ability of injecting CO_2 underground into geologic crude oil and natural gas deposits to enhance extraction of oil and gas from these deposits. These benefits can be quantified by multiplying the price of the additional crude oil or natural gas extracted as a result of CO_2 injection by the going market price of oil or gas. As this chapter is currently being written in early 2016, the real prices of crude oil and natural gas resources received by oil and gas producers are at record lows worldwide. These relatively low prices have a negative impact on the private benefits of CO_2 injection projects for enhancing oil and gas projects.

Thus, a c ritical component of whether or not such projects will be economically feasible to oil and gas companies is the expected price path of future oil and gas prices. Such price paths are difficult to estimate empirically [2, 8, 11]. However, based on economic theory and Hotelling's rule in particular, we expect theoretically that the market price of any exhaustible, non-renewable natural resource, including crude oil and natural gas, to follow an upward-sloping price path in the long run as the resource becomes scarcer.

[3]For consistency, in the chapter the units of MtC and US $/tC are being used to describe economic values for marginal costs and benefits on average, assuming MC ≈ AC in all long-run CCS operations. We will note where the units of MtCO_2 and US $/$CO_2$ are applied as alternative measures. All are equivalent: (1) US $27.3/t$CO_2$ (= US $100/tC) [26]; (2) US $10/t$CO_2$ is approximately equivalent to US $37/tC [15].

[4]Because the atomic weights of carbon are 12 atomic mass units and carbon dioxide is 44 atomic mass units, a ratio factor of 3.67, or 44/12, is used, meaning one ton of carbon equals 3.67 tons of carbon dioxide, which can also approximately equal 1 tC = 3.7 tCO_2 (as computed from (US $37/tC)÷(US $10/t$CO_2$), or 3.66 t$CO_2$ = (US $100/tC) ÷ (US $27.3/t$CO_2$)). However, only the factor of 3.67 is applied for computation of all estimates in this chapter.

[5]These costs can be easily and quickly observed in Anderson and Newell (2004) (please see Table 3 in Ref. [8]).

[6]The later measures may be slightly higher after having been adjusting for inflation over time. Assuming a gas price of US $3 per million Btu (MBtu), which was the average price over the past decade, transport and storage costs of $37/tC stored were reported in [8]. Moreover, one can apply the following formulas to see how adjusted/expected benefits and costs are affected by inflation rates over time, that is, adjusted benefits in current-year = dollars in base-year × ($CPI_{Current\text{-}year}$/$CPI_{Base\text{-}year}$), and adjusted costs in current-year = dollars in base-year × ($PPI_{Current\text{-}year}$/$PPI_{Base\text{-}year}$), where CPI is the consumer price index, and PPI is the producer price index.

http://dx.doi.org/10.5772/67000

From economic theory, we can also predict that increasing market prices of exhaustible, non-renewable energy resources such as crude oil and natural gas will eventually lead to the substitution of these relatively high-cost energy sources by relatively cheaper energy sources. For example, sometime in the future it may be economically feasible and desirable to shift completely over to some "backstop technology" for producing energy including solar and wind power and the "holy grail" of virtually unlimited energy production—nuclear fusion ([5], Chapter 3).

In addition to the private benefits to oil and gas companies of CCS projects that enhance oil and gas production, private benefits of CCS projects as a whole also include private benefits of global warming mitigation such as reduced health costs to individuals, reduced damages to agricultural crops, and reduced damages to human-built structures in flood-prone areas. These private benefits can be quantified using private health-care expenditures, the market value of agricultural crops, and the costs of replacing or repairing human-built structures [16].

There are also many public benefits of CCS projects associated with global warming mitigation. These public benefits include economic values associated with protecting fish and wildlife habitat (e.g., Polar Bear habitat in Artic regions) and human cultures (e.g., Indigenous, Native or First-Peoples in Artic regions). Non-market economic valuation techniques including contingent valuation and choice experiments can be used to quantify these types of nonmarket benefits ([5], Chapter 13).

The total benefits of carbon capture and storage (TB_{ccs})can be expressed in the equation form as

$$TB_{ccs} = \left(TPB_{ccs}^{s}\right) + \left(TSB_{ccs}^{s}\right) \tag{3}$$

where,

TPB_{ccs}^{s} is the total private benefits of carbon captured and stored

TSB_{ccs}^{s} is the total social benefits of carbon captured and stored.

For economic efficiency purposes, we must also measure the marginal benefits of human-engineered CCS. The short-run marginal costs (MBCCS) of human-engineered CCS are defined as

$$MB_{CCS} = \frac{\partial TB_{CCS}}{\partial Q_{CO_2}} \tag{4}$$

where Q_{CO_2} is the quantity of CO_2 captured and stored.

4.2. Measures of total marginal benefits of CSS

In Section 4.1, we describe that the total benefits of carbon capture and storage technologies are received by both public and private entities. For economic efficiency analyses, we use the total marginal benefits of human-engineered CCS (MBccs) given by the private marginal benefits of carbon captured and stored (PMBccs) plus the social marginal benefits (SMBs) of carbon captured and stored (SMBccs). In the following sections, we discuss quantitative estimates of the private marginal benefits of CCS (PMBccs) and the social marginal benefits of CCS (SMBccs).

4.2.1. Private marginal benefits and carbon capture and utilization

A Canadian Pembina Institute Publication [17] reported post-CO_2 capture diverging into two pathways—carbon sequestrations (CCS) (already discussed so far) and carbon capture and utilization (CCU) (discuss in this subsection). CCU applications fall under two main approaches: the conversion approach and nonconversion approach.[7] Since the twenty-first century, technological advances have made various CCU applications under these two main approaches more practical and profitable [18–20].

CCU conversion approach applications range from mineralization (e.g. varied utilized forms of carbonate applications), biological transformation (e.g. algae cultivation applications), and chemical transformation (e.g. liquid fuel applications including methanol, polymer/chemical feedstock, and urea yield boosting[8]). CCU non-conversion approach applications are generally aimed for the purposes of desalination and enhanced techniques including enhanced oil recovery (EOR), enhanced geothermal systems, and enhanced coal-bed methane [17].

Thus, the economics of CCU technologies then lies in potential net benefits received from reutilizations of the captured CO_2. Within 5–10 years, CCU conversion approaches including mineralization (considered as permanent-based performance) and biological and chemical innovations (considered as non-permanent-based performance) have been estimated to be utilized in a range of 5 to more than 300 $MtCO_2$ per year [17]. Within the same time frame, CCU nonconversion approaches will yield, in both permanent and non-permanent potential performance, an estimated 5–300 $MtCO_2$ in enhanced techniques and between 30 and 300 $MtCO_2$ in desalination [17].[9]

According to Refs. [21, 22], it is estimated that each year about 80%, or 9 million metric tons (MtC) of captured CO_2 used by commercial industry, are in EOR operations. The net marginal benefits (PMBccs) of stored carbon to EOR and enhanced coal-bed methane recovery operations have been estimated in the range of US \$15/tC to \$30/tC [23].[10] There certainly exist

[7]CCU is also called carbon capture and reuse or carbon capture and recycling (CCR) [17].

[8]Urea, also known as carbamide, is an organic compound with the chemical formula CO(NH2)2 and one of the most common forms of solid nitrogen fertilizer. Urea is produced by the reaction between ammonia and CO2. See ([24], Appendix B).

[9]Permanent and non-permanent potential performances are referred to permanent and non-permanent storage. According to the Global CCS institute, reuse technologies that permanently store CO2 are considered to be an alternative form of CCS and referred to as "alternative CCS." EOR, ECBM, EGS, carbonate mineralization, concrete curing, bauxite residue carbonation, and potentially algae cultivation (depending on the end product) are considered to be alternative forms of CCS. See ([24] Part I: Section 3.2).

[10]In recent work [25], it was estimated that EOR storage of CO2 could generate net benefits as high as \$335/tC stored, or cost as much as \$270/tC stored. In a base-case calculation, EOR generates average net benefits of about \$45/tC stored [8].

http://dx.doi.org/10.5772/67000

additional net benefits from other applications described above, but empirical estimates of these benefits are not yet available.

4.2.2. Social marginal benefits

Unlike quantifying direct total private benefits, an attempt to measure public or social benefits can be quite a challenging task for researchers since there involve concepts of types of costs and nonmarket valuations of public goods and services provided to the population.

Before it is attempted to explain how social marginal benefits arrive in this subsection of the chapter, there are three concepts needed to explain since we simply use the reported range (not derived explicitly) of the estimates for SMB from various sources. First of all, we simply define that private costs are the costs that individual decision makers are facing given actual established market prices. Second, social costs are the private costs plus the costs of economic externalities on society. These social costs are the prices derived from market prices, where opportunity costs are taken into account. Finally, social cost of carbon (SCC) is the discounted monetized sum of the annual net losses from impacts caused by an additional unit of carbon emitted presently and is measured in US \$/tC or US \$/tCO_2 ([26] Chapter 3, p. 135).

According to the economic theory, at an economically efficient mitigation level the marginal social benefits of carbon reduction (SMB) are equal to the social costs of carbon, where SCC is defined as avoided total damages for an additional ton of carbon abated ([26], Chapter 3, p. 233). Thus, using estimates of SCC ([26], Chapter 20) and the assumption that SCC = SMB at an economically efficient carbon price, we can infer estimates of SMB_{CCS} currently in the range US \$14/tC to \$350/tC (or US \$4/tCO_2 to \$95/tCO_2).[11] By assuming a 2.4% per year increase in emissions, the estimated range for SMB_{CCS} in the year 2030 is between US \$29/tC and \$694/tC (or US \$8/tCO_2 to 189/tCO_2).[12] By adding private marginal benefits (PMB_{CCS}) from the previous section to social marginal benefits (SMBccs) from the current section, we estimate total marginal benefits of CCS (MBccs) to fall in a range of US \$29/tC–\$380/tC currently to US \$49/tC–\$735/tC in 2030.[13]

5. Optimal CCS provision

5.1. Concept of economically efficient level of CSS size

According to economic efficiency, the optimal level of carbon capture and storage is where the marginal benefits and marginal costs of CO_2 captured and stored are equal. In **Figure 1**, we show the marginal benefit curve for CCS (MB_{CCS}), and the marginal cost curve for CCS (MC_{CCS}). The marginal benefit curve is downward sloping because, following the law of diminishing

[11]Median and 95th percentile estimates reported in [27].

[12]The estimated social cost of carbon reported by [28] including uncertainty, equity weighting, and risk aversion is \$44 per ton of carbon (or \$12 per ton CO2) in 2005 US\$. Second, including uncertainty increases the expected value of the SCC by approximately 8%. Finally, equity weighting generally tends to reduce the SCC.

[13]For consistency, we assume there is also a 2.4% per year increase in the PMBCCS reported in [23]. Thus, for 2030 the estimated range for PMBCCS is between US \$20/tC and \$41/tC.

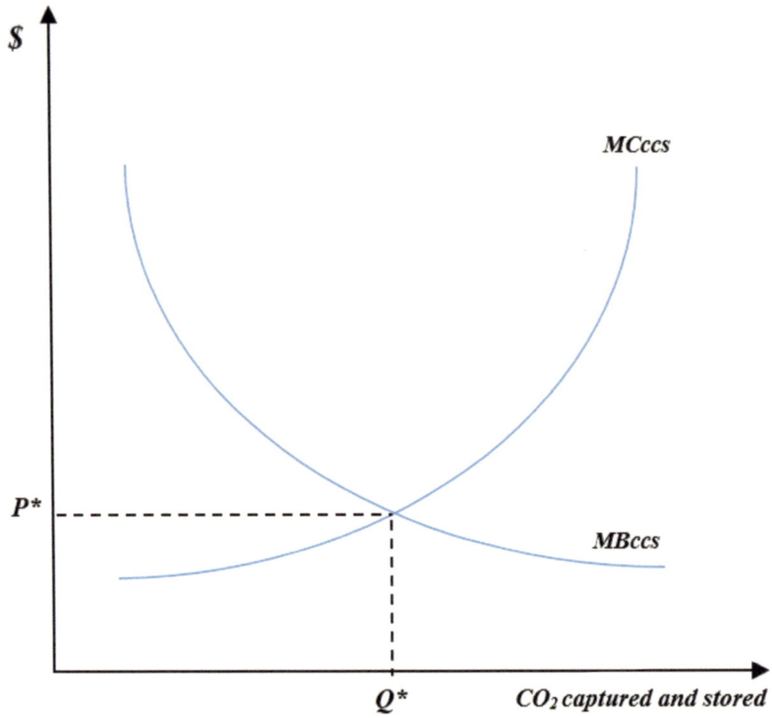

Figure 1. Economically efficient level of CCS.

returns, each additional unit of CO_2 captured and stored provides less private and social benefits. The marginal cost curve is upward sloping because both the private and social costs of CCS go up with each additional unit of CO_2 captured and stored. The upward-sloping nature of the marginal cost curve indicates that it would be very expensive (and likely cost prohibitive) to capture and store 100% of all CO_2 found in emissions from a point source such as a coal-fired power plant or industrial factory.

The economically efficient level of CCS (Q*) is shown graphically in **Figure 1** where the marginal benefit curve and marginal cost curve for CCS cross; at this point,

$$MC_{CCS} = \frac{\partial TC_{CCS}}{\partial Q_{CO_2}} = MB_{CCS} = \frac{\partial TB_{CCS}}{\partial Q_{CO_2}} \quad (5)$$

If all private and social benefits and costs of CCS could be "internalized" into economic markets, transactions between buyers and sellers could lead automatically to an economically efficient level of CCS, given certain conditions (e.g., perfect competition). It is notoriously difficult, however, to "internalize" all social benefits and costs because of the public good (or "bad") characteristics of these benefits and costs such as nonexclusiveness and nonrivalry. Thus, achieving an economically efficient level of CCS would most likely require some degree of government intervention into markets such as economic incentives (e.g., taxes and subsidies) and/or direct regulation ([5], Chapter 10).

http://dx.doi.org/10.5772/67000

5.2. Estimates of CSS optimal level

As previously described in this chapter, under the condition where marginal benefits and marginal costs of CO_2 captured and stored are equal, there exists a relationship between the optimal carbon price and the optimal level of carbon capture and storage. For a given carbon price range of US \$146–\$257/tC (or US \$40–\$70/tCO_2), the optimal level of CO_2 captured and stored is in the estimated range of 0–8MtC (or 0–29.48MtCO_2) per year [29, 30].

6. Summary and conclusions

From a public policy perspective, since the general public also benefits from carbon dioxide being captured, stored, and prevented from entering the atmosphere, there is economic justification for public policies targeted at providing economic incentives for private companies to invest in CCS technology, such as direct subsidies or tax breaks. Whether or not CCS technology will prove to be one of the "tools" in the global warming, mitigation "tool box" in the long run is yet to be seen.

In addition to the Petra Nova project in the United States, private companies in Canada, Germany, and China are investing in large-scale CCS projects, with mixed economic feasibility results from a private firm perspective. Scaling-up from the private firm level to the society level where public benefits from global warming mitigation are taken into account, the private and public economic benefits of CCS projects seem likely to outweigh the private costs. Thus, public polices, which help private companies to defray the high costs of large-scale CCS projects, may be justified from an overall benefit-cost analysis perspective.

Author details

John C. Bergstrom* and Dyna Ty

*Address all correspondence to: jberg@uga.edu

University of Georgia, Athens, GA, USA

References

[1] Rhodes J.S. and Keith D.W. Engineering economic analysis of biomass igcc with carbon capture and storage. Biomass and Bioenergy. 2005;29:440–450.

[2] Haszeldine R.S. Carbon capture and storage: how green can black be? Science. 2009;325(25):1647–1652.

[3] House K.Z., Baclig A.C., Ranjan M., van Nierop E.A., Wilcox J., and Herzog H. J. Economic and energetic analysis of capturing CO_2 from ambient air. Proceedings of the National Academy of Sciences of the United States of America. 2011;108(51):20428–20433.

[4] Pires J.C.M., Martins F.G., Alvim-Ferraz M.C.M., and Semoes M. Recent developments on carbon capture and storage: an overview. Chemical Engineering Research and Design. 2011;89:1446–1460.

[5] Bergstrom J.C. and Randall A. Resource economics: an economic approach to natural resource and environmental policy, 4th edition. Cheltenham, UK and Northampton, MA, USA: Edward Elgar; 2016.

[6] Nichani K., Bapat S., and Kumari A. A study of combating climate change with carbon capture and sequestration. International Journal of Engineering and Technical Research. 2014;2(9):269–272.

[7] Jenkins J. Financing mega-scale energy projects: a case study of the petra nova carbon capture project. University of Chicago, Chicago, IL, USA: Paulson Institute; 2015.

[8] Anderson S. and Newell R. Prospects for carbon capture and storage. Annual Review of Environmental Resources. 2004;29:109–142.

[9] Herzog H.J. Scaling up carbon dioxide capture and storage: from megatons to gigatons. Energy Economics. 2011;33(4):597–604.

[10] Azar C., Lindgren K., Larson E., and Mollersten K. Carbon capture and storage from fossil fuels and biomass—costs and potential role in stabilizing the atmosphere. Climate Change. 2006;74:47–79.

[11] Rubin E.S., Chen C., and Rao A.B. Cost and performance of fossil fuel power plants with CO_2 capture and storage. Energy Policy. 2007;35:4444–4454.

[12] Lackner K.S. A guide to CO_2 sequestration. Science. 2003;300(13):1677–1678.

[13] Rhodes C.J. Carbon capture and storage. Science Progress. 2012;95(4):473–483.

[14] Herzog H.J., Drake E., and Adams E. CO_2 capture, reuse, and storage technologies for mitigating global climate change: a white paper. Cambridge, MA: Massachusetts Institute of Technology Energy Laboratory; 1997.

[15] David J. and Herzog H. J. The cost of carbon capture. Presented Paper, Fifth International Conference on Greenhouse Gas Control Technologies; 13–16 August 2000; Cairns, Australia.

[16] Baker E., Chon H., and Keisler J. Carbon capture and storage: combining economic analysis with expert elicitations to inform climate policy. Climate Change. 2009;96:379–408.

[17] Kenyon D. and Jeyakumar B. Carbon capture and utilization. Calgary, Canada: The Pembina Institute; March 2015. [Cited September 2016]. Available from https://www.pembina.org/pub/carbon-capture-and-utilization

[18] International Energy Agency (IEA). Tracking clean energy progress 2015: energy technology perspectives 2015. Excerpt IEA Input to the Clean Energy Ministerial, IEA Publications; 2015.

[19] Kenyon D. The technologies behind carbon utilization. Calgary, Canada: The Pembina Institute; 2015.

[20] U.S. Department of Energy. Carbon dioxide enhanced oil recovery: untapped domestic energy supply and long term carbon storage solution. National Energy Technology Laboratory (NETL); March 2010. [cited September 2016]. Available from https://www.netl.doe.gov/file%20library/research/oil-gas/CO2_EOR_Primer.pdf

[21] U.S. Department of Energy. A climate change solution beneath our feet? Office of Fossil Energy; November 2002 [cited September 2016]. Available from http://www.fe.doe.gov/techline/tl_sequestration_aep.shtml

[22] Chargin A. and Socolow R. Fuels decarbonization and carbon sequestration: report of a workshop. Princeton, NJ: Princeton University, Center for Energy and Environmental Studies, School of Engineering and Applied Science; 1997.

[23] Stevens S.H., Kuuskraa V.A., and Gale J. Sequestration of CO_2 in depleted oil and gas fields; global capacity, costs and barriers. Presented Paper, Fifth International Conference on Greenhouse Gas Control Technologies; 13–16 August, 2000; Cairns, Australia.

[24] The Global CCS Institute. Accelerating the uptake of CCS: industrial use of captured carbon dioxide. Docklands, Australia: The Global CCS Institute; 2011.

[25] Bock B., Rhudy R., Herzog H., Klett M., Davison J., De La Torre Ugarte D. G., and Simbeck D.R. Economic evaluation of CO_2 storage and sink enhancement options. Washington, DC: U.S. Department of Energy & Tennessee Valley Authority Public Power Institute; 2002.

[26] The Intergovernmental Panel on Climate Change (IPCC). Climate change 2007: mitigation of climate change. Contribution of working group III to the fourth assessment report of the intergovernmental panel on climate change published for the intergovernmental panel on climate change. Cambridge, UK and New York: Cambridge University Press; 2007.

[27] Tol R.S.J. The marginal damage costs of climate change: an assessment of the uncertainties. Energy Policy. 2005;33(16):2064–2074.

[28] Nordhaus W. Estimates of the social cost of carbon: background and results from the rice-2011 model. NBER Working Paper 17540. Cambridge, Massachusetts: National Bureau of Economic Research; 2011.

[29] Middleton R.S. and Bielicki J.M. A scalable infrastructure model for carbon capture and storage: *simccs*. Energy Policy. 2009;37(3):1052–1060.

[30] Kuby M.J., Bielicki J.M., and Middleton R.S. Optimal spatial deployment of CO_2 capture and storage given a price on carbon. International Regional Science Review. 2011;000(00):1–21.